Edited by
Klaus K. Unger,
Nobuo Tanaka, and
Egidijus Machtejevas

Monolithic Silicas in Separation Science

Related Titles

Fritz, J. S., Gjerde, D. T.

Ion Chromatography

2009
ISBN: 978-3-527-32052-3

Miller, J. M.

Chromatography

Concepts and Contrasts, 2nd Ed.

2009
ISBN: 978-0-470-53025-2

Cohen, S. A., Schure, M. R. (eds.)

Multidimensional Liquid Chromatography

Theory and Applications in Industrial Chemistry and the Life Sciences

2008
ISBN: 978-0-471-73847-3

Forciniti, D.

Industrial Bioseparations

Principles and Practice

2007
ISBN: 978-0-8138-2085-9

Barry, E. F., Grob, R. L.

Columns for Gas Chromatography

Performance and Selection

2007
ISBN: 978-0-471-74043-8

Hahn-Deinstrop, E.

Applied Thin-Layer Chromatography

Best Practice and Avoidance of Mistakes

2007
ISBN: 978-3-527-31553-6

Pyell, U. (ed.)

Electrokinetic Chromatography

Theory, Instrumentation and Applications

2006
ISBN: 978-0-470-87102-7

Cox, G. B. (ed.)

Preparative Enantioselective Chromatography

2005
ISBN: 978-1-4051-1870-5

Edited by
Klaus K. Unger, Nobuo Tanaka, and Egidijus Machtejevas

Monolithic Silicas in Separation Science

Concepts, Syntheses, Characterization, Modeling and Applications

WILEY-VCH Verlag GmbH & Co. KGaA

The Editors

Prof. Klaus K. Unger
Am Alten Berg 40
64342 Seeheim-Jugenheim
Germany

Prof. Nobuo Tanaka
School Science & Technology
Kyoto Inst. of Technology
Matsugasaki, Sakyo-ku
Kyoto 606-8585
Japan

Dr. Egidijus Machtejevas
Merck KGaA
Perf. & Life Science Chemicals
Frankfurter Str. 250
64293 Darmstadt
Germany

All books published by Wiley-VCH are carefully produced. Nevertheless, authors, editors, and publisher do not warrant the information contained in these books, including this book, to be free of errors. Readers are advised to keep in mind that statements, data, illustrations, procedural details or other items may inadvertently be inaccurate.

Library of Congress Card No.: applied for

British Library Cataloguing-in-Publication Data
A catalogue record for this book is available from the British Library.

Bibliographic information published by the Deutsche Nationalbibliothek
The Deutsche Nationalbibliothek lists this publication in the Deutsche Nationalbibliografie; detailed bibliographic data are available on the Internet at http://dnb.d-nb.de.

© 2011 WILEY-VCH Verlag GmbH & Co. KGaA, Weinheim

All rights reserved (including those of translation into other languages). No part of this book may be reproduced in any form – by photoprinting, microfilm, or any other means – nor transmitted or translated into a machine language without written permission from the publishers. Registered names, trademarks, etc. used in this book, even when not specifically marked as such, are not to be considered unprotected by law.

Typesetting   Toppan Best-set Premedia Limited, Hong Kong
Printing and Binding   Fabulous Printers Pte Ltd.

Cover Design   Formgeber, Eppelheim

Printed in Singapore
Printed on acid-free paper

ISBN: 978-3-527-32575-7

Contents

Preface *XIII*
List of Contributors *XV*

1 The Basic Idea and the Drivers *1*
Nobuo Tanaka and Klaus K. Unger
1.1 Definitions *1*
1.2 Monoliths as Heterogeneous Catalysts *1*
1.3 Monoliths in Chromatographic Separations *2*
1.4 Conclusion and Perspectives *4*
References *5*

Part One Preparation *9*

2 Synthesis Concepts and Preparation of Silica Monoliths *11*
Kazuki Nakanishi
2.1 Introduction *11*
2.2 Background and Concepts *12*
2.2.1 Sol-Gel Reactions of Silica *12*
2.2.2 Polymerization-Induced Phase Separation *14*
2.2.3 Domain Formation by Phase Separation *15*
2.2.4 Arresting Transient Structure within a Solidifying Network *17*
2.2.5 Macropore Control *19*
2.2.6 Mesopore Control *21*
2.3 Synthesis of Silica Monoliths *21*
2.3.1 Silica Source and Catalyst *21*
2.3.2 Additives to Induce Phase Separation *22*
2.3.3 Preparation Procedure for Bulk Monolith *24*
2.3.3.1 Dissolution of PEO in Acidic Water *25*
2.3.3.2 Addition of TMOS for Hydrolysis *25*
2.3.3.3 Transferring the Solution to a Mold *25*
2.3.3.4 Gelation and Aging *26*

Monolithic Silicas in Separation Science. Edited by K.K. Unger, N. Tanaka, and E. Machtejevas
© 2011 WILEY-VCH Verlag GmbH & Co. KGaA, Weinheim
ISBN: 978-3-527-32575-7

2.3.4	Preparation Procedure for Capillary Monolith	26
2.3.5	Aging of Wet Monolith to Tailor Mesopores	27
2.3.6	Drying and Heat Treatment	28
2.3.7	Miscellaneous Factors for Better Reproducibility	28
2.4	Monolithic Columns Prepared in the Laboratory	29
2.4.1	Epoxy-Clad Columns	29
2.4.2	Capillary Columns	30
2.4.3	Other Types	31
2.5	Summary	31
	References	31

3 Preparation and Properties of Various Types of Monolithic Silica Stationary Phases for Reversed-Phase, Hydrophilic Interaction, and Ion-Exchange Chromatography Based on Polymer-Coated Materials 35
Oscar Núñez and Tohru Ikegami

3.1 Stationary Phases for Reversed-Phase Chromatography 35
3.2 Stationary Phases for Hydrophilic-Interaction Chromatography Separations 38
3.3 Stationary Phases for Ion-Exchange Separations 42
3.4 Advantages of Polymer-Coated Monolithic Silica Columns 43
References 45

Part Two   Characterization and Modeling 47

4 Characterization of the Pore Structure of Monolithic Silicas 49
Romas Skudas, Matthias Thommes, and Klaus K. Unger

4.1 Monolithic Silicas 49
4.2 General Aspects Describing Porous Materials 50
4.3 Characterization Methods of the Pore Structure of Monolithic Silicas 53
4.3.1 Mercury Intrusion-Extrusion 53
4.3.2 Inverse Size-Exclusion Chromatography 60
4.3.3 Nitrogen Sorption 64
4.3.4 Liquid Permeation 67
4.3.5 Microscopy and Image Analysis 70
4.4 Comparison of the Silica Monolith Mesopore-Characterization Data 73
4.5 Comparison of the Silica Monolith Flow-Through Pore-Characterization Data 74
References 75

5 Microscopic Characterizations 81
Haruko Saito, Kazuyoshi Kanamori, and Kazuki Nakanishi

5.1 Introduction 81

5.2	Preparation of Macroporous Silica Monolith *82*	
5.3	Laser Scanning Confocal Microscope Observation *83*	
5.4	Image Processing *84*	
5.5	Fundamental Parameters *84*	
5.5.1	Porosity *85*	
5.5.2	Surface Area *86*	
5.5.3	Characteristic Wavelength *86*	
5.5.4	Macropore Size and Skeleton Thickness *89*	
5.5.5	Chord Length *90*	
5.5.6	Mean Curvature and Gaussian Curvature *91*	
5.5.7	Curvature Distributions *93*	
5.5.7.1	Comparison between Different Porosities *93*	
5.5.7.2	Comparison between Different Macropore Sizes *94*	
5.6	Three-Dimensional Observation of Deformations in Confined Geometry *95*	
5.6.1	Synthesis and Characters of Organic–Inorganic Hybrid Monoliths *95*	
5.6.2	Deformed Macroporous Structures between Plates in MF System *96*	
5.6.3	Deformed Macroporous Structures between Plates in MM System *99*	
	References *102*	

6 Modeling Chromatographic Band Broadening in Monolithic Columns *105*
Frederik Detobel and Gert Desmet

6.1	Introduction *105*	
6.2	The General Plate-Height Model *106*	
6.2.1	Meaning of k'', u_i and K_p *107*	
6.2.2	Expressions for H_{ax} *108*	
6.2.3	Estimation of D_{mol} and D_{skel} *109*	
6.2.4	Expressions for $k_{f,tp}$ and $k_{f,skel}$ *110*	
6.2.5	Selection of the Characteristic Reference Lengths *111*	
6.2.6	Complete Plate-Height Equation *113*	
6.3	Use of the General Plate-Height Model to Predict Band Broadening in TSM Structures *114*	
6.3.1	Nonporous Skeleton Case *115*	
6.3.2	Porous Skeleton Case *116*	
6.4	Conclusion *120*	
	Acknowledgments *121*	
	Symbols *121*	
	Greek Symbols *122*	
	Subscripts *122*	
	References *122*	

7 Comparison of the Performance of Particle-Packed and Monolithic Columns in High-Performance Liquid Chromatography 127
Georges Guiochon

7.1 Introduction 127
7.2 Basic Columns Properties 128
7.2.1 Total, External and Internal Column Porosity 129
7.2.1.1 Definition of the Total Column Porosity 129
7.2.1.2 Measurement of the Total Column Porosity 129
7.2.1.3 Definitions of the Column External and Internal Porosities 131
7.2.1.4 Measurement of the Column External and Internal Porosities 131
7.2.2 Column Permeability 132
7.2.2.1 Permeability of Packed Columns 132
7.2.2.2 Permeability of Monolithic Columns 132
7.2.3 Column Efficiency 135
7.2.3.1 The HETP Equation 136
7.2.3.2 Reduced HETP and Reduced Velocity 137
7.2.4 Column Impedance 137
7.3 Comparison of the Through-Pore Structures and Related Properties 138
7.3.1 Porosity and Through-Pore Structure 138
7.3.1.1 External Porosity of Packed and Monolithic Columns 138
7.3.1.2 Importance of the Size of the Through-Pores 139
7.3.1.3 Average Size of the Through-Pores in Packed Columns 139
7.3.1.4 Average Size of the Through-Pores of Monolithic Columns 140
7.3.2 Column Permeability 140
7.3.2.1 Permeability of Packed Columns 140
7.3.2.2 Permeability of Monolithic Columns 140
7.3.3 Column Radial Homogeneity 142
7.3.3.1 Radial Heterogeneity of Packed Columns 142
7.3.3.2 Radial Heterogeneity of Monolithic Columns 143
7.3.3.3 Consequences of Column Radial Heterogeneity 144
7.4 Thermodynamic Properties 144
7.4.1 Retention 144
7.4.1.1 Retention Factors 145
7.4.1.2 Reproducibility of Retention Factors and Isotherms 146
7.4.2 Column Loadability 146
7.5 Kinetic Properties and Column Efficiency 147
7.5.1 Axial Dispersion 147
7.5.2 Mass-Transfer Kinetics 148
7.5.3 Column Impedance 150
7.5.4 Kinetic Properties 151
7.6 Conclusions 151
Acknowledgments 153

Symbols *153*
Greek Symbols *154*
References *154*

Part Three Applications *157*

8 Quantitative Structure–Retention Relationships in Studies of Monolithic Materials *159*
Roman Kaliszan and Michał J. Markuszewski
8.1 Fundamentals of Quantitative Structure–Retention Relationships (QSRR) *159*
8.2 Quantitative Relationships between Analyte Hydrophobicity and Retention on Monolithic Columns *163*
8.3 QSRR Based on Structural Descriptors from Calculation Chemistry *166*
8.4 LSER on Monolithic Columns *169*
8.5 Concluding Remarks *171*
References *171*

9 Performance of Silica Monoliths for Basic Compounds. Silanol Activity *173*
David V. McCalley
9.1 Introduction *173*
9.2 Reproducibility of Commercial Monoliths for Analysis of Bases *174*
9.3 Activity of Monoliths towards Basic Solutes *175*
9.4 Contribution of Overload to Peak Shapes of Basic Solutes *180*
9.5 Van Deemter Plots for Commercial Monoliths *180*
9.6 Performance of Hybrid Capillary Silica Monoliths for Basic Compounds *183*
9.7 Conclusions *186*
References *187*

10 Quality Control of Drugs *189*
Mohammed Taha, Abdelkarem Abed, and Sami El Deeb
10.1 Introduction *189*
10.2 Analysis of Pharmaceutics *189*
10.3 Natural Products Analysis *190*
10.4 Analysis Speed and Performance *191*
10.5 Method Transfer *193*
10.6 Separation of Complex Mixtures *196*
10.7 Monolith Derivatives and Versatile Application *198*
10.8 Summary and Conclusions *201*
References *202*

11	**Monolithic Stationary Phases for Fast Ion Chromatography** *207*
	Pavel N. Nesterenko and Paul R. Haddad
11.1	Fast Ion Chromatography *207*
11.2	Historical Development of Fast Ion Chromatography *207*
11.3	3 Advantages of the Bimodal Porous Structure of the Silica Monolith Matrix *210*
11.4	Type and Properties of Silica Monolithic Columns Used in IC *212*
11.5	Modification of Silica Monoliths for IC Separations *216*
11.5.1	Bare-Silica Monoliths as Ion Exchangers *216*
11.5.2	Coated Reversed-Phase Silica Monolithic Ion-Exchange Columns *217*
11.5.3	Silica Monoliths with Covalently Bonded Ion-Exchange Groups *221*
11.6	Operational Parameters *221*
11.7	Analytical Applications *223*
11.8	Future Work *225*
	References *226*

12	**Monolithic Chiral Stationary Phases for Liquid-Phase Enantioseparation Techniques** *231*
	Bezhan Chankvetadze
12.1	Introduction *231*
12.2	Organic Monolithic Materials for the Separation of Enantiomers *233*
12.3	Silica-Based Monolithic Materials for the Separation of Enantiomers *235*
12.3.1	Monolithic Silica Columns with Physically Adsorbed Chiral Selector *236*
12.3.2	Monolithic Silica Columns with Covalently Attached Chiral Selector *237*
12.4	Summary of the Present State-of-the-Art and Problems to be Solved in the Future *245*
	References *245*

13	**High-Speed and High-Efficiency Separations by Utilizing Monolithic Silica Capillary Columns** *249*
	Takeshi Hara, Kosuke Miyamoto, Satoshi Makino, Shohei Miwa, Tohru Ikegami, Masayoshi Ohira, and Nobuo Tanaka
13.1	Introduction *249*
13.2	Preparation of Monolithic Silica Capillary Columns *250*
13.3	Properties of Monolithic Silica Capillary Columns *252*
13.4	Monolithic Silica Capillary Columns for High-Efficiency Separations *254*
13.4.1	Performance of Long Monolithic Silica Capillary Columns *254*
13.4.2	Examples of High-Efficiency Separations *256*
13.4.2.1	Isocratic Mode *256*
13.4.3	Performance of Long Capillary Columns for Peptides in Gradient Mode *259*

13.5	Monolithic Silica Capillary Columns for High-Speed Separations *261*	
13.5.1	Monolithic Silica Columns Having Increased Phase Ratios *261*	
13.5.2	Performance of High-Speed Monolithic Silica Columns *265*	
13.5.3	Comparison of Performance with a Particle-Packed Column *266*	
13.6	Future Considerations *267*	
13.7	Conclusion *268*	
	References *269*	
14	**Silica Monolithic Columns and Mass Spectrometry** *273*	
	Keith Ashman	
14.1	Introduction *273*	
14.2	Offline Chromatography, LC MALDI MS *274*	
14.3	Online ESI LC/MS/MS for Proteomics and Selected Reaction Monitoring (SRM) *275*	
14.4	Online Reactors and Affinity Columns Coupled to Mass Spectrometry *278*	
14.5	Conclusion *279*	
	References *280*	
15	**Silica Monoliths for Small-Scale Purification of Drug-Discovery Compounds** *285*	
	Alfonso Espada, Cristina Anta, and Manuel Molina-Martín	
15.1	Introduction *285*	
15.2	Instrumental and Operating Considerations *286*	
15.2.1	Analytical Conditions *286*	
15.2.2	Preparative Conditions *287*	
15.3	Preparative Separations and Sample Loading *288*	
15.3.1	Semipreparative Monolithic 10×100 mm Column *288*	
15.3.2	Preparative Monolithic 25×100 mm Column *290*	
15.4	Purification of Drug-Discovery Compounds *292*	
15.5	Conclusions *294*	
	Acknowledgment *295*	
	References *295*	
16	**Monolithic Silica Columns in Multidimensional LC-MS for Proteomics and Peptidomics** *297*	
	Egidijus Machtejevas and Eglė Machtejevienė	
16.1	Introduction *297*	
16.2	Liquid Chromatography as a Tool Box for Proteomics *300*	
16.3	Selectivity of Columns for MD-LC *303*	
16.4	Dimensions of Columns in MD-LC *305*	
16.5	Monolithic Silica Columns *307*	
16.6	Applications of Monolithic Silica in Proteomics – A Brief Survey *310*	
16.7	Summary and Conclusions *314*	
	References *314*	

17	**Silica Monoliths in Solid-Phase Extraction and Solid-Phase Microextraction** *319*
	Zhi-Guo Shi, Li Xu, and Hian Kee Lee
17.1	Introduction *319*
17.2	Extraction Process *320*
17.3	Extraction Platforms *321*
17.3.1	Online Extraction *321*
17.3.2	Offline Extraction *322*
17.4	Applications *322*
17.4.1	SPE and SPME *322*
17.4.1.1	Silica Monolith from Entrapped Particles *322*
17.4.1.2	Silica Monolith from Direct Sol-Gel Strategy *327*
17.4.2	Other Applications of Silica Monolith *332*
17.5	Conclusion and Outlook *332*
	References *333*

Index *335*

Preface

One of the most prominent drivers in the field of separation science and technology is the search for novel and efficient materials as adsorbents to improve the mass transfer kinetics and to allow fast separations. While the major attention was directed to provide particle packed columns with smaller and smaller particles the idea to develop continuous beds based on silica monoliths was pioneered by Professor N. Soga and K. Nakanishi from Kyoto University, Japan, utilizing the template approach. It was a milestone in the development of silica monoliths when both researchers (NS and KN) had the splendid idea to introduce them as continuous beds in High Performance Liquid Chromatography (HPLC). In close collaboration with Professor K. Nakanishi, Dr Minakuchi and one of our co-editors (NT) performed the synthesis of such columns for HPLC. However, there was a serious limitation to apply monolithic silicas in 4 and 4.6 mm I.D. column format as the shrinkage of silica calling for a leak-tight and pressure stable cladding. This problem was finally solved by researchers (Dr K. Cabrera and Dr D. Lubda) from Merck KGaA, Darmstadt, Germany.

When my research group (KKU) became access to research samples from Merck, Darmstadt, at the mid of 1990s we were fascinated by the potential of silica monoliths as continuous beds in HPLC due to their flexibility in adjusting and controlling the morphology , pore structure and surface chemistry and thus enabling to optimize the chromatographic performance parameters. The second-generation monolithic silica columns just appearing seem to provide much higher performance than the first-generation columns commercialized in 2000.

The focus of Professor Tanaka's group was the preparation and improvement of monolithic fused silica capillaries to generate high efficiency columns and to compare them with particle packed fused silica columns.

Professor Tanaka and Dr E. Machtejevas both could demonstrate the potential of such columns in various life science applications such as proteomics and peptidomics using multidimensional HPLC/MS.

After almost twenty years of extensive research and development in the field the three authors (KKU, NT and EM) became convinced that it is time to review the work under the various aspects of separation science.

The authors are jointly indebted to the division of performance and life science chemicals of Merck KGaA, Darmstadt, Germany, for the generous support and

Monolithic Silicas in Separation Science. Edited by K.K. Unger, N. Tanaka, and E. Machtejevas
© 2011 WILEY-VCH Verlag GmbH & Co. KGaA, Weinheim
ISBN: 978-3-527-32575-7

for supplying numerous micrographs of the material. The authors want to express their gratitude to the team of Wiley-VCH, Weinheim, Germany, in particular to Ms W. Wüst, for their excellent cooperation and assistance in editing the book. We would appreciate the comments of the readers for the improvement of a further edition.

Klaus K. Unger, Nobuo Tanaka and Egidijus Machtejevas
October 2010

List of Contributors

Abdelkarem Abed
Al-Azhar University-Gaza
Department of Pharmaceutical
Chemistry
Gaza, P.O. Box 1277
Palestinian Territories

Cristina Anta
Centro de Investigación Lilly S.A.
Analytical Technologies Department
Avda. de la Industria 30
28108 Alcobendas, Madrid
Spain

Keith Ashman
Centro Nacional de Investigaciones
Oncológicas (CNIO)
C/ Melchor Fernández Almagro, 3
28029 Madrid
Spain

Bezhan Chankvetadze
Tbilisi State University
School of Exact and Natural Sciences
Department of Physical and
Analytical Chemistry and Molecular
Recognition and Separation Science
Laboratory
Chavchavadze Ave 3
0179 Tbilisi
Georgia

Gert Desmet
Vrije Universiteit Brussel
Department of Chemical Engineering
Pleinlaan 2
1050 Brussels
Belgium

Frederik Detobel
Vrije Universiteit Brussel
Department of Chemical Engineering
Pleinlaan 2
1050 Brussels
Belgium

Sami El Deeb
Al-Azhar University-Gaza
Department of Pharmaceutical
Chemistry
Gaza, P.O. Box 1277
Palestinian Territories

Alfonso Espada
Centro de Investigación Lilly S.A.
Analytical Technologies Department
Avda. de la Industria 30
28108 Alcobendas, Madrid
Spain

Georges Guiochon
University of Tennessee
Department of Chemistry
Knoxville, TN 37996-1600
USA

Monolithic Silicas in Separation Science. Edited by K.K. Unger, N. Tanaka, and E. Machtejevas
© 2011 WILEY-VCH Verlag GmbH & Co. KGaA, Weinheim
ISBN: 978-3-527-32575-7

Paul R. Haddad
University of Tasmania
School of Chemistry, Australian
Centre for Research on Separation
Science, School of Chemistry
Private Bag 75, Hobart
7001, Tasmania
Australia

Takeshi Hara
Justus Liebig University of Giessen
Department of Physical Chemistry
Heinrich Buff Ring 58
35392 Giessen
Germany

Tohru Ikegami
Kyoto Institute of Technology
Graduate School of Science and
Technology
Department of Biomolecular
Engineering
Gosho-Kaido-cho, Matsugasaki,
Sakyo-ku
Kyoto 606-8585
Japan

Roman Kaliszan
Medical University of Gdańsk
Department of Biopharmaceutics and
Pharmacodynamics
ul. Gen. J. Hallera 107
80-416 Gdańsk
Poland

Kazuyoshi Kanamori
Kyoto University
Graduate School of Science,
Department of Chemistry
Kitashirakawa, Sakyo-ku
Kyoto 606-8502
Japan

Hian Kee Lee
National University of Singapore
Department of Chemistry
3 Science Drive 3
Singapore 117543
Singapore

Egidijus Machtejevas
Merck KGaA
Performance & Life Science
Chemicals / Laboratory Business,
Product Manager Analytical
Chromatography
Frankfurter 250
64293 Darmstadt
Germany

Eglė Machtejevienė
Kaunas Medical University
Department of Obstetrics and
Gynecology
Eivenių 2
50009 Kaunas
Lithuania

Satoshi Makino
Kyoto Institute of Technology
Matsugasaki, Sakyo-ku
Kyoto 606-8585
Japan

Michał J. Markuszewski
Medical University of Gdańsk
Department of Biopharmaceutics and
Pharmacodynamics
ul. Gen. J. Hallera 107
80-416 Gdańsk
Poland
Nicolaus Copernicus University
Ludwik Rydygier Collegium
Medicum, Department of Toxicology
ul. M. Skłodowskiej-Curie 9
85-094 Bydgoszcz
Poland

David V. McCalley
University of the West of England
Centre for Research in Biomedicine
Frenchay
Bristol BS16 1QY
UK

Shohei Miwa
Kyoto Institute of Technology
Matsugasaki, Sakyo-ku
Kyoto 606-8585
Japan

Kosuke Miyamoto
Kyoto Institute of Technology
Matsugasaki, Sakyo-ku
Kyoto 606-8585
Japan

Manuel Molina-Martín
Centro de Investigación Lilly S.A.
Analytical Technologies Department
Avda. de la Industria 30
28108 Alcobendas, Madrid
Spain

Kazuki Nakanishi
Kyoto University
Graduate School of Science,
Department of Chemistry
Kitashirakawa, Sakyo-ku
Kyoto 606-8502
Japan

Pavel N. Nesterenko
University of Tasmania
School of Chemistry
Australian Centre for Research on
Separation Science
School of Chemistry
Private Bag 75, Hobart
7001
Tasmania, Australia

Oscar Núñez
University of Barcelona
Department of Analytical Chemistry
Av. Diagonal, 647
08901 Barcelona
Spain

Masayoshi Ohira
GL Sciences, Inc.
237-2 Sayamagahara
Iruma 358-0032
Japan

Haruko Saito
Kyoto University
Graduate School of Science,
Department of Chemistry
Kitashirakawa, Sakyo-ku
Kyoto 606-8502
Japan

Zhi-Guo Shi
National University of Singapore
Department of Chemistry
3 Science Drive 3
Singapore 117543
Singapore

Romas Skudas
Merck KGaA
PC-RLP-Polymer Materials
Frankfurter str. 250
64293 Darmstadt
Germany

Mohammed Taha
Al-Azhar University-Gaza
Department of Pharmaceutical
Chemistry
Gaza, P.O. Box 1277
Palestinian Territories

Nobuo Tanaka
Kyoto Institute of Technology
Matsugasaki, Sakyo-ku
Kyoto 606-8585
GL Sciences, Inc.
237-2 Sayamagahara
Iruma 358-0032
Japan

Matthias Thommes
Quantachrome Instruments
1900 Corporate Drive
Boynton Beach, FL 33426
USA

Klaus K. Unger
Johannes Gutenberg University
Institut fuer Anorganische Chemie
und Analytische Chemie
Duesberweg 10-14, DE-55099
Germany

Li Xu
National University of Singapore
Department of Chemistry
3 Science Drive 3
Singapore 117543
Singapore

1
The Basic Idea and the Drivers
Nobuo Tanaka and Klaus K. Unger

1.1
Definitions

The term "monolith" means a single stone and is used to describe a large stone of special origin. Ayer's rock in the middle of Australia is such an object. A detailed description of the history of monoliths is given by Svec and Tennikova [1]. We would like to define a monolith in this chapter as a continuous porous object whose morphology and pore structure can be varied in a wide range. A monolith can be a foam, a ceramic piece, it can be made of a cross-linked polymer or composed of an oxide or even carbon. The morphology can change between a membrane, a thin disk, a cylindrical rod, and a ceramic carrier of substantial size. The pores range from millimeter size down to mesopores and micropores. Monoliths can have a broad monomodal distribution, a distinct bimodal distribution and even a trimodal distribution with a hierarchical order of the pores.

1.2
Monoliths as Heterogeneous Catalysts

The control of morphology and pore structure of monoliths has made them ideal support materials for specific heterogeneous catalysis in environmental applications where fast reactions take place at high temperatures. The catalytic reduction of exhaust of automobiles [2] and the removal of pollutants from stationary sources for example, the selective reduction of NO_x, the removal of volatile organic compounds (VOC), the decomposition of NO and the oxidation of CO from exhaust gas [3] are typical examples.

For the first-mentioned application, a variety of ceramic monolithic supports are offered on the market that have mostly square channels and a honeycomb structure. The monoliths are made of cordierite and other high-temperature materials. They differ in the channel width and the thickness of channels. Generally, they are manufactured by extrusion from pastes that contain an oxide hydrate precursor and additives [4]. The shaped green bodies are dried and subjected to

Monolithic Silicas in Separation Science. Edited by K.K. Unger, N. Tanaka, and E. Machtejevas
© 2011 WILEY-VCH Verlag GmbH & Co. KGaA, Weinheim
ISBN: 978-3-527-32575-7

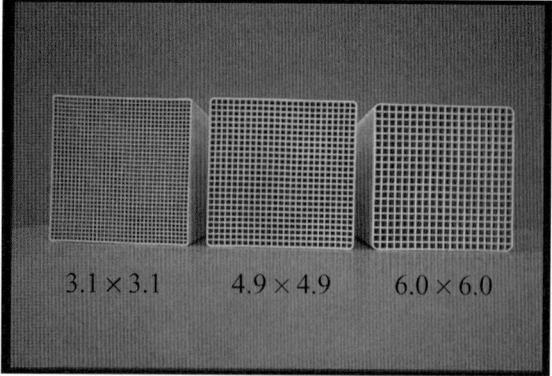

Figure 1.1 View of a ceramic monolith DeNO$_x$ catalysts for reduction of nitrogen oxides at power plants (By courtesy of Steuler Anlagenbau GmbH & Co, KG, Höhr-Grenzhausen, Germany).

calcination. The monolith itself is nonporous. A porous layer at the channel surface is deposited through a wash-coat procedure by which a layer of finely divided oxide for example, gamma-alumina is formed. Catalytic compounds are precious metals such as platinum, palladium and rhodium that are deposited by a wet impregnation procedure. Typically, a monolithic catalyst has a loading of 2 to 5 g of precious metal per liter volume. Due to the complex composition of the components of the exhaust gas, the variation of gas velocity and reaction temperature, so-called three-way catalyst systems have been designed and developed [5]. The catalytic converters for automobiles usually have an guaranteed average lifetime of 100 000 miles.

The monolithic catalysts for stationary sources for example, at power plants have much larger dimensions [3]. For instance, monoliths for selective catalytic reduction (=SCR catalysts) in NO$_x$ reduction have cells of 6 mm in diameter, a length of 1 meter and a wall thickness of 1 mm as shown in Figure 1.1. The advantages of such monoliths are: a low pressure drop, a low axial dispersion and a low radial heat flow.

1.3
Monoliths in Chromatographic Separations

The first materials to be used as continuous columns for gas and liquid chromatography were polyurethane foams [6–8]. However, the columns made offered a low column performance and exhibited a limited pressure stability.

Polymer-based monolithic materials were introduced by Hjerten et al. [9].

The pioneering work in the field of silica monoliths was made by Soga and Nakanishi [10, 11]. They combined the sol-gel chemistry of silica with a phase-separation mechanism of a system of slow dynamics resulting from the addition of hydrophilic polymers. The formation processes comprise a spinodal decomposition that produces a cocontinuous domain structure before fragmentation leading to a two-phase system. The transient cocontinuous domain structure was frozen by gelation to form the sponge-like morphology consisting of the silica hydrogel phase and the solvent phase. Interestingly, the morphology and pore structure of the resulting monolith can be changed over a wide range depending on the composition of tetramethoxysilane as the silica source, the solvent and the polymer in the phase diagram [12].

First, a cocontinuous monolith structure of silica skeletons and macropores in the size range between 1 and 10 µm is formed by a competitive process between a sol-gel transition and phase separation. A subsequent treatment of the monolith with an alkaline aqueous solution yielded mesoporous silica structures with a pore-size range of 5–30 nm. The macropores are termed as through-pores enabling a convective flow through the continuous column bed, while the mesopore (diffusional pores) are responsible for the selective separation by adsorption/desorption. It was the merit of the group in Kyoto to discover the potential and added value of silica monoliths for high-performance liquid chromatography (HPLC) [13].

When the monolithic silicas were manufactured as rods of 4 and 4.6 mm inner diameter, respectively, the task remained to clad the single pieces into leak-tight pressure-stable cylindrical columns with corresponding fittings. At the beginning the problem was solved by adapting the radial-compression column cartridge system developed by Waters where silica particles are filled in a plastic tube and the filled tube is then compressed by applying and maintaining an external pressure [13]. This device, however, was not a practical solution for monolithic silicas. Researchers from Merck KGaA, Darmstadt, solved the problem by a cladding process using poly(ether-ether-ketone) (PEEK) [14]. Monolithic silica columns are now commercially available as Chromoliths with varying column ID down to fused-silica capillaries. In the case of MonoClad columns available from GL Science, a silica monolith is covered by two layers of polymers and encased in a stainless steel tubing. This structure has higher pressure resistance than PEEK cladding [15]. It should be emphasized that the capillary silica monoliths do not require cladding as they are prepared *in situ* in fused-silica capillaries.

The major advantages of monolithic silica columns as compared to particle-packed columns were seen in the fast separation, short analysis time, and the low column pressure drop [16]. The first family of Chromolith columns of 4.6 mm ID exhibited a column performance comparable to a column packed with 3–4 µm particles and a pressure drop equivalent to those packed with 8–13 µm particles [17, 18]. The macropores of the Chromolith columns were approx. 2 µm and the mesopores approx. 12 nm. The surface functionality was an octadecyl bonding. Most of the applications were with low molecular weight analytes and with peptides.

In the years after the commercial introduction of the first monolithic columns, the direction for further development has become clear. Following the UPLC concept introduced by Waters [19], development work has focused on two areas:

1. Developing short monolithic silica columns capable of efficiencies up to 200 000 plates per meter and high-speed separations in a minute or less – performance parameters like a particulate column packed with sub-2-µm particles, but achieved with monolithic silica columns at much lower backpressure.

2. Developing monolithic columns producing 100 000 theoretical plates or more with a single column for high-resolution liquid chromatography without requiring very high operating pressures.

The primary quest is how can we optimize the design of a monolithic silica in order to achieve these goals.

Ultimately, the expectations led to research activities directed towards a better characterization of silica monoliths and to modeling and simulation studies correlating the properties of monoliths with the performance parameters in HPLC [20–22].

The primary questions were:

a) How can one characterize reliably the pore structure parameters, e.g., in monolithic silica capillaries?
b) Which pore structural parameters are relevant to improving the column performance and which values should be achieved?
c) Can one verify the proposed structures by a proper synthesis?

We are not yet in a position to answer these questions immediately.

However, substantial ground-breaking work in the named areas has been performed by Gzil [23], Grimes and Skudas [20, 22], Guiochon and coworkers [24, 25], and by Tallarek [18, 26].

In essence, the results and predictions need an experimental verification to develop much better materials. So far, column efficiency similar to that obtainable with 2.5 µm particles was reported [15, 27]. While the generation of 1 000 000 theoretical plates was shown to be possible with a silica monolith prepared in a capillary [28], the preparation of a homogeneous domain structure remains problematic. Moreover, the development of silica monoliths with regard to chromatographic selectivity by designing appropriate surface functionalities and their application to biopolymer separations is still in its infancy [29].

1.4
Conclusion and Perspectives

Monolithic silica columns, as compared to particle-packed columns, offer a number of advantages [30–32]:

1. Monolithic columns do not require a frit system, which is particularly advantageous in the case of a column in a fused-silica capillary.

2. Monolithic columns are robust and pressure stable and allow one to vary the flow rate in a wide range.

3. The pressure drop can be optimized as a function of the column performance by adjusting the pore structural parameters.
4. Silica monoliths can be designed according to the analyte targets: small-size pharmaceutical molecules, peptidic analytes and high molecular weight biopolymers.
5. Silica monolithic columns can be developed to provide a very large number of theoretical plates, e.g., more than 1 000 000, though accompanied by a long separation time, or a fast separation comparable with the performance of a particulate column packed with sub-2.5-µm particles.

Monolithic columns do have disadvantages with respect to their preparation:

1. Rod-type columns need cladding and this seems to be the biggest problem in column production. The cladding must be pressure-tight, but in addition may not disturb the flow or adsorption/desorption characteristics of the monolithic silica in the area where the monolithic silica is in contact with the cladding material.
2. Relatively large through-pores that contribute to a high permeability and low separation impedance may eventually limit the performance at high speed due to the large A-term contribution to the band broadening [25], unless monolithic silicas are prepared having very small-sized domains and high homogeneity.

Much work still has to be done to realize the expected performance of monolithic silica columns that can, at least in some aspects, exceed that of particulate columns and to achieve this with column formats suitable for practical applications.

References

1. Svec, F., and Tennikova, T.B. (2003) *Historical Review in Monolithic Materials: Preparation, Properties and Applications* (eds F. Svec, T.B. Tennikova, and Z. Deyl), Elsevier, Amsterdam, pp. 1–14.
2. Lox, E.S.J., and Engler, B.H. (1997) *Environmental Catalysis – Mobile Sources* in Handbook of Heterogeneous Catalysis, vol. 4 (eds G. Ertl, H. Knözinger, and J. Weitkamp), Wiley-VCH Verlag GmbH, Weinheim, pp. 1559–1633.
3. Janssen, F.J. (1997) *Environmental Catalysis – Stationary Sources* in Handbook of Heterogeneous Catalysis, vol. 4 (eds G. Ertl, H. Knözinger, and J. Weitkamp), Wiley-VCH Verlag GmbH, Weinheim, pp. 1633–1668.
4. Howitt, J.S. (1987) *Advances in Automotive Catalysis Supports* in Catalysis and Automotive Pollution Control (eds A. Crucq and A. Frennet), Elsevier, Amsterdam, pp. 301–313.
5. Weigert, W., and Koberstein, E. (1976) Autoabgasreinigung mit multifunktionellen Katalysatoren. *Angew. Chem. Int. Ed. Engl.*, **88**, 657–663.
6. Ross, W.D., and Jefferson, R.T. (1970) In situ-formed open-pore polyurethane as chromatography supports. *J. Chromatogr. Sci.*, **8**, 386–389.
7. Schnecko, H., and Bieber, O. (1971) Foam filled columns in gas chromatography. *Chromatographia*, **4**, 109–112.

8. Lynn, T.R., Rushneck, D.R., and Cooper, A.R. (1974) High resolution-low pressure liquid chromatography. *J. Chromatogr. Sci.*, **12**, 76–79.
9. Svec, F., Tennikova, T.B., and Deyl, Z. (eds) (2003) *Monolithic Materials: Preparation, Properties and Applications*, Elsevier, Amsterdam.
10. Nakanishi, K., Motowaki, S., and Soga, N. (1992) Preparation of SiO_2-TiO_2 gels with controlled pore structure via sol-gel route. *Bull. Inst. Chem. Res.*, **70** (2), 144–151.
11. Nakanishi, K. (1997) Pore structure control of silica gels based on phase separation. *J. Porous Mater.*, **4**, 67–112.
12. Nakanishi, K., Shikata, H., Ishizuka, N., Koheiya, N., and Soga, N. (2000) Tailoring mesopores in monolithic macroporous silica for HPLC. *J. High Resolut. Chromatogr.*, **23**, 106–110.
13. Minakuchi, H., Nakanishi, K., Soga, N., Ishizuka, N., and Tanaka, N. (1996) Octadecylsilylated porous silica rods as separation media for reversed-phase liquid chromatography. *Anal. Chem.*, **68**, 3498–3501.
14. Cabrera, K., Lubda, D., Eggenweiler, H.M., Minakuchi, H., and Nakanishi, K. (2000) A new monolithic-type HPLC column for fast separations. *J. High Resolut. Chromatogr.*, **23**, 93–99.
15. Miyazaki, S., Takahashi, M., Ohira, M., Terashima, H., Morisato, K., Nakanishi, K., Ikegami, T., Miyabe, K., and Tanaka, N. (2010) Monolithic silica rod columns for high-efficiency reversed-phase liquid chromatography. *J. Chromatogr. A*, Submitted.
16. Siouffi, A.M. (2003) Silica-gel based monoliths prepared by the sol-gel method: facts and figures. *J. Chromatogr. A*, **1000**, 801–818.
17. Bidlingmaier, B., von Doehren, N., and Unger, K.K. (1999) Comparative Study on the Column Performance of Microparticulate and Monolithic C 18 Bonded Reversed Phase Columns. *J. Chromatogr. A*, **832**, 11–16.
18. Leinweber, F.C., Lubda, D., Cabrera, K., and Tallarek, U. (2002) Characterization of silica-based monoliths with bimodal pore size distribution. *Anal. Chem.*, **74**, 2470–2424.
19. Mazzeo, J.R., Neue, U.D., Kele, M., and Plumb, R.S. (2005) A new separation technique takes advantage of sub-2-µm porous particles. *Anal. Chem.*, **77**, 460 A–465 A.
20. Grimes, B.A., Skudas, R., Unger, K.K., and Lubda, D. (2007) Pore Structural characterization of monolithic silica columns by inverse size exclusion chromatography. *J. Chromatogr.*, **1144**, 14–29.
21. Thommes, M., Skudas, R., Unger, K.K., and Lubda, D. (2008) Textural characterization of native and N-alkyl bonded silica monoliths by mercury intrusion/extrusion, nitrogen sorption and inverse size exclusion chromatography. *J. Chromatogr. A*, **1191**, 57–66.
22. Skudas, R., Grimes, B.A., Thomes, M., Lubda, D., and Unger, K.K. (2009) Flow-through pore characteristics of silica monoliths and their impact on chromatographic separation performance. *J. Chromatogr. A*, **1191**, 2625–2636.
23. Gzil, P. (2004) General rules for the optimal external porosity of LC supports. *Anal. Chem.*, **76**, 6707–6718.
24. Mriziq, K.S., Abia, J.A., Lee, Y., and Guiochon, G. (2008) Structural radial heterogeneity of a silica-based wide-bore monolithic column. *J. Chromatogr. A*, **1193**, 97–103.
25. Gritti, F., and Guiochon, G. (2009) Mass transfer kinetic mechanism in monolithic columns and application to the characterization of new research monolithic samples with different average pore sizes. *J. Chromatogr. A*, **1216**, 4752–4767.
26. Bruns, S., Mllner, T., Kollmann, M., Schachtner, J., Hltzel A., and Tallarek U. (2010) Confocal Laser Scanning Microscopy Method for Quantitative Characterization of Silica Monolith Morphology. *Anal. Chem.*, **82**, 6569–6575.
27. Hara, T., Kobayashi, H., Ikegami, T., Nakanishi, K., and Tanaka, N. (2006) Performance of monolithic silica capillary columns with increased phase ratios and small-sized domains. *Anal. Chem.*, **78**, 7632–7642.

28. Miyamoto, K., Hara, T., Kobayashi, H., Morisaka, H., Tokuda, D., Horie, K., Koduki, K., Makino, S., Núñez, O., Yang, C., Kawabe, T., Ikegami, T., Takubo, H., Ishihama, Y., and Tanaka, N. (2008) High-efficiency liquid chromatographic separation utilizing long monolithic silica capillary columns. *Anal. Chem.*, **80**, 8741–8750.
29. Núñez, O., Nakanishi, K., and Tanaka, N. (2008) Preparation of monolithic silica columns for high-performance liquid chromatography. *J. Chromatogr. A*, **1191**, 231–252.
30. Tanaka, N., Nakanishi, K., Minakuchi, H., Ishizuka, N., and Kobayashi, H. (2001) Monolithic columns – a new type of chromatographic support for liquid chromatography. *Anal. Chem.*, **73**, 420a–429a.
31. Guiochon, G. (2007) Monolithic Columns in High Performance Liquid Chromatography. *J. Chromatogr. A*, **1168**, 101–168.
32. Unger, K.K., Skudas, R., and Schulte, M.M. (2008) Particle Packed Columns and Monolithic Columns in High Performance Liquid Chromatography – Comparison and Critical Appraisal. *J. Chromatogr.*, **1184**, 393–415.

Part One
Preparation

2
Synthesis Concepts and Preparation of Silica Monoliths
Kazuki Nakanishi

2.1
Introduction

Historically, the synthesis of monolithic silica (silicate) having controlled macropores dates back to the middle of the twentieth century, when researchers in Corning Incorporated developed the so-called Vycor® process [1, 2]. Their extensive experimental study synchronized with the development of electron microscopy [3] as well as theoretical interpretation [4, 5] of the phase-separation process that occurs under a thermodynamically unstable condition, termed spinodal decomposition. Since the Vycor® process utilizes the phase separation of a melt-quenched metastable glass reheated near its glass-transition temperature, the development of heterogeneous structure in the glass could be adequately controlled by temperature programs. A critical drawback of the process as a method of fabricating porous materials was that subsequent leaching of a pore-forming phase was inevitable. Leaching one of finely interconnected glass phases in the length scale of few nanometers to submicrometers requires a substantial time depending on the sample thickness. As the sample specimen becomes thicker, the stress accumulated by the leaching process easily surpasses the strength of partly leached silicate glass, which makes it difficult to fabricate thick or bulky pieces without cracking or fracture. As a result, plates and hollow tubes with a thickness of a few millimeters were the main products. The method has been further improved by other researchers [6, 7], among others the invention of Shirasu porous glass (SPG) based on alumino-borosilicate composition using fly ash as a part of starting materials explored a novel application field of porous glasses as a membrane emulsifier [8].

In the early era of sol-gel synthesis, porous silica was regarded as an intermediate to fully sintered pore-free silica glass, as was exactly the case with Vycor® and other silica gels based on colloidal processes [9]. The size of continuous pores controlled was as large as a few hundred nanometers, which was too small to efficiently transport liquids through the specimen. In 1991, the first paper that described the structure control method of macroporous silica was published [10]. An incorporation of water-soluble polymer into an alkoxysilane-based sol-gel

Monolithic Silicas in Separation Science. Edited by K.K. Unger, N. Tanaka, and E. Machtejevas
© 2011 WILEY-VCH Verlag GmbH & Co. KGaA, Weinheim
ISBN: 978-3-527-32575-7

process made it possible to fabricate pure silica gels having well-defined interconnected macropores in the size range of micrometers. This method has been gradually extended using various kinds of water-soluble polymers, surfactants or other additives, to siloxane-based organic–inorganic hybrids and metal oxides such as titania, zirconia and alumina [11]. As early as 1993, the first presentation was made by Tanaka *et al.* that the silica monolith with well-defined macropores and tailored mesopores can be favorably applied to HPLC separation medium [12]. Continuous efforts have been made in the latter 1990s by Minakuchi *et al.* to fabricate monolithic columns with comparable efficiency and pronounced advantage over those of particle-packed columns [13–15]. The starting composition of silica gels used in commercialized monolithic silica columns includes only four substances, namely tetraalkoxysilane, water-soluble polymer, acid catalyst and water. Although the underlying science is somewhat complicated, the synthesis of macroporous silica monolith can be reproduced by anyone who keeps some critical points in mind. This chapter is devoted to concisely explain scientific background of synthesizing silica monoliths with controlled macro- and mesopores, as well as to offer a practical guide to synthesizing them in a laboratory.

2.2
Background and Concepts

2.2.1
Sol-Gel Reactions of Silica

The use of water-glass as a silica source has long been involved in a procedure in industry to manufacture silica gels especially in the form of particles [16, 17]. Recent requirements for higher purity have shifted the silica source from water-glass to alkoxide. As described later, water-glass as well as colloidal dispersions can favorably be used to fabricate macroporous silica monoliths when the high purity is not specifically required [9, 18]. In the following section, the sol-gel reaction of silica is described based on the hydrolysis and polycondensation of alkoxides as a typical example of gel formation in an aqueous-alcoholic silica system.

The sol-gel transition of a metal-alkoxide solution is generally caused by two kinds of reactions, that is, hydrolysis and polycondensation. Although the former one is necessary for the latter to take place, these reactions are known to proceed in parallel from a very early stage of the whole reaction. In normal cases, the hydrolysis is initiated by mixing water with alkoxide in the presence of alcohol as a cosolvent. The hydrolysis of M-OR produces M-OH that subsequently condenses with another M-OH to produce polycondensed species containing an M-O-M linkage and water. A successive condensation leads to the growth of metalloxane (in the case of Si, siloxane) oligomers that subsequently link together to form a gel network. In the presence of a limited amount of water, the alcohol producing condensation is also possible. The overall reactions for silicon alkoxide, $Si(OR)_4$ are expressed as follows;

$$Si(OR)_4 + H_2O \rightarrow Si(OH)(OR)_3 + ROH \quad (2.1)$$

$$\equiv Si-OH + \equiv Si-OH \rightarrow \equiv Si-O-Si\equiv + H_2O \quad (2.2)$$

$$\equiv Si-OH + \equiv Si-OR \rightarrow \equiv Si-O-Si\equiv + ROH \quad (2.3)$$

Silicon alkoxides exhibit extraordinarily slow hydrolysis and polycondensation kinetics compared with alkoxides of other kind of metals such as Ti, Zr, B and Al. In contrast to the fact that most alkoxides readily convert to a dispersion or aggregates of hydroxide or oxide powder on contacting a large amount of water, a controlled hydrolysis of silicon alkoxides is much easier and the silica gel materials can be processed into various shapes such as a fiber by spinning a viscous sol, a film by dipping a substrate in a less viscous sol and a bulk by casting a sol and allowing it to gel in a mold.

The overall kinetics of reactions (2.1–2.3) depends strongly on the water concentration, the kind of alkoxy groups and also on the kind and concentration of the catalyst. In the experimental systems described below, acid catalysts are used to promote the reaction. Under acidic conditions, the property of resultant gel is known to vary with the water concentration particularly across the range where the mole ratio of water/alkoxide becomes 2–4. The differences in the property of resultant gels are as follows;

1. Acid-catalyzed, low water (water/alkoxide ≤2)
   Gels with lower cross-linking density are formed, which are suitable to form gel fibers and coating films. A substantial amount of unreacted alkoxy groups tend to remain in the gel network.

2. Acid-catalyzed, high water (water/alkoxide ≥4)
   Gels with higher cross-linking density are formed, which exhibit higher tendency to remain monolithic when dried carefully. Hydroxy groups dominate in the interior of the gel network, and pores are distributed mostly below a few nanometers.

The gelation time, usually defined as the time between the start of hydrolysis and the sol-gel transition where the bulk fluidity of the sample is lost, depends strongly on the pH of the reacting solution. For silica, it is the longest, that is, the polycondensation is the slowest, around the isoelectric point of amorphous silica, pH = 2–3, and becomes shorter toward both directions of the pH axis [19]. Since the macropore formation of the silica sol-gel system described below is a competitive process of the phase separation and the sol-gel transition, its dynamic behavior is influenced by factors such as the molecular weight and its distribution of silica oligomers, the affinity between the solvent mixture and the oligomers, the rate of development of the gel network, etc. Therefore, the composition and catalyst concentration of the starting solution are key parameters that determine the resultant macropore morphology.

2.2.2
Polymerization-Induced Phase Separation

The compatibility in a system containing at least one kind of polymeric species can be estimated by the thermodynamic treatment known as the Flory–Huggins formulation [20]. The Gibbs free energy change of mixing for a binary system can be expressed as;

$$\Delta G = -T\Delta S + \Delta H = RT\left\{\left(\frac{\Phi_1}{P_1}\right)\ln\Phi_1 + \left(\frac{\Phi_2}{P_2}\right)\ln\Phi_2 + \chi_{12}\Phi_1\Phi_2\right\} \quad (2.4)$$

Here, the Φ_is and P_is (i = 1, 2) denote the volume fraction and the degree of polymerization of each component, χ_{12} is the interaction parameter. The former two terms in the bracket express the entropic contribution, and the last term the enthalpic contribution. Since the decrease in absolute values of the negative entropic terms destabilizes the system, it is evident that an increase of the degree of polymerization of either component makes the mixture less compatible. When the sign of ΔG turns from negative to positive, a driving force of phase separation arises. In other words, an initially single-phase solution containing a polymerizable component becomes less stable with the progress of polymerization reaction, finally results in the separation into different phases. Exactly the same occurs if the positive enthalpic contribution increases as the polymerization proceeds. A polycondensation reaction that consumes polar parts of molecules, for example, that between silanol groups, mediated in a polar solvent is a possible case for the substantial change in the enthalpic term during the polymerization.

Reviewing Equation 2.4, the decrease either in T or ΔS results in an increase of ΔG, that is, the system becomes destabilized against homogeneous mixing. The decrease in T corresponds to the ordinary cooling, while that in ΔS corresponds to the polymerization that decreases the degree of freedom among the polymerizing components. The equation indicates that decreases in T and ΔS equally contribute to cause phase separation of a mixture. These decreases in T and ΔS are termed "physical cooling" and "chemical cooling", respectively (Figure 2.1). Irrespective of the mode of "cooling", once the phase separation is induced, the process of domain formation follows an identical path, described in the next section. In the rest of this chapter, the phase separation induced by polymerization is extensively described in relation to the principle of macropore control of silica monoliths. Phenomenologically, the macropore formation is closely related to the formation of porous granules of instant coffee where a drop of liquid coffee is instantaneously frozen into a solid composed of concentrated coffee extracts and ice crystals followed by the sublimation of ice to form interconnected macropores. An important difference between physical and chemical cooling is that the former is reversible and can be easily controlled artificially, but the latter is often irreversible and only the rate of cooling can be adjusted by experimental parameters.

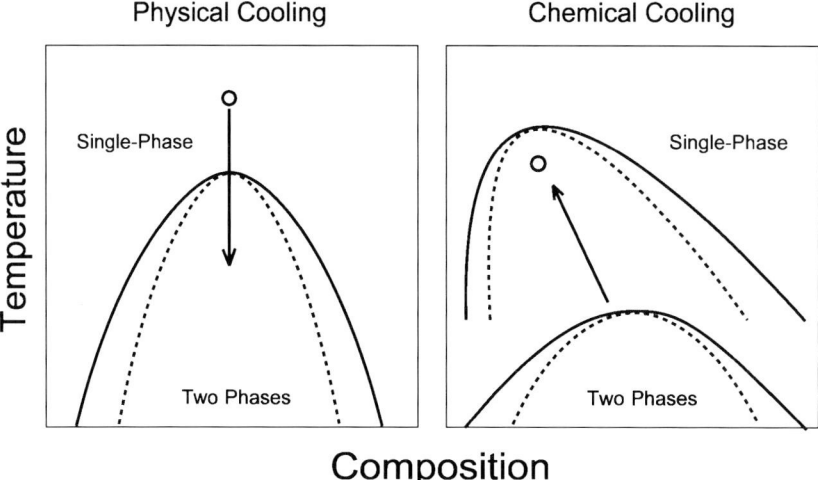

Figure 2.1 Comparison between physical and chemical coolings of the systems with miscibility windows. In the chemical cooling, the composition and temperature of an initially homogeneous mixture becomes included in the miscibility gap with the progress of the polymerization reaction.

2.2.3
Domain Formation by Phase Separation

Phase separation in a binary system is experimentally observed as a formation of two conjugate phase regions having, respectively, different chemical compositions generally called "phase domains". In most physical cooling processes, a one-phase mixture is thrust into the two-phase region by a temperature jump. In the two-phase region surrounded by the binodal line where the first derivative of Gibbs free energy of the system with respect to the composition becomes zero (solid curves in Figure 2.1), exists an internal region surrounded by the spinodal line on which the second derivative becomes zero (broken curves in Figure 2.1). The two lines coincide with each other at the critical consolute point where both the first and second derivatives become zero. Corresponding to the composition–temperature region where the phase separation is initiated, two modes of phase separation are known; one is a nucleation and growth (NG) that takes place between binodal and spinodal lines, and the other a spinodal decomposition (SD) that takes place within the spinodal line.

In the early stage of the NG process, a small fraction of dispersed domains with equilibrium conjugate composition, called nuclei, are generated. The nucleation is classified into homogeneous and heterogeneous origins. The former is believed to take place by thermally induced microscopic composition fluctuation, and the latter by impurity or inclusions that offer higher-energy sites favorable for the

1. Nucleation and Growth (Diffusion limited)

Small but discrete phase regions (nuclei) are generated followed by their individual growth into dispersed and matrix phases.

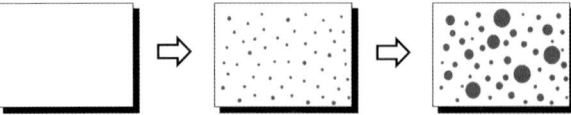

2. Spinodal Decomposition (Spontaneous)

Phase regions that initially have diffuse interface and small concentration difference develop into those with well-defined interface including cocontinuous domain structure.

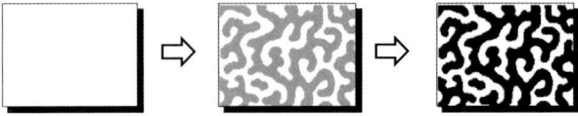

Figure 2.2 Formation mechanisms of phase-separated structures; 1. Nucleation and Growth, 2. Spinodal Decomposition

nucleation. Additional atoms or molecules subsequently diffuse toward the nuclei to increase the individual size of the nucleated domains. Since the growth process is the activated diffusion of chemical species driven by the concentration gradient, any kind of fluctuation, for example, temperature, composition, impurity or inclusion, etc., will influence the local rate of material transfer and leads to the inhomogeneous growth of the nuclei. Therefore, the precise control over the domain size with a sharp distribution is generally difficult. The domain morphology consisting of a matrix phase and a dispersed phase is a normal result of the NG mechanism (Figure 2.2-1).

The spinodal decomposition, on the other hand, starts with an infinitesimal composition fluctuation with a characteristic wavelength and its amplitude continuously grows to make higher contrast between the phase domains. The material transfer is characterized by an "uphill diffusion" where chemical species move against the concentration gradient without an activation barrier. When the mixture has the critical composition of the system as well as comparable volume fractions, a bicontinuous domain structure tends to develop and remains unbroken for a substantial period of time. In the early stage described by the linearized theory of Cahn [21], the composition fluctuation with a fixed wavelength increases its amplitude exponentially with time. Subsequently, while the composition difference between the domains continues to increase, the domain size also starts to increase (intermediate stage), and eventually the domains reach respective equilibrium compositions, while their sizes continue to increase (late stage)

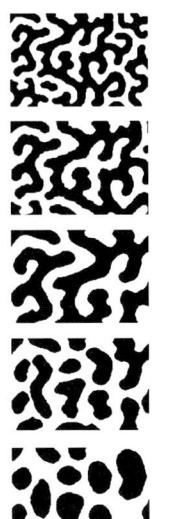

Development of cocontinuous structure

Self-similar coarsening 1

Self-similar coarsening 2

Fragmentation of domains

Spheroidization and sedimentation

Figure 2.3 Time evolution of spinodally decomposed isotropic phase domains driven by surface energy. After the domains grow only in characteristic size while maintaining the connectivity (self-similar growth), fragmentation and spheroidization follow to minimize the interfaces with energetically unfavorable curvatures.

driven by the surface energy [22, 23] (Figure 2.2-2). From the morphological point of view, the intermediate and late stages are regarded as the coarsening stage where the domains grow in size (Figure 2.3). The fragmentation of the domains often occurs during the coarsening, leading to the formation of morphologies similar to those obtained as a result of NG mechanism.

In polymeric systems, due to their extraordinarily slow and molecular-weight-dependent kinetics of molecular motion and viscoelastic behavior of phase domains, various anomalies in the coarsening process of the spinodal decomposition are reported [24, 25]. In the present silica system, there exists a certain time region where the gelling solution behaves similarly to a viscoelastic network with a finite relaxation time. The deformation of developing domains prior to gelation significantly influences the homogeneity of macroporous structure formed by the sol-gel reaction accompanied by the spinodal decomposition.

2.2.4
Arresting Transient Structure within a Solidifying Network

The domain formation induced by the spinodal decomposition includes the coarsening process in which the characteristic size of the bicontinuous structure grows from a shorter to a longer length scale with time. In the gel-forming polymerization reaction, on the other hand, the mobility within the network becomes restricted from a longer length scale to a shorter one. If the two processes of phase separation and gel formation take place competitively, what is expected? If we interpret the phenomenon as a freezing process of the transient structure of phase separation, several examples can be found in the preceding studies.

A simple example of a physically cooled polymer–solvent system to form porous structure was first shown experimentally by Keller *et al.* [26] for atactic polystyrene, investigated in more detail by Berghmans *et al.* [27], where the transient structure developed by the phase separation is frozen-in by the glass transition or crystallization of the polymer-rich phase, and the whole sample forms a macroscopic gel. A variety of morphologies could be obtained by changing the polymer concentration and the cooling rate.

A simple principle is applicable to the morphology control of the above examples as well as chemically cooled silica sol-gel systems exemplified below; the final morphology is determined by the timing of structure freezing relative to the onset and development process of the phase-separating domains. The higher the adopted cooling rate becomes, the shorter time is allowed for the coarsening of domains, resulting in the morphology with a finer structure. Finer domains can be obtained also by delaying the onset of phase separation without altering the freezing (gelling) process (Figure 2.4). Since in an isothermally polymerizing

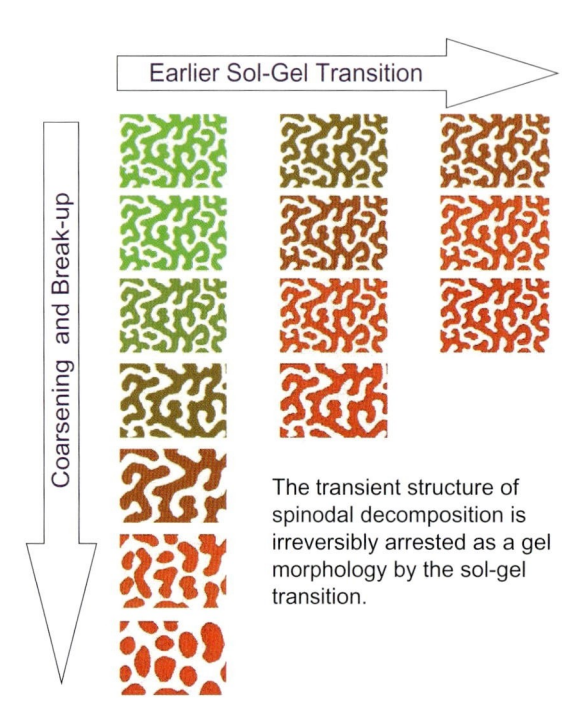

Figure 2.4 Relation between coarsening phase-separated domains and structure arrested by the sol-gel transition. The structures observed in resultant gels correspond to those at the bottoms of each row. Earlier sol-gel transition relative to the onset of phase separation results in finer structure.

system, both the phase-separation tendency and polycondensation rate are predetermined by the starting composition, all the pore characteristics, including pore volume, can be controlled by precisely adjusting the starting composition under a fixed temperature throughout the reaction. Regarding the catalyst, a base-catalyzed alkoxy-derived silica generally exhibits an inhomogeneous network structure and is not suitable to fix the well-defined domains developed in the length scale of micrometers.

2.2.5
Macropore Control [28]

As described above, the diameter of macropores and the thickness of silica skeletons are closely related to the characteristic sizes of phase-separated domains that are developed in the course of a sol-gel reaction. Here, the term "domain" denotes a pair of neighboring conjugate phases corresponding to the minimum characteristic wavelength of the heterogeneous structure. The term "domain size" usually denotes the sum of sizes of a gel-phase and its neighboring fluid-phase averaged over the sample. Even if a phase-separated wet gel is dried by removing the fluid-phase and the solvent, often accompanied by shrinkage, one can define the domain size as the sum of the diameter of the shrunk pore and its neighboring shrunk skeleton. Most silica sol-gel systems that phase separate in the course of a sol-gel transition contain three major components, silica, solvent and a phase-separation inducer, and they separate into two phases. The distribution of the three major components to the two conjugate phases, one rich in gelling silica and the other rich in nongelling component, hereafter denoted as gel-phase and fluid-phase, determines the volume fractions of macropores and silica skeletons.

Silanol groups that are abundantly present on the surface of siloxane oligomers often strongly interact with the ligands of polymers or surfactants added as a phase-separation inducer. Hydrogen-accepting ligands such as ether or carbonyl oxygens attractively interact with silanols to form hydrogen bonds. Moreover, in the case of vinyl polymers having such ligands as side chains or polyoxyethylenes, the close "lattice match" between the spacing of the ligands and that of silanols makes the hydrogen bonds concertedly stronger. Many kinds of water-soluble polymers, therefore, adsorb on the surface of silica oligomers rather than being homogeneously dissolved in a water–alcohol solvent. There exist other kinds of water-soluble polymers that do not specifically interact with silanols, but still can be used as a phase-separation inducer. Poly(sodium styrenesulfonate), poly(acrylic acid), poly(ethyleneimine) and poly(allylamine) under acidic conditions are examples of such weakly interacting polymers. Since their interaction with silanols largely depend on the surface charge on silica oligomers, the phase-separation tendency is significantly influenced by the solution pH.

With either type of strongly or weakly interacting phase separation inducer, the phase-separation tendency itself depends on the concentration of the polymer or surfactant, although the details of the dependence differ from system to system. In principle, the sizes of resultant gel- and fluid-phases are controlled by the concentrations of additive polymer or surfactant. The pore volume, which is determined by the volume fractions of gel- and fluid-phases, however, is determined very differently between the strongly and weakly interacting phase-separation inducers. With the additives that strongly interact (e.g., form hydrogen bonds) with silanols, the silica oligomers grown in the solution is tightly adsorbed by the additive molecules. On phase separation, the conjugate phases become one composed mainly of silica and adsorbed additive, and the other containing a majority of solvent mixture. The volume fraction of the gel-phase is determined by the volume fractions occupied by silica and additive, and that of the fluid-phase by those of majority of solvents. In this case, the volume fraction of macropores formed after the removal of the fluid-phase can be controlled mainly by the volume fraction of solvent mixture. Since the domain size is controlled by the additive/Si ratio in principle, the pore size (domain size) and pore volume can be independently controlled (Figure 2.5).

With the additives that only weakly interact with silanols, on the other hand, the gel-phase contains a majority of the silica and the fluid phase holds a majority of the additives, with the solvent mixture being distributed evenly as a common solvent. In this case, the controls over pore size by the concentration of additives necessarily influence the volume fraction of the fluid-phase, which becomes macropores afterwards. The amount and composition of the solvent plays only a minor role in determining the pore size and pore volume. Since the pore size and

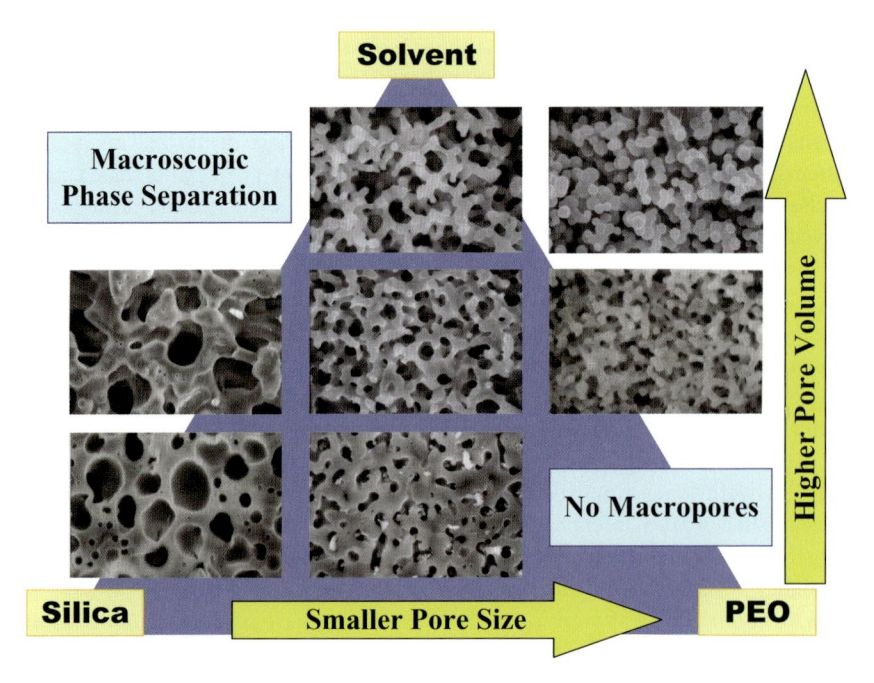

Figure 2.5 Relation between starting composition and resultant gel morphology in TMOS-PEO-solvent pseudoternary system. Pore size is controlled by PEO/Si ratio and pore volume by the fraction of solvent.

pore volume become interdependent through the additive concentration, the starting compositions where desired phase-separated structure can be obtained becomes limited compared with the case of strongly interacting additives.

2.2.6
Mesopore Control [29]

The sol-gel process accompanied by phase separation provides silica gels having phase-separated bicontinuous microdomains, one already solidified as a wet silica gel and the other still remains as fluid. Simple evaporative solvent removal to some extent preserves the shape and size of the wet silica gel domains to give a macroporous silica gel. The monolithicity, however, is hard to preserve even if the wet gel has well-connected micrometer-range open pores. Since the microscopic network of silica is determined by the acid-catalyzed polymerization, a substantial capillary stress corresponding to the pore size in the range of nanometers is exerted by the evaporation of solvents. To avoid or minimize undesired shrinkage and cracking during drying, an aging process under weakly basic condition is known to be effective in the case of pure silica gels.

The mechanism of pore coarsening by the aging under a basic condition is explained by a classical Ostwald ripening theory based on the difference in solubility of solid (hydrated silica in this case) as a function of the surface roughness. That is, the dissolution is most enhanced on the sharp points with the smallest positive curvature, whereas the reprecipitation is most pronounced at the cavities with the smallest negative curvature. As a result, with an elapse of aging, roughness is removed and the whole surface is reorganized into that with only larger points and cavities. If this process occurs in the three-dimensional network of silica gels, smaller pores are eliminated and the whole pore system is reorganized into that with larger pores.

In the case of less-water-soluble solids such as titania, zirconia and alkyl-modified silsesquioxanes, aging under more severe conditions, such as hydrothermal conditions, is required to tailor the mesopore structure. In these cases, the classical explanation based on an appreciable solubility of solid into water seems to be inadequate. Additional mechanisms including cooperative reorganization of partially cross-linked metalloxane network under a strong hydrothermal condition should be considered. In any case, the rate of pore coarsening is accelerated by increasing the temperature.

2.3
Synthesis of Silica Monoliths

2.3.1
Silica Source and Catalyst

In many exercises of synthesizing macroporous silica monoliths, alkoxysilanes, especially tetramethoxysilane, TMOS, and tetraethoxysilane, TEOS, are most convenient choices as silica sources with respect to purity, availability and price. For the purpose of preparing a rigid silica network (as frameworks of porous products), alcohols are usually not involved in the starting compositions. In the case of hydrolyzing TEOS under a weakly acidic condition, it sometimes becomes difficult to obtain a homogeneous solution by simply mixing the starting materials. In the standard procedures containing dilute acetic acid and urea described below, TMOS is used mainly for the sake of easier hydrolysis. Methyl-modified alkoxides such as methyltrimethoxysilane, MTMS, or methyltriethoxysilane, MTES, are usually used as minor components up to ~30% especially in the case of synthesizing capillary monoliths. In such copolymerization of TM(E)OS-MTM(E)S monoliths, the order and timing of adding respective alkoxides into an aqueous solution for hydrolysis should be strictly controlled as described below. For laboratory-scale uses, high-purity alkoxysilanes are available from many of the manufacturers that produce silicone compounds. In purchasing separate large lots of alkoxysilane for production, one should be careful about the impurity level and alcohol content. An inclusion of appreciable amount of alcohol corresponds to decreased silica content and will retard the gel formation, and thus changes the macroporous structure, compared with the case of neat alkoxysilane. Although it is generally possible to reproduce a specific macroporous structure by adjusting concentrations of other components, the content of initial alcohol, if any, should be strictly controlled. Any acid or base impurity may cause partial hydrolysis and polycondensation of alkoxysilanes depending on the duration and conditions of storage, which makes reproducible production of silica monoliths difficult.

The use of water-glass (aqueous alkaline silicate) is preferred in the case of large-scale productions of moderate purity demand. Water-glass is available both on laboratory and industry scales. Since the effective silica content of water-glass is usually higher than that finally obtained from alkoxysilane solutions, dilution with distilled water can be included in the process to mix viscous components better. Detailed laboratory-scale procedures of synthesizing macroporous silica monoliths from water-glass can be found in the literature [18].

As described above, acid-catalyzed hydrolysis is mandatory for the synthesis of silica monoliths with well-defined macropore structure. The concentration of catalyst always plays an important role in determining the final macroporous morphology of silica monoliths. The kind of catalyst, however, has a minor effect on morphology, because hydrolysis and polycondensation of alkoxysilane mostly depend on the effective pH of the reaction solution. In this regard, many of the mineral acids such as hydrochloric acid and nitric acid (with a special exception of hydrofluoric acid) can be used similarly to each other. The need for precisely controlling weakly acidic conditions, acetic acid is favorably used in the concentration range between 1 M to 1 mM.

For the purpose of obtaining robust porous silica monoliths, in many practices, the addition of external solvents other than water is avoided to keep as high an oxide content as possible. That is, into an acidic aqueous solution of a phase-separation inducer, neat alkoxide is added under vigorous stirring for hydrolysis.

As a result, the solvent is composed simply of water, alcohol and an acid catalyst. If an external solvent other than the parent alcohol of the corresponding alkoxide is intentionally added, the rate of gel formation and the phase-separation tendency are influenced by the kind and concentration of the solvent. Most polar solvents are compatible with the parent water–alcohol mixtures and simply work as diluents of the polymerizing system. The general results are that the domain size becomes coarser with an addition of external solvent. Amides derived from formic acid and acetic acid are important exceptions, especially used under strongly acidic conditions. The acid-catalyzed hydrolysis of amides produces ammonia or amine that raises the pH of the polymerizing solution and leads to an accelerated gel formation. With these external solvents, finer domain sizes than that of the parent composition can be obtained.

2.3.2
Additives to Induce Phase Separation

Polymerization-induced phase separation seldom occurs in simple silica sol-gel systems based on the hydrolysis of tetraalkoxysilanes because of the high compatibility among the constituents; silica oligomers, water and alcohol. Except in special cases associated with a limited amount of water for hydrolysis, polymeric or surfactant additives are mandatory to induce phase separation in polymerizing silica systems. In Table 2.1, experimentally confirmed combinations of such additives are listed together with specific oxide or inorganic–organic hybrid gel compositions. Water-soluble polymers and surfactants are convenient additives in that stable supplies can be expected from major chemical manufacturers. In most cases, the phase separation is monotonously enhanced with increased concentration of additives, of which excessive addition results in the macroscopic phase separation seen as sedimented precipitates and a supernatant. Important and apparent exceptions, however, are seen in the case of poly(ethylene oxide), PEO, with the average molecular weight being higher than ~50 000. In the case of PEO with relatively low molecular weight <20 000, the phase-separation tendency gradually increases from null addition, passes through a maximum, and then decrease again with further additions. With an increase of the molecular weight, the concentration corresponding to the maximum phase-separation tendency shifts to a very low concentration, less than a few per cent by weight, and in most practical concentrations the phase-separation tendency monotonously decreases with the additive concentration.

2.3.3
Preparation Procedure for Bulk Monolith

Monoliths for ordinary (inner diameter of 4.6 mm) HPLC columns are usually prepared in pure silica composition, where variations in the kind of additives and catalysts are possible. Taking the PEO-TMOS system, which has been industrially most successful as an example, detailed steps are described below. Figure 2.6

Table 2.1 Combinations of precursors and additives that are used to fabricate macroporous gels.

Precursors	Hydrolysis/Polycondensation Conditions (mol/mol)	Additives (excl. Water, Catal.)	Gel Phase	Fluid Phase
Tetraalkoxysilane	Water/Si <2	Polar Solvent	Siloxane	Solvent Mixture
Alkyltrialkoxysilane	Water/Si <3	Alcohols	Alkyl-terminated Siloxane	Solvent Mixture
Bis(trialkoxysilyl)alkane	Water/Si >10	None	Alkylene-bridged Siloxane	Solvent Mixture
Tetraalkoxysilane	Water/Si >4	Strongly Hydrogen-bonding Polymer, Nonionic Surfactant, Cationic Surfactant	Siloxane + Polymer	Solvent Mixture
Tetraalkoxysilane	Water/Si >4	Weakly Hydrogen-bonding Polymer, Anionic Surfactant	Siloxane	Solvent Mixture + Polymer
Water Glass	Destabilization by acid	Weakly Hydrogen-bonding Polymer Anionic Surfactant	Polysilicate	Sovent Mixture + Polymer
Titania Colloid	Destabilization by base	Strongly Hydrogen-bonding Polymer	Titanoxane + Polymer	Solvent Mixture
Titanium Alkoxide	Water/Si >4		Titanoxane + Polymer	Solvent Mixture
Zirconium Alkoxide	Water/Si >4		Zirconoxane + Polymer	Solvent Mixture
Aluminum Chloride	Water/Si >4	Strongly Hydrogen-bonding Polymer	Aluminoxane	Solvent Mixture + Polymer

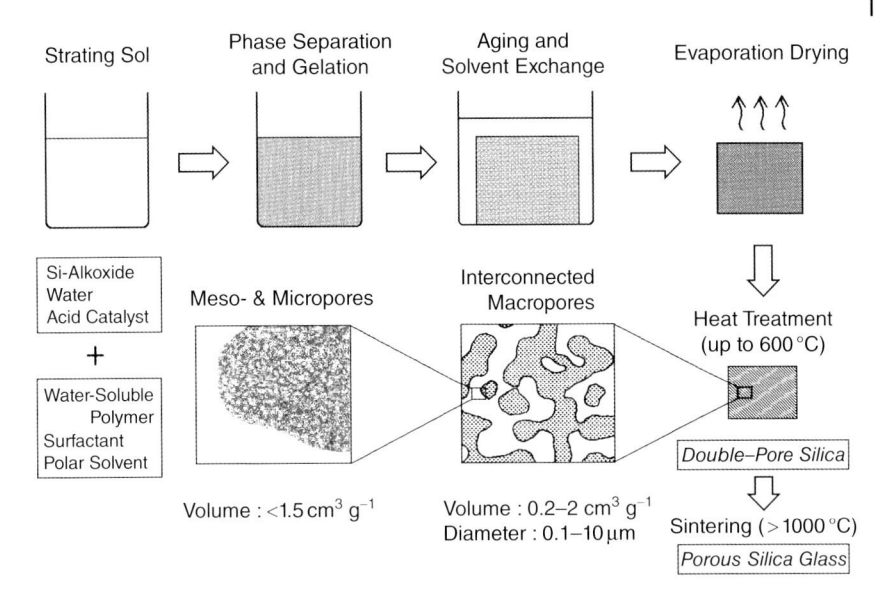

Figure 2.6 Overall process of fabricating macro/mesoporous silica monolith.

shows the overall process of preparing macro/mesoporous silica monoliths. Although no change in the process necessarily does harm to the quality and reproducibility of the products, several critical points highlighted below should be fully taken into account.

2.3.3.1 Dissolution of PEO in Acidic Water

PEO can be used as a phase-separation inducer with a broad range of average molecular weight between ~8000 and over 1 000 000. Practically, the dissolution of PEO solid requires longer times as the molecular weight increases. The use of PEO with the average molecular weight of 10 000 is recommended in view of easier dissolution into water, moderate viscosity, as well as commercial availability. Acid catalyst and other additives can be premixed with or dissolved in water before adding PEO. In the present example, 10 mM acetic acid is used as a catalyst together with urea corresponding to 1 M ammonia upon complete hydrolysis. To this homogeneously dissolved aqueous solution of acid-urea, a predetermined amount of PEO in the form of powder or flake is added under stirring. A few minutes are usually enough to dissolve PEO 10 000 homogeneously.

2.3.3.2 Addition of TMOS for Hydrolysis

To the aqueous solution of acid-urea-PEO described above, TMOS is added under stirring at a controlled temperature. Ice cooling is convenient for lab-scale or pilot-scale synthesis up to a few liters of reaction solution in a single batch, but somewhat higher temperatures, such as 30 °C, works fine. Care should be taken not to release volatile components during the hydrolysis reaction in order for the solution composition not to be deviated from the start. One should also be careful to keep the temperature of the reaction solution as constant as possible and not to locally heat it by insufficient or inhomogeneous mixing. With 10 mM acetic acid, the hydrolysis of TMOS takes 30 min under ice-cooled conditions. Under stirring, an initially heterogeneous liquid mixture composed of droplets of TMOS dispersed in an aqueous medium gradually becomes uniformly turbid. Then, after 10 to 15 min (depending on the volume of solution and conditions of stirring) the mixture turns into a clear solution. The reaction heat of hydrolysis is generated in this stage. If, for some reason, one uses a higher concentration of acid for hydrolysis, the heat evolution becomes intensive so that the solution temperature easily reaches the boiling point of the alcohols in the system. Again, for better reproducibility, reproducible temperature control throughout the above hydrolysis process is critical.

2.3.3.3 Transferring the Solution to a Mold

After obtaining a transparent solution containing hydrolyzed TMOS, it will be transferred into a mold where a columnar-shaped gel monolith is solidified. In most practices, tubes made of polycarbonate or polypropylene resin are conveniently used. Glass tubes with a hydrophilic inner surface are not suitable because the silica gel forms a tight bond with the surface. The formation of tight bonds between the mold and gel piece makes it difficult to take out the solidified specimen without damaging its surface. Hydrophobized glass tubes can be favorably used when precisely controlled inner and outer diameters of the mold are required. The inner diameter of the mold tube is chosen so as to result in an appropriate target diameter of dried and heat-treated gel monolith. For example, gels for 4.6 mm ID column is usually prepared in a mold with 6.0 mm ID. The mold should be tightly sealed so as not to allow the compositional change of the solution inside mainly due to the evaporation of alcohols.

2.3.3.4 Gelation and Aging

During the gelation process, the mold should be placed still, avoiding any possible mechanical disturbance, especially around the sol-gel transition point. Although it is sometimes useful for better column production to homogenize the solution by gently swinging the mold, it should be kept still at least 15–30 min before the gelation time. In order to minimize the deviation from the targeted temperature, the use of a water bath is recommended with regard to the heat capacity. With a transparent mold, one can observe the onset of phase separation as the transparent solution becomes translucent to turbid. Under optimized conditions, the gelation is nearly completed a few minutes after the solution turned turbid. The wet gel monolith thus solidified should be kept in a closed condition for an extended duration typically a half day to overnight. This "aging" period is important to stabilize the macropore structure as well as to obtain a sharper pore-size distribution after aging at higher pH and temperature conditions.

2.3.4
Preparation Procedure for Capillary Monolith

Capillary monoliths are prepared from the starting compositions similar to those used in bulk monoliths. With capillaries less than 100 μm ID, pure silica compositions can be prepared without serious defects to influence the performance of the column. With thicker capillaries, however, the partial substitution of TMOS with MTMS is effective to avoid undesired shrinkage within the capillaries. In preparing TMOS-MTMS monoliths, premixed alkoxides can be used similarly to the case of pure alkoxide in the hydrolysis step. Alternatively, TMOS is added to the aqueous solution FIRST, and after some time for the hydrolysis of TMOS, MTMS is mixed in addition. Under strongly acidic conditions, MTMS is hydrolyzed much faster than TMOS. The sequential hydrolysis in the order of MTMS $\to$ TMOS therefore results in preferential hydrolysis and subsequent polycondensation of MTMS before the addition of TMOS, which leads to enhanced heterogeneity in the resultant partially methyl-modified network.

Precursor solutions are introduced into capillaries typically 300–500 mm in length using a syringe or an HPLC pump, then both ends are put through the air-tight septum into the precursor solution in the vials (Figure 2.7). Placed in an air-circulating oven, the precursor solution in the capillary and vials are allowed to gel. Alternatively, both ends of the capillaries can be sealed by using appropriate adhesives. Even with these precautions to tightly seal the capillary, limited parts near both ends perform much worse than those out of the central parts

Monolithic Capillary Column

Fused Silica Capillary filled with precursor solution

Vials with septa
Precursor solution

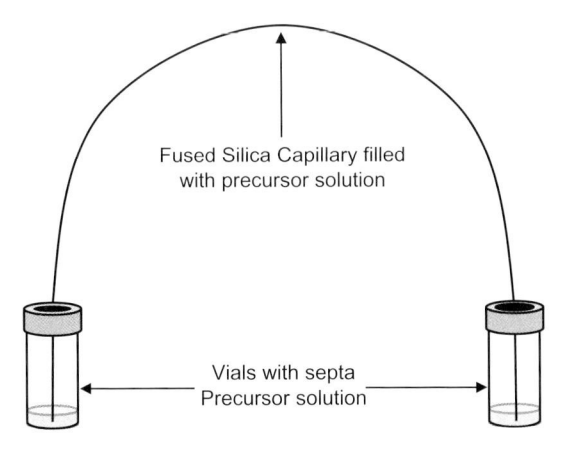

Figure 2.7 Schematic presentation of preparation of capillary monolith column. The whole system is placed in a thermostat for gelation and aging.

(confirmed by cutting long capillaries). It is therefore recommended to prepare capillary columns longer than desired, so that both end parts can be cut off for better performance.

2.3.5
Aging of Wet Monolith to Tailor Mesopores

For both bulk and capillary columns, the aging process in addition to that carried out directly after gelation is required to enlarge the mesopore size suitable for the separation of targeted substances. In the case of bulk monoliths, the wet gel specimens are carefully taken out of the cylindrical mold, immersed in a sufficient amount of approximately equivolume mixture of parent alcohol and water. This step is important to avoid any unexpected deformation or cracking due to the difference in osmotic pressure between the inside and outside of the wet gels. After thoroughly renewing the alcohol–water several times, the external solution is replaced either with dilute aqueous ammonia or aqueous urea solution. In using ammonia solution for aging, similar care must be taken not to abruptly expose the wet gel to the large difference in pH. In the case of urea solution, the concentration is usually set so that the complete hydrolysis of added urea to correspond to 1 M aqueous ammonia in the given amount of water. Aging conditions to obtain a specific value of median mesopore diameter differ from case to case. As a rough guide, at 80–100 °C overnight produces 10 nm, 110–120 °C for 1–2 days gives 20 nm. Higher temperature and pressure using an autoclave can produce mesopores larger than 50 nm. In the last case, however, a significant modification in the inner structure of the gel skeletons usually occurs, which makes it difficult to precisely define the pore diameter effective for chromatographic separations.

In the case of capillaries, the aging to tailor mesopores is carried out all within the capillary. The mother liquor contained in the as-gelled capillary is thoroughly replaced with aqueous ammonia or urea similarly to the case described above. Then, the whole capillary with its both ends tightly sealed is placed in an oven at the designated temperature. Monolithic silicas formed within capillaries are hard to directly characterize due to the limited amount of samples. In practice, the bulk monolith prepared under identical conditions is used to represent the pore properties of the capillary monolith.

2.3.6
Drying and Heat Treatment

A typical bulk silica monolith having 2 μm macropores and 10 nm mesopores can be dried in an air-circulating oven at 40 °C for a day or two. The actual time required for drying mainly depends on the thickness of the columnar-shaped specimen. Care should be taken not to rapidly evaporate the solvent out of the specimen. Ideally the drying and heat treatment should be carried out in a single oven without transferring the dried gels into another. In transferring the dried gels to the other oven or furnace for heat treatment, their temperature should be controlled so as to avoid exerting heat-shock to the specimens.

As far as the aging to expand mesopores is complete before drying, the drying and heat-treatment conditions have minimal effect on the properties of the final silica monolith. A linear program to heat the gels to 600–700 °C for a few hours can be favorably used. The surface area, as well as micropore volume, may vary with the final holding temperature and duration. In many silica monoliths tailored to retain 12-nm mesopores (similar to the most popular commercially available Chromolith®), the surface area value is approximately 300–350 m^2 g^{-1} after heating at 700 °C for 2 h. If monoliths are prepared with additional organosilanes such as MTMS, the highest possible heat treatment temperature becomes lower than 400 °C, to avoid damaging organic ligands.

2.3.7
Miscellaneous Factors for Better Reproducibility

As described above, the preparation of macroporous silica monolith with tailored mesopores consists of many steps. Each step requires specific attention for better performance with good reproducibility. The starting mixture should be free of any solid inclusion such as dust or particles as large as 1 µm. Hydrophobic inclusions such as fragmented fabric sometimes become a cause of defects larger than a few micrometers in size, which become crucial for performance of HPLC separations. Filtration of precursor solutions before transferring to the mold is effective to minimize the effect of these potential inclusions. The temperature profile of the solutions after the addition of TMOS until gelation should be kept identical from batch to batch. If the solution containing hydrolyzed TMOS experiences a longer time than usual at lower temperature, the resultant macropore structure becomes coarser due to delayed gelation relative to the onset of phase separation. It is also important to avoid any loss of volatile components during the process. If any small amount of loss is inevitable, for example, in transferring the solution to the mold, the loss should be controlled to be identical from batch to batch.

2.4
Monolithic Columns Prepared in the Laboratory

2.4.1
Epoxy-Clad Columns

A bulk columnar-shaped silica monolith can be fabricated into an HPLC column by encasing it into a pressure resistant vessel with both its ends exposed so as to be connected with LC tubings. The side surface of the silica monolith and the inner wall of cylindrical vessel (often termed "clad") must be coherently attached to perform reasonable separations in HPLC. For the purpose of tightly sealing the side wall of a silica monolith, a tubular film made of thermoshrinking PTFE

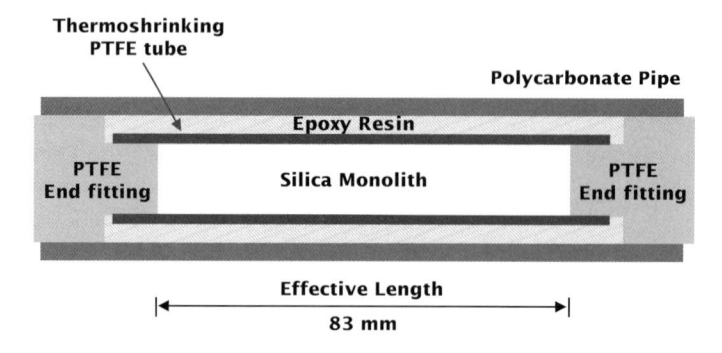

Monolithic Column Fabricated in Laboratory

2.0–4.6 mmϕ silica monolith

Figure 2.8 Schematic presentation of assembly of the parts for a column with bulk monolithic silica.

film is covered over the monolith, then heated under appropriate conditions to shrink the tube to be tightly fitted to the monolith. Then, the monolith covered with thin PTFE is cut into an appropriate length. The ends of the silica monolith are then polished flat to be attached with PTFE end-fittings that are designed to minimize dead volumes. Together with the end-fittings, the monolith is placed along the central axis of the outer pipe made of polycarbonate resin, then an epoxy resin mixture is poured into the space between the pipe and the monolith. After the solidification of the epoxy resin, sometimes curing at higher temperature is required, the column becomes ready for use (Figure 2.8). Due to the relatively low temperatures required for the above epoxy-clad formation, chemically modified silica monoliths can be fabricated into HPLC columns similarly. The column length and the shape of end-fittings are designed so as to be fitted in the radial compression module (RCM 8X10, Waters Corp., Milford MA), with which HPLC tubings are connected.

2.4.2
Capillary Columns

Capillary columns can be used as soon as the final heat treatment is completed. In the common cases of TMOS–MTMS hybrid monoliths prepared in fused silica capillaries with polyimide coating, the heat-treatment temperature should be limited to 330 °C in air. At this temperature, the removal of volatile components as well as PEO is complete, and the damage on the polyimide coating can be minimized. Additional surface modifications such as octadecylsilylation are to be carried out on-column. Due to significantly low pressure drop compared with ordinary particle-packed columns, several capillary monoliths can be connected to perform higher separation. The connection can be realized using a stainless steel union using graphite vespel ferrules (MVSU/004 and MGVF-004, GL Sciences) as connectors. Capillaries with the thickness from 50 to 200 μm ID can be connected to endure pressures as high as 50 MPa.

2.4.3
Other Types

Silica monoliths and their derivative materials with relatively large macropores are now used for several kinds of extraction, purification and protein-digestion tools [30–32]. As these kinds of application require no chromatographic separation efficiency, porous monoliths prepared in the laboratory can be easily used for the same purposes. Since a heat-treated macroporous silica monolith can be sanded or carved without serious fracture, thin disks can be fabricated by cutting the columnar shapes in the radial direction. By placing appropriately modified macroporous silica monoliths in pipette tips or spin columns for centrifugal separations, simple and efficient purification, concentration or enzymatic reactions can be realized in the laboratory.

2.5
Summary

Concepts and preparation methods of macroporous silica monoliths are described in some practical details. Although the underlying science is somewhat complicated and may not be familiar to many analytical scientists, the sol-gel synthesis in general is a simple and easy way to prepare materials readily used for efficient separations. The exploration of ultrahigh performance, however, is limited to capillary columns in laboratories. Due to the lack of accessible technology to place bulk silica monolith into an appropriate casing that can assure high performance in HPLC separations, fabrication of long columns with larger bores currently seems difficult. In the case that users need highly efficient accessibility to the solid surface modified by any desired ligand, silica monoliths can be convenient vessels to realize a broad variations of chemical operations.

References

1. Nordberg, M.E. (1944) Properties of some Vycor-brand glasses. *J. Am. Ceram. Soc.*, **27**, 299–304.
2. Hood, H.P., and Nordberg, M.E. (1940) Method of treating borosilicate glasses. US Patent 2215039.
3. Vogel, W. (1985) *Chemistry of Glass*, The American Ceramic Society, Inc., Ohio.
4. Cahn, J.W. (1961) On spinodal decomposition. *Acta Metall.*, **9**, 795–801.

5. Cahn, J.W., and Charles, R.J. (1965) The initial stages of phase separation in glasses. *Phys. Chem. Glasses*, **6**, 181–191.
6. Tanaka, H., Yazawa, T., Eguchi, H., Nagasawa, H., Matsuda, N., and Einishi, T. (1984) Precipitation of colloidal silica and pore-size distribution in high silica porous glass. *J. Non-Cryst. Solids*, **65**, 301–309.
7. Nakashima, T., Shimizu, M., and Kawano, M. (1987) Articles of porous glass and process for preparing the same. US Patent 4,657,875.
8. Vladisavljevic, G.T., Kobayashi, I., Nakajima, M., Williams, R.A., Shimizu, M., and Nakashima, T. (2007) Shirasu Porous Glass membrane emulsification: characterisation of membrane structure by high-resolution X-ray microtomography and microscopic observation of droplet formation in real time. *J. Membr. Sci.*, **302**, 243–253.
9. Shoup, R.D. (1976) Controlled pore silica bodies gelled from silica-alkali silicate mixtures, in *Colloid and Interface Science*, vol. 3 (ed. M. Kerker), Academic Press, New York, pp. 63–69.
10. Nakanishi, K., and Soga, N. (1991) Phase separation in gelling silica–organic polymer solution: Systems containing poly[sodium styrenesulfonate]. *J. Am. Ceram. Soc.*, **74**, 2518–2530.
11. Nakanishi, K. (2006) Sol-gel process of oxides accompanied by phase separation. *Bull. Chem. Soc. Jpn.*, **79**, 673–691.
12. Tanaka, N., Ishizuka, N., Hosoya, K., Kimata, K., Minakuchi, H., Nakanishi, K., and Soga, N. (1993) *Kuromatogurafi*, **14**, 50–51.
13. Minakuchi, H., Nakanishi, K., Soga, N., Ishizuka, N., and Tanaka, N. (1996) Octadecylsililated porous silica rods as separation media for reversed-phase liquid chromatography. *Anal. Chem.*, **68**, 3498–3501.
14. Minakuchi, H., Nakanishi, K., Soga, N., Ishizuka, N., and Tanaka, N. (1997) Effect of skeleton size on the performance of octadecylsilylated continuous porous silica columns in reversed-phase liquid chromatography. *J. Chromatogr. A*, **762**, 135–146.
15. Minakuchi, H., Nakanishi, K., Soga, N., Ishizuka, N., and Tanaka, N. (1998) Effect of domain size on the performance of octadecylsilylated continuous porous silica columns in reversed-phase liquid chromatography. *J. Chromatogr. A*, **797**, 121–131.
16. Iler, R.K. (1976) *The Chemistry of Silica*, John Wiley & Sons, Inc., New York.
17. Unger, K.K. (1979) *Porous Silica its Properties and Use as Support in Column Liquid Chromatography*, Journal of Chromatography Library, Elsevier, Amsterdam.
18. Takahashi, R., Sato, S., Sodesawa, T., and Azuma, T. (2004) Silica with bimodal pores for solid catalysts prepared from water glass. *J. Sol-Gel Sci. Technol.*, **31**, 373–376.
19. Brinker, C.J., and Scherer, G.W. (1990) *Sol-Gel Science; The Physics and Chemistry of Sol-Gel Processing*, Academic Press, New York.
20. Flory, P.J. (1971) *Principles of Polymer Chemistry*, Cornell University Press, Ithaca, New York.
21. Cahn, J.W. (1965) Phase separation by spinodal decomposition in isotropic systems. *J. Chem. Phys.*, **42**, 93–99.
22. Hashimoto, T., Itakura, M., and Hasegawa, H. (1986) Late stage spinodal decomposition of a binary polymer mixture. I. Critical test of dynamical scaling on scattering function. *J. Chem. Phys.*, **85**, 6118–6128.
23. Hashimoto, T., Itakura, M., and Shimidzu, N. (1986) Late stage spinodal decomposition of a binary polymer mixture. II. Scaling analyses on $Q_m(\tau)$ and $I_m(\tau)$. *J. Chem. Phys.*, **85**, 6773–6786.
24. Tanaka, H., and Nishi, T. (1988) Anomalous phase separation behavior in a binary mixture of poly(vinyl methyl ether) and water under deep quench conditions. *Jpn. J. Appl. Phys.*, **27**, L1787–L1790.
25. Tanaka, H. (2000) Viscoelastic phase separation. *J. Phys. Condens. Matter*, **12**, R207–R264.
26. Hikmet, R.M., Callister, S., and Keller, A. (1988) Thermoreversible gelation of atactic polystyrene: phase transformation and morphology. *Polymer*, **29**, 1378–1388.

27. De Rudder, J., Berghmans, H., and Arnauts, J. (1999) Phase behaviour and structure formation in the system syndiotactic polystyrene/cyclohexanol. *Polymer*, **40**, 5919–5928.
28. Nakanishi, K. (1997) Pore structure control of silica gels based on phase separation. *J. Porous Mater.*, **4**, 67–112.
29. Nakanishi, K., Takahashi, R., Nagakane, T., Kitayama, K., Koheiya, N., Shikata, H., and Soga, N. (2000) Formation of hierarchical pore structure in silica gel. *J. Sol-Gel Sci. Technol.*, **17**, 191–210.
30. Shintani, Y., Zhou, X., Furuno, M., Minakuchi, H., and Nakanishi, K. (2003) Monolithic silica column for in-tube solid-phase microextraction coupled to high-performance liquid chromatography. *J. Chromatogr. A*, **985**, 351–357.
31. Miyazaki, S., Morisato, K., Ishizuka, N., Minakuchi, H., Shintani, Y., Furuno, M., and Nakanishi, K. (2004) Development of a monolithic silica extraction tip for the analysis of proteins. *J. Chromatogr. A*, **1043**, 19–25.
32. Ota, S., Miyazaki, S., Matsuoka, H., Morisato, K., Shintani, Y., and Nakanishi, K. (2007) High-throughput protein digestion by trypsin-immobilized monolithic silica with pipette-tip formula. *J. Biochem. Biophys. Methods*, **70**, 57–62.

3
Preparation and Properties of Various Types of Monolithic Silica Stationary Phases for Reversed-Phase, Hydrophilic Interaction, and Ion-Exchange Chromatography Based on Polymer-Coated Materials

Oscar Núñez and Tohru Ikegami

Once the monolithic silica column is formed it can be chemically functionalized in order to obtain the desired stationary phase. In general, the modification can be performed by following either a silylation or a polymer-coating procedure as shown in Figure 3.1. On the former, the modification can be carried out by the direct reaction between the silica and a wide variety of silane reagents with the desired functional moieties, for example, octadecyldimethyl-N,N-diethylaminosilane and alkyltriethoxysilanes, etc. On the latter, the introduction of a spacer is carried out as a first step, and then the chemical reaction with a ligand that will introduce the desired functional group. Recently, polymerization methods are also being used to functionalize silica monoliths. In these cases either a spacer or an anchor group is first bonded to the silica before the chemical modification is performed. Additionally, dynamic coating of anion-exchange or cation-exchange functionalities onto reversed-phase monolithic columns is also a pathway to prepare monolithic columns. In this chapter, the chemical modification methods on monolithic silica columns after the silica preparation based on polymer-coated materials to obtain various types of stationary phases (reversed phase, hydrophilic interaction, ion exchange) are discussed, and some properties of these columns are compared to those of particle-packed columns.

3.1
Stationary Phases for Reversed-Phase Chromatography

Reversed-phase (RP) separations is one of the most widely used chromatographic modes in HPLC today, and most publications dealing with monolithic silica columns are focused on this chromatographic mode. The bonding of any alkyl chain or other chemical groups to prepare stationary phases in HPLC can be performed with almost the same methods as those used to bind these groups to the surface of silica particles. Most of the publications dealing with monolithic silica columns for RP separations employed the chemical modification using octadecyldimethyl-N,N-diethylaminosilane (ODS-DEA) [1] to obtain the so-called ODS silica monoliths [2–8]. With this procedure, monolithic silica columns, both

Monolithic Silicas in Separation Science. Edited by K.K. Unger, N. Tanaka, and E. Machtejevas
© 2011 WILEY-VCH Verlag GmbH & Co. KGaA, Weinheim
ISBN: 978-3-527-32575-7

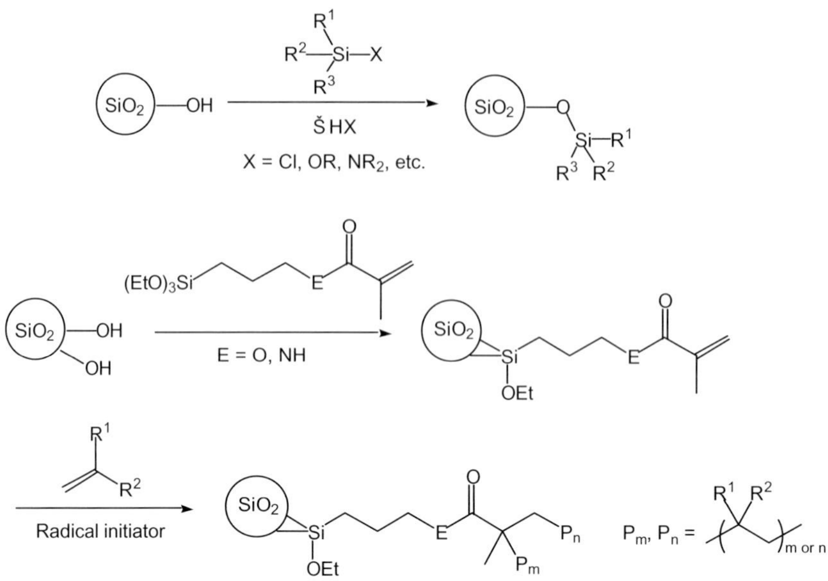

Figure 3.1 Chemical modification processes to functionalize monolithic silica columns.

rod and capillary, can be modified by continuously feeding 20% ODS-DEA (v/v) solution in toluene to the columns at 60 °C for 3 h, followed with washes with dry toluene and dry THF. These columns keep high permeability after the modification process, because of their higher porosity and the large through-pore-size–skeleton-size ratios, although in general small phase ratios (volume of the stationary phase divided by volume of the mobile phase in a column) are obtained. A high porosity means that a small amount of silica is present in the column, which will represent a small phase ratio after the chemical modification. This fact will be more pronounced in monolithic silica capillary columns. For instance, the phase ratio in a monolithic silica capillary column prepared from tetramethoxysilane (TMOS) estimated on the basis of retention factors (k) of alkylbenzenes was 0.02–0.03 [9] or 0.04–0.05 [5] compared to 0.06 for a silica rod and 0.19 for a particle-packed column [10]. So, in general, smaller retention factors, in many cases, 1/3 or less will be obtained with monolithic silica capillary columns or monolithic silica rod columns modified with C_{18} phases, compared to C_{18} particle-packed columns [9].

In general, high retention factors are preferred if high column efficiency is to be maintained under various conditions. This high retentivity is not expected for monolithic silica columns having high porosity resulting in a small phase ratio. One way to increase retention is by increasing the phase ratio, which can be obtained by raising the silica content in the capillary column [5] or by increasing the amount of stationary phase bonded to silica support in a column, which can be achieved by using polymerization reactions. These reactions can increase the phase ratio of monolithic silica columns, for instance, by grafting polymer chains to the surface of the silica monolith in a similar way as has been performed on silica particles [11, 12]. For instance, high efficiency and highly retentive monolithic silica capillary columns were reported by a polymerization of octadecyl methacrylate using α,α'-azobisisobutyronitrile (AIBN) as a free-radical initiator [13]. The first step in this chemical modification method consisted of a bonding reaction with 3-methacryloxypropyltrimethoxysilane (MOP) in order to anchor a monomer with a double bond that will undergo polymerization with octadecyl methacrylate. This step was common for other chemical modifications by polymerization and other anchor groups such as 3-(methacrylamidopropyl)triethoxysilane (MAS) can be used by applying a similar bonding procedure. After washing the monolithic silica columns with toluene and methanol for 24 h, the MOP bonding was performed by feeding the columns with a MOP:pyridine 1:1 (v/v) solution for 48 h at 80 °C, followed by washings using toluene and methanol. To perform the polymer-bonding reaction, the columns were first rinsed with toluene for 3 h and then the polymerization reaction mixture, which contained 250 µl of octadecyl methacrylate (ODM) monomer and 250 µl toluene solution of AIBN (38.6 mg/ml), was charged into the column and allowed to react at 80 °C for 3 h. After polymerization, the columns were washed with toluene (24 h) followed by methanol. By performing the polymerization at higher monomer and higher initiator concentrations, columns with higher retentivity than ODS columns were obtained ($k_{amylbenzene}$ up to 4.2 compared to 1.7 for ODS in 80% methanol mobile phases), while keeping a high efficiency similar to that of an ODS column (H_{uracil} 10 µm) [13]. In fact, it was observed that the higher the amount of monomer available in the polymerization reaction mixture, the higher the surface coverage achieved. This would raise the phase ratio of the stationary phase as was observed by a considerably increase in retention properties of alkylbenzenes ($k_{amylbenzene}$ values of 1.7 to 4.2 for 100 and 300 µl monomer, respectively). In contrast, the column retention did not change much with the variation of initiator concentration in the polymerization reaction mixture, for instance, a variation of $k_{amylbenzene}$ from 3.20 to 3.51 (80% methanol mobile phase) was obtained for AIBN concentrations between 1.2 to 19.3 mg/l [13]. Moreover, although an increase in monomer concentration produced a loss in column efficiency (due to the increase in phase ratio and the corresponding decrease in total porosity), by increasing the amount of initiator in the polymerization reaction an enhancement in column efficiency was observed. At the separation employing 80/20 (v/v) acetonitrile/water, the degree of loss in column efficiency was around 3 µm, however, in 80/20 (v/v) methanol/water, it was around 20 µm in the range from 100 to 300 µl of ODM in the reaction mixture.

Figure 3.2 shows scanning electron micrograph images of the monolithic silica columns before and after the polymerization of octadecyl methacrylate. Apparently, the morphology of the columns does not change by the polymerization. Before the polymerization, the total porosity and the external porosity were 92 and 77%, respectively. After the polymerization, they were 84 and 73%, respectively, as determined by size-exclusion chromatography [13].

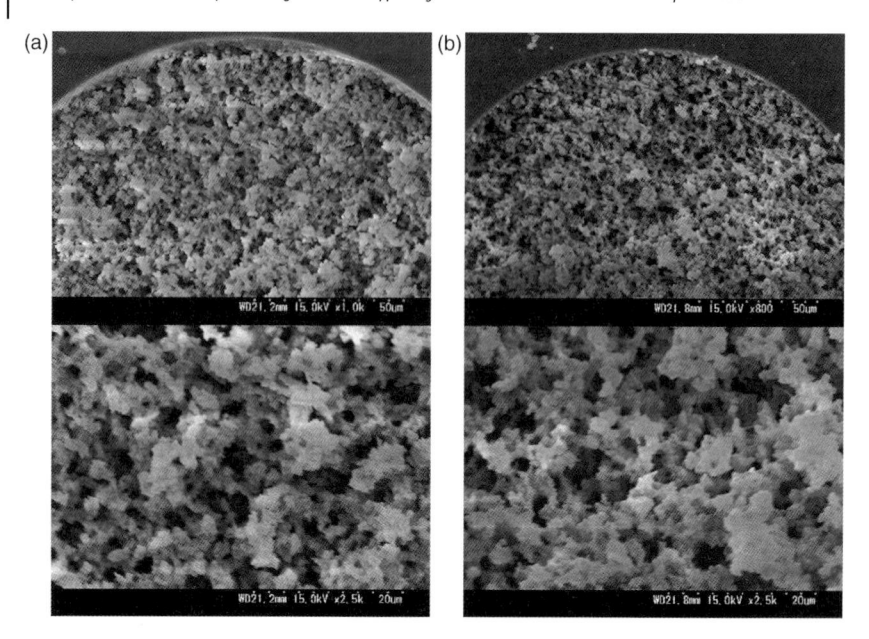

Figure 3.2 Scanning electron micrograph images of the monolithic silica capillary column obtained before (a) and after (b) polymerization with octadecyl methacrylate. Reproduced from Ref. [13] with the permission of Elsevier.

In a recent report, the selectivity of these ODM columns for some polar and nonpolar compounds was studied [14]. In general, the polymer-coated monolithic silica stationary phase ODM seems to have the greater preference for compounds with aromatic character, rigid and planar structures, and smaller length-to-breadth ratios (more compact structures), compared to the conventional C_{18} ODS stationary phases. As an example, Figure 3.3 shows the separation of tocopherol isomers using a 25 cm × 200 μm I.D. ODM monolithic silica capillary column. As can be seen, separation of the four tocopherol isomers was achieved with 95% methanol mobile phase, especially for β- and γ-tocopherol isomers, while no separation was observed with an ODS column under the same conditions, even by using a long ODS monolithic silica capillary column (440 cm × 100 μm I.D.), or other conventional C_{18} stationary phases [15, 16]. This polymeric stationary phase maintained the efficiency of the monolithic silica support.

3.2
Stationary Phases for Hydrophilic-Interaction Chromatography Separations

Hydrophilic interaction chromatography (HILIC) separations using monolithic silica columns have been studied recently in order to prepare new materials for the separation of highly polar compounds that showed limited retention in the

3.2 Stationary Phases for Hydrophilic-Interaction Chromatography Separations

(a)

$R_1=R_2=R_3=Me$ α-tocopherol
$R_1=R_3=Me$ $R_2=H$ β-tocopherol
$R_1=H$ $R_2=R_3=Me$ γ-tocopherol
$R_1=R_2=H$ $R_3=Me$ δ-tocopherol

(b) ODM column
$u = 1.09$ mm/s
$L = 25$ cm

ODS column
$u = 1.01$ mm/s
$L = 25$ cm

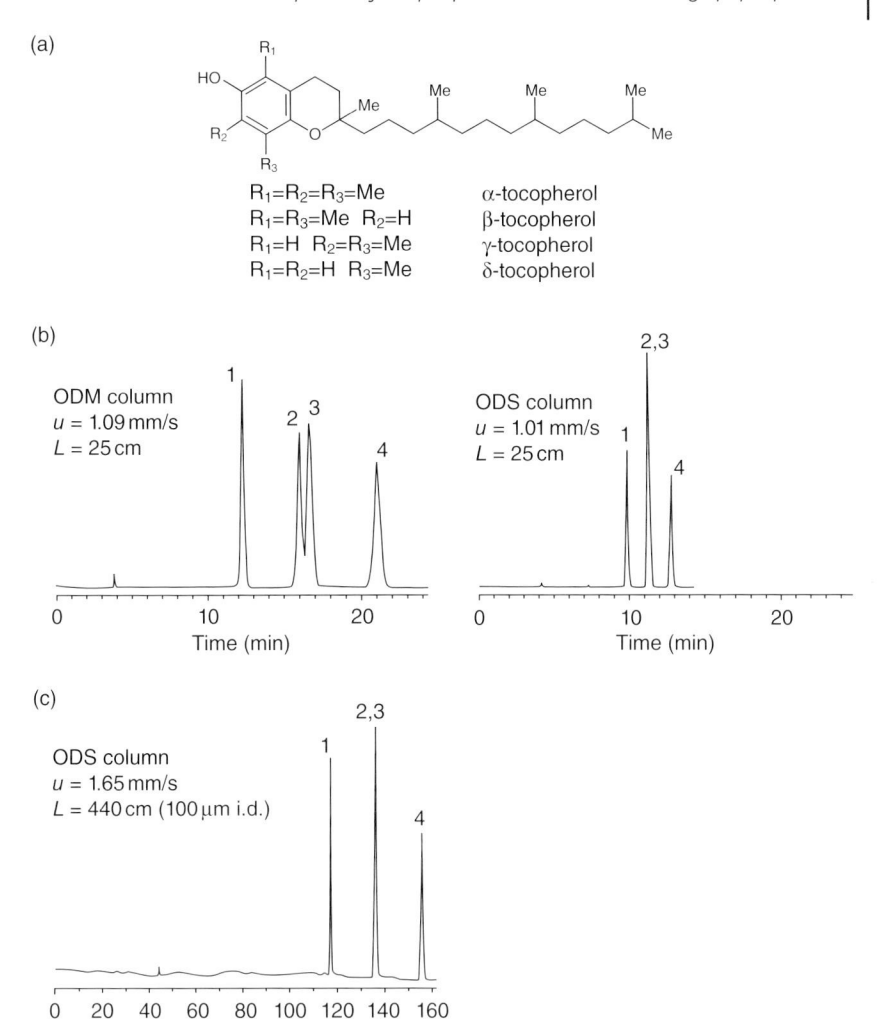

(c) ODS column
$u = 1.65$ mm/s
$L = 440$ cm (100 μm i.d.)

Figure 3.3 (a) Chemical structures of tocopherol homologues. (b) Separation of tocopherols with ODM and ODS columns (25 cm × 200 μm I.D.) using a methanol:water 95 : 5 (v/v) solution as mobile phase. Thiourea was used as the t_o standard. Detection performed at 295 nm and 30 °C. (c) Separation of tocopherols using a long ODS monolithic silica capillary column (440 cm × 100 μm I.D.). Other conditions as in (b). Peak identification: 1, δ-tocopherol; 2, β-tocopherol; 3, γ-tocopherol; 4, α-tocopherol. Reproduced from Ref. [14] with the permission of Elsevier.

RP mode. Until now, only a few publications have dealt with the use of monolithic silica columns in HILIC mode, and only some of them based on polymer-coated materials. This approach to prepare HILIC monolithic silica columns utilizes polymerization of vinyl monomers having functional groups, such as amides, carboxylic acids, sulfonic acids, amines, and ammonium salts, to immobilize

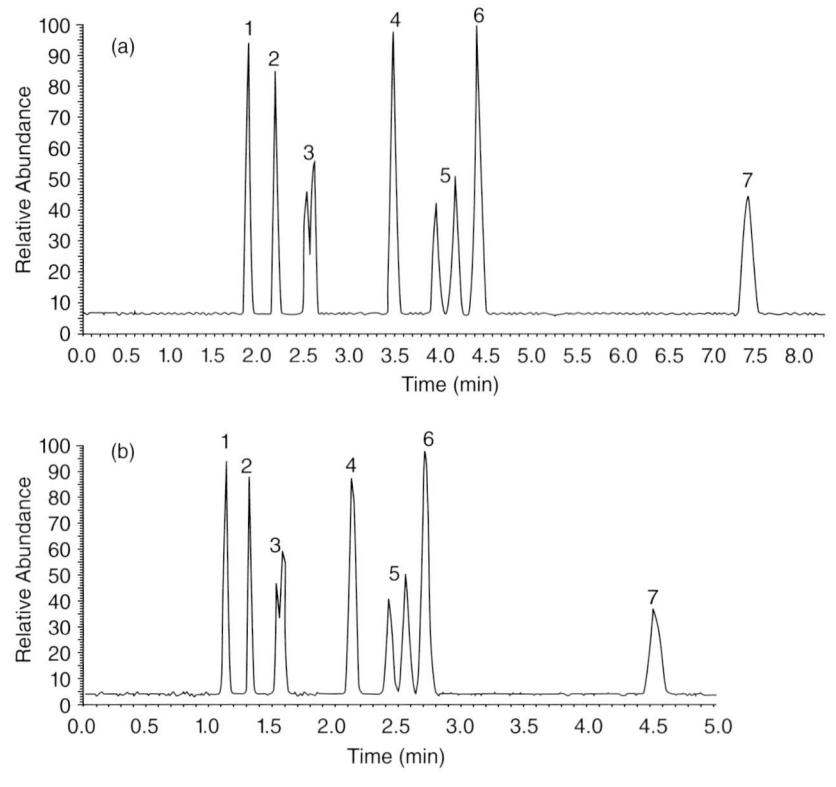

Figure 3.4 LC-ESI-MS base peak chromatograms of a mixture of mono-, di-, and trisaccharides obtained in 80% acetonitrile (13 mM ammonium acetate), flow rate 0.2 ml min^{-1}; linear velocities (a) 3.9 mm s^{-1} (split ratio 34:1), and (b) 7.6 mm s^{-1} (split ratio 17:1); polyacrylamide-bonded (PAAm) monolithic silica capillary column 267 mm × 200 μm I.D.; detection, 2.2 kV negative ESI-ion-trap MS. Peak identification: 1, ribose; 2, sedoheptulose; 3, glucose; 4, sucrose; 5, maltose; 6, trehalose; 7, raffinose. Reproduced from Ref. [21] with the permission of Springer.

various functionalities onto silica surfaces modified with methacryloyl groups [17]. In this procedure, monolithic silica capillary columns modified with 3-(methacryloxypropyl)trialkoxysilanes, or 3-(methacrylamidopropyl)trialkoxysilanes can be used as "universal platforms". This anchoring process can be achieved by passing through the capillary column with a 1:1 (v/v) mixture of anchor group: pyridine at 80 °C for 24 h, followed by washes with toluene and methanol for 24 h (the procedure was repeated twice to obtain the monolithic silica capillary columns modified with the corresponding functionality). The polymerization reaction is then quite simple: a solution containing the vinyl monomer with the adequate functional group, and a radical initiator such as AIBN or ammonium peroxydisulfate (APS), was charged into the column, and allowed to react at 60–80 °C. Then, the polymers that did not bind to the column are conveniently washed out, and

3.2 Stationary Phases for Hydrophilic-Interaction Chromatography Separations | 41

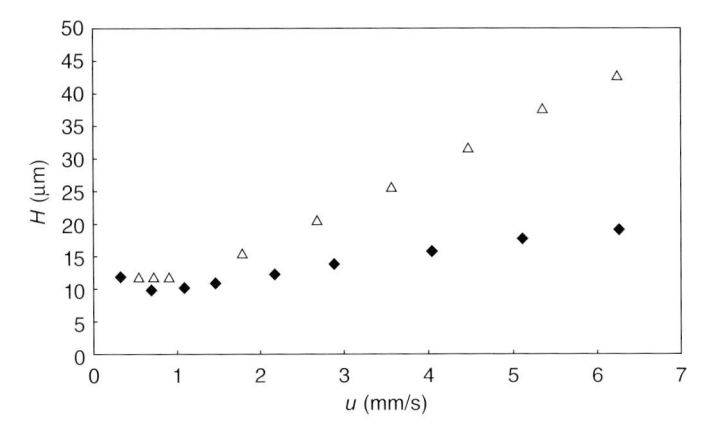

Figure 3.5 The van Deemter plots obtained for HILIC mode elution. (♦) Poly(acrylic acid)-bonded-monolithic silica capillary column (200 mm × 200 μm I.D.) in 90% acetonitrile/(0.2% formic acid) at 25 °C. Solute: adenosine ($k = 3.8$). Injection volume: 2 μl split injection. (△) ZIC-HILIC column, 150 mm × 4.6 mm I.D., 5 μm particles in 80% acetonitrile/buffer at 25 °C. Solute: cytosine ($k = 1.3$). Reproduced from Ref. [19] with the permission of Elsevier.

a HILIC-mode polymer-coated monolithic silica capillary column was obtained [18–21]. As in the case of reversed-phase chromatography stationary phases, one advantage of the modification by on-column polymerization on the monolithic silica is that the amount of the polymer (the stationary phase) can be controlled by changing the concentration of monomers in the reaction mixture.

Starting from the modification of silica using MOP or MAS, some vinyl monomers such as acrylamide [18, 21] and acrylic acid [19, 20] were used to obtain HILIC-type monolithic silica capillary columns. In an early step of this research topic, Ikegami et al. prepared 100 μm I.D. × 38 cm polyacrylamide monolithic silica capillary columns that showed HILIC-mode retention characteristics with three times greater permeability and slightly higher column efficiency compared to a commercially available amide-type HILIC column packed with 5 μm particles [18]. This polyacrylamide column generated a column efficiency of $H = 16$ μm for uridine in 90% methanol mobile phase (the bare silica showed $H = 10$ μm). After the improvement of the silica support to increase phase ratio [5], this polyacrylamide-type column could generate $H = 7$ μm for an optimal flow rate of 1 mm s^{-1} [21]. Figure 3.4 shows as an example the chromatographic separation of oligosaccharides in HILIC mode using the above-mentioned polyacrylamide monolithic silica capillary column by liquid chromatography–electrospray ionization–mass spectrometry (LC-ESI-MS), using an ion-trap MS analyzer [21].

Horie et al. prepared 200 μm I.D. poly(acrylic acid) monolithic silica capillary columns with a total performance much higher than the conventional particle-packed HILIC columns currently available [19]. For instance, Figure 3.5 shows the van Deemter plot obtained for adenosine ($k = 3.8$) with a 20-cm poly(acrylic acid)-modified monolithic silica capillary column in 90% acetonitrile at 25 °C. At the optimum linear velocity, the theoretical plate height H was less than 10 μm, and the H value was kept below 20 μm even at $u = 6$ mm s^{-1}, demonstrating that the column is suitable for fast separations, and showing much better results than those observed for a well-known commercially available HILIC column packed with 5-μm particles, ZIC-HILIC (150 mm × 4.6 mm I.D.). The high performance observed with the monolithic silica capillary column modified with poly(acrylic acid) suggests that the HILIC mode can be an alternative to the reversed-phase chromatography mode for a wide range of compounds, especially for those of high polarity in isocratic as well as gradient elution. This type of stationary phase could be used for a weak cation-exchange phase and a HILIC phase to separate proteins, peptides, nucleosides, and oligosaccharides [20]. Some monolithic silica columns modified with cation- and anion-exchange stationary phases [22, 23], that will be discussed in the next section, can also be used as HILIC phases, because they will easily form water-rich phases for partitioning equilibrium between mobile and stationary phases. HILIC mode separations using the above-mentioned monolithic columns modified with polar polymers will provide more efficient and faster separation than particle-packed columns.

3.3
Stationary Phases for Ion-Exchange Separations

The use of porous monolithic stationary-phase media for high-performance separations of inorganic and organic ions has increased in 2007, as it can be reflected by the number of reviews based upon the application of monolithic stationary phase for ion-exchange chromatography in both HPLC and capillary electrochromatography that have recently appeared [22, 24]. An area of emerging interest is the modification of monolithic silica columns to make them useful for ion analysis, although quite a few publications are based on polymer-coated materials.

As commented before, Ikegami *et al.* prepared an HILIC monolithic silica capillary column by on-column polymerization of acrylic acid [20]. This column was also reported to be used in weak cation-exchange (WCX) mode for the separation of proteins and nucleosides even at high linear velocity. The column provided fair permeability after the polymer-coating step. A high-speed separation of proteins at $u = 4.66$ mm s^{-1} with high column efficiency was possible, they achieved a high-speed separation of nucleosides within one minute using the column at $u = 8.67$ mm s^{-1}, a fact that suggests the suitability of this column to be used as a second separation dimension in multidimensional HPLC systems.

Following a similar modification procedure, highly efficiency anion- and cation-exchange micro-HPLC columns modified by on-column polymerizations of methacrylates possessing amine or sulfonic acid functional groups to prepare strong cation exchange (SCX), strong anion exchange (SAX), and weak anion exchange (WAX) modes were reported [23]. The authors also described the preparation of a mixed-mode anion/cation exchange column prepared from a step-by-step functionalization using polymerization reactions after obtaining a monolithic silica capillary column bonded with the MAS anchor. First, polymerization was performed with a mixture of *p*-styrenesulfonic acid (pSSA) monomer and ammonium persulfate as initiator at 50 °C. After washing out the remaining reaction mixture with water and methanol for 24 h, the column was filled with the second polymerization mixture, 3-diethylamino-2-hydroxypropyl methacrylate (DAHMA) monomer and AIBN as initiator, and allowed to react at 70 °C. The performance of this new mixed-mode pSSA/DAHMA polymer-coated column was compared to that of two monolithic capillary columns (a pSSA column and a DAHMA column) connected in series to prepare a new column, by studying the separation of four nucleotides and four nucleic bases. When the two polymer-coated columns were connected in series the column efficiency was kept high; the total plate heights of each peak were in the range of 18–36 μm (Figure 3.6a). With the mixed-mode pSSA/DAHMA polymer-coated column plate heights for each peak in the range of 27–50 μm were obtained (Figure 3.6b), although retention factors of the employed samples were not the same as with the connected column. It is obvious that improvement of the column preparation in this way is needed, but this result shows a potential of the column preparation by the step-by-step polymer coating using different monomer to obtain several mixed-mode columns, for example, RP-SCX, RP-WCX, RP-SAX, or RP-WAX, combinations that were effective to separate complex mixtures [25, 26].

Recently, Jafariah *et al.* reported the preparation of a strong anion exchange stationary phase by on-column copolymerization of *N*-[3-(dimethylamino)propyl] acrylamide methyl chloride-quaternary salt (DMAPAA-Q) with MOP bonded monolithic silica column [27]. The polymerization was carried out using APS as an initiator at 60 °C for 2 h. The performance of this DMAPAA-Q polymer-coated column resulted to be much better than that of a *N*-[2-(dimethylamino)ethyl] acrylate methyl chloride-quaternary salt (DMAEA-Q) bonded to monolithic silica reported by Ikegami *et al.* [23]. The column showed plate heights in the range between 9 and 14 μm for different anions, being column efficiencies considerably higher than those of other types of ion exchange capillary columns reported at the literature [28]. Under a capillary electrochromatography (CEC) mode, the column provided far better separation efficiency of $H = 4\,\mu m$.

3.4
Advantages of Polymer-Coated Monolithic Silica Columns

One of the advantages of the polymerization methods to functionalize porous silica is that these procedures are easily applicable for monolithic silica columns as the modification is performed *in situ* in most of the cases. This modification allows one to maintain the original efficiency after the polymer-coated bonding, unless polymers caused a significant reduction in mass transfer.

In general, this method for the stationary phase preparation has several advantages over the preparation of particle-packed columns as follows: (1) the MOP or

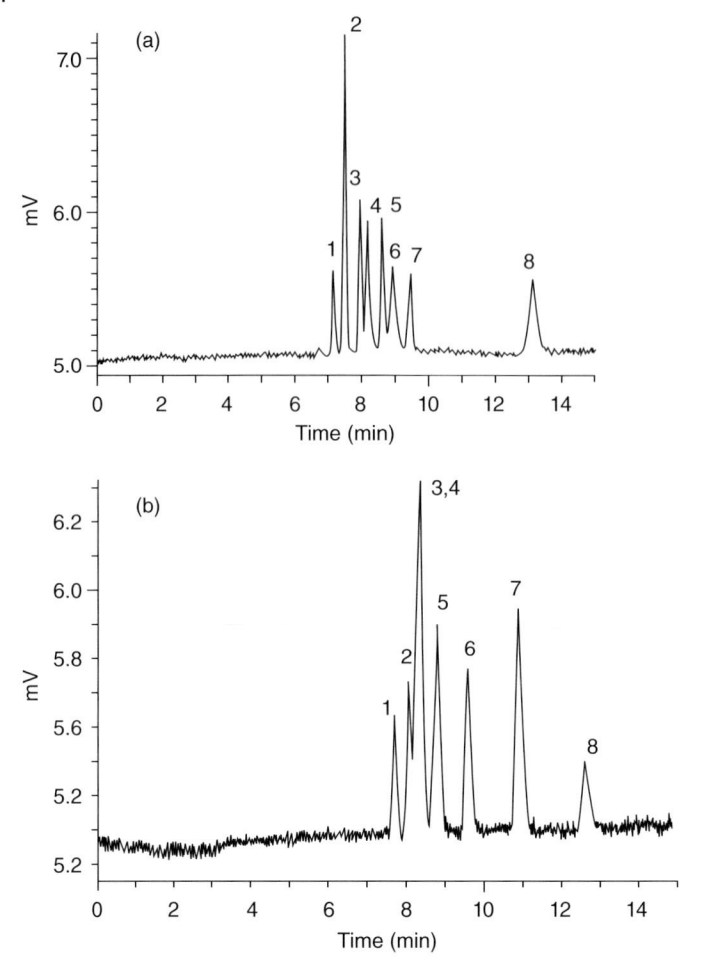

Figure 3.6 Separation of nucleic bases and nucleotides using mixed-mode capillary columns. Chromatographic conditions: mobile phase, 0.05 M $(NH_4)_2HPO_4$ (pH 3.0); detection, $\lambda = 254$ nm; temperature, ambient. (a) DAHMA polymer-coated monolithic silica capillary column (200 μm I.D. × 300 mm) + pSSA polymer-coated monolithic silica capillary column (200 μm I.D. × 300 mm), connected using microtight union; pressure, 6.6 MPa. (b) Mixed-mode pSSA/DAHMA polymer-coated monolithic silica capillary column (200 μm I.D. × 600 mm); pressure, 5.9 MPa. Peak identification: 1, 5′-CMP; 2, uracil; 3, thyamine; 4, cytosine; 5, adenine; 6, 5′-AMP; 7, 5′-UMP; 8, 5′-GMP. Reproduced from Ref. [23] with the permission of the Japan Society of Analytical Chemistry.

MAS modified monolithic silica column can be used as universal platforms, as it has been described before, and various stationary phases with different functionalities can be prepared by changing only the monomer; (2) there is no need to prepare silanes possessing highly reactive functional groups; (3) a good permeability and column efficiency can be maintained after the polymerization, and if a reduction in column efficiency is observed with polymers of high molecular weight, it can be improved by increasing initiator concentration during preparation [13]; (4) it is free from column clogging, preventing the agglomeration of particles that could occur when monomers are polymerized in the presence of silica particles; (5) the amount of bonded phase can be controlled by the polymerization conditions, either monomer or initiator concentration, modeling in this way the retention properties of the column; (6) the phase ratio can be increased compared to the silica modification using silylating reagents.

References

1. Tanaka, N., Konishita, H., Araki, M., and Tsuda, T. (1985) On-column preparation of chemically bonded stationary phase with maximum surface coverage and high reproducibility, and its application to packed microcapillary columns. *J. Chromatogr.*, **332**, 57.
2. Ishizuka, N., Kobayashi, K., Minakuchi, M., Nakanishi, K., Hirao, K., Hosoya, K., Ikegami, T., and Tanaka, N. (2002) Monolithic silica columns for high-efficiency separations by high-performance liquid chromatography. *J. Chromatogr. A*, **960**, 85.
3. Motokawa, M., Kobayashi, H., Ishizuka, N., Minakuchi, M., Nakanishi, K., Jinnai, H., Hosoya, K., Ikegami, T., and Tanaka, N. (2002) Monolithic silica columns with various skeleton sizes and through-pore sizes for capillary liquid chromatography. *J. Chromatogr. A*, **961**, 53.
4. Motokawa, M., Ohira, M., Minakuchi, H., Nakanishi, K., and Tanaka, N. (2006) Performance of octadecylsilylated monolithic silica capillary columns of 530μm inner diameter in HPLC. *J. Sep. Sci.*, **29**, 2471.
5. Hara, T., Kobayashi, H., Ikegami, T., Nakanishi, K., and Tanaka, N. (2006) Performance of monolithic silica capillary columns with increased phase ratios and small-sized domains. *Anal. Chem.*, **78**, 7632.
6. Rieux, L., Lubda, D., Niederlander, H.A.G., Verpoorte, E., and Bischoff, R. (2006) Fast, high-efficiency peptide separations on a 50-μm reversed-phase silica monolith in a nanoLC–MS set-up. *J. Chromatogr. A*, **1120**, 165.
7. Cabrera, K., Lubda, D., Eggenweiler, H.-M., Minakuchi, H., and Nakanishi, K. (2000) A new monolithic-type HPLC column for fast separations. *J. High Resolut. Chromatogr.*, **23**, 93.
8. Yang, C., Ikegami, T., Hara, T., and Tanaka, N. (2006) Improved endcapping method of monolithic silica columns. *J. Chromatogr. A*, **1130**, 175.
9. Ishizuka, N., Minakuchi, H., Nakanishi, K., Soga, N., Hosoya, K., and Tanaka, N. (2000) Performance of a monolithic silica column in a capillary under pressure-driven and electrodriven conditions. *Anal. Chem.*, **72**, 1275.
10. Minakuchi, H., Nakanishi, K., Soga, N., Ishizuka, N., and Tanaka, N. (1997) Effect of skeleton size on the performance of octadecylsilylated continuous porous silica columns in reversed-phase liquid chromatography. *J. Chromatogr. A*, **762**, 135.
11. Ansarian, H.R., Derakhshan, M., Rahman, M.M., Sakurai, T., Takafuji, M., Taniguchi, T., and Ihara, H. (2005) Evaluation of microstructural features of a new polymeric organic stationary phase grafted on silica surface: a paradigm of characterization of HPLC-stationary phases by a combination of suspension-state 1H NMR and solid-state 13C-CP/MAS-NMR. *Anal. Chim. Acta*, **547**, 179.
12. Derakhshan, M., Ansarian, H.R., Rahman, M.M., Sakurai, T., Takafuji, M., and Ihara, H. (2005) A new method for evaluation of the mobility of silica-grafted alkyl chains by suspension-state ^{1}H NMR. *Can. J. Chem.*, **83**, 1792.

13. Núñez, O., Ikegami, T., Kajiwara, W., Miyamoto, K., Horie, S., and Tanaka, N. (2007) Preparation of high efficiency and highly retentiva monolithic silica capillary columns for reversed-phase chromatography by chemical modification by polymerization of octadecyl methacrylate. *J. Chromatogr. A*, **1156**, 35.

14. Núñez, O., Ikegami, T., Miyamoto, N., and Tanaka, N. (2007) Study of monolithic silica capillary column coated with poly(octadecyl methacrylate) for the reversed-phase liquid chromatographic separation of some polar and non-polar compounds. *J. Chromatogr. A*, **1175**, 7.

15. Stecher, G., Huck, C.W., Stöggl, W.M., and Bonn, G.K. (2003) Phytoanalysis: a challenge in phytomics. *Trends Anal. Chem.*, **22**, 1.

16. Szabó, Z., Ohmacht, R., Huck, C.W., Stöggl, W.M., and Bonn, G.K. (2005) Influence of the pore structure on the properties of silica based reversed phase packings for LC. *J. Sep. Sci.*, **28**, 313.

17. Ikegami, T., Tomomatsu, K., Takubo, H., Horie, K., and Tanaka, N. (2008) Separation efficiencies in hydrophilic interaction chromatography. *J. Chromatogr. A*, **1184**, 474.

18. Ikegami, T., Fujita, H., Horie, K., Hosoya, K., and Tanaka, N. (2006) HILIC mode separation of polar compounds by monolithic silica capillary columns coated with polyacrylamide. *Anal. Bioanal. Chem.*, **386**, 578.

19. Horie, K., Ikegami, T., Hosoya, K., Saad, N., Fiehn, O., and Tanaka, N. (2007) Highly efficient monolithic silica capillary columns modified with poly(acrylic acid) for hydrophilic interaction chromatography. *J. Chromatogr. A*, **1164**, 198.

20. Ikegami, T., Horie, K., Jaafar, J., Hosoya, K., and Tanaka, N. (2007) Preparation of highly efficient monolithic silica capillary columns for the separations in weak cation-exchange and HILIC modes. *J. Biochem. Biophys. Methods*, **70**, 31.

21. Ikegami, T., Horie, K., Saad, N., Hosoya, K., Fiehn, O., and Tanaka, N. (2008) Highly efficient analysis of underivatized carbohydrates using monolithic-silica-based capillary hydrophilic interaction (HILIC) HPLC. *Anal. Bioanal. Chem.*, **391**, 2533.

22. Chambers, S.D., Glenn, K.M., and Lucy, C.A. (2007) Developments in ion chromatography using monolithic columns. *J. Sep. Sci.*, **30**, 1628.

23. Ikegami, T., Ichimaru, J., Kajiwara, W., Nagasawa, N., Hosoya, K., and Tanaka, N. (2007) Anion- and cation-exchange MicroHPLC utilizing poly(methacrylates)-coated monolithic silica capillary columns. *Anal. Sci.*, **23**, 109.

24. Schaller, D., Hilder, E.F., and Haddad, P.R. (2006) Monolithic stationary phases for fast ion chromatography and capillary electrochromatography of inorganic ions. *J. Sep. Sci.*, **29**, 1705.

25. Link, A.J., Eng, J., Schieltz, D.M., Carmack, E., Mize, G.J., Morris, D.R., Garvik, B.M., and Yates, J.R., III (1999) Direct analysis of protein complexes using mass spectrometry. *Nat. Biotechnol.*, **17**, 676.

26. Nogueira, R., Laemmerhofer, M., and Lindner, W. (2005) Alternative high-performance liquid chromatographic peptide separation and purification concept using a new mixed-mode reversed-phase/weak anion-exchange type stationary phase. *J. Chromatogr. A*, **1089**, 158.

27. Jaafar, J., Watanabe, Y., Ikegami, T., Miyamoto, K., and Tanaka, N. (2008) Anion exchange silica monolith for capillary liquid chromatography. *Anal. Bioanal. Chem.*, **391**, 2551.

28. Zakaria, P., Hutchinson, J.P., Avdalovic, N., Liu, Y., and Haddad, P.R. (2005) Latex-coated polymeric monolithic ion-exchange stationary phases. 2. micro-ion chromatography. *Anal. Chem.*, **77**, 417.

Part Two
Characterization and Modeling

4
Characterization of the Pore Structure of Monolithic Silicas
Romas Skudas, Matthias Thommes, and Klaus K. Unger

4.1
Monolithic Silicas

The silica-based monoliths are composed of a single porous piece and exhibit two types of pores: flow-through pores in the micrometer (μm) size range and mesopores in the nanometer (nm) size range (see Figure 4.1). This structure allows high accessibility of the stationary surface through the micrometer-size flow-through pores (the spaces between the skeleton of the stationary phase) and the rapid mass transfer of the molecules in the diffusional mesopores being embedded in the skeleton of the monoliths. This concept was already applied in heterogeneous catalysis, where monoliths are used as catalyst supports, enabling fast diffusion of reactants to the active catalyst surface, binding the reactants quickly and effectively, stabilizing the activated complex, and releasing the products of the reaction, while assuring a high reactivity through a high accessibility. In liquid chromatography (LC), where monoliths are used as stationary phases, this structure enables fast mass transfer of molecules to the functionalized surface, where the separation takes place.

The ability to control and to adjust both pore-size regimes permits monolithic silicas to be obtained with desirable flow-through pore sizes (as chromatographic supports in the range of 0,5 μm to 6 μm), and mesopore sizes (as chromatographic supports in the range of 10 to 30 nm) [1–13]. Further, in LC the mesopores must be adapted to the size and shape of the analytes to ensure and to facilitate the diffusion. For example, favorable LC separation of peptide samples is achieved on increasing the mesopore diameter from 12 nm to 25 nm [13]. The adjustment of pore structural parameters of the monolithic silicas provides a unique opportunity to design stationary phases for targeted liquid-separation applications with optimum hydrodynamic properties and mass-transfer kinetics [14]. The major challenge is to reliably characterize the pore structural parameters of monolithic silicas and to link these parameters with the hydrodynamic and kinetic properties, for example, the column permeability, the column efficiency, the speed of analysis and other target parameters. Primarily, it is necessary to address the major methods used at present to characterize the bimodal pore structure of

Monolithic Silicas in Separation Science. Edited by K.K. Unger, N. Tanaka, and E. Machtejevas
© 2011 WILEY-VCH Verlag GmbH & Co. KGaA, Weinheim
ISBN: 978-3-527-32575-7

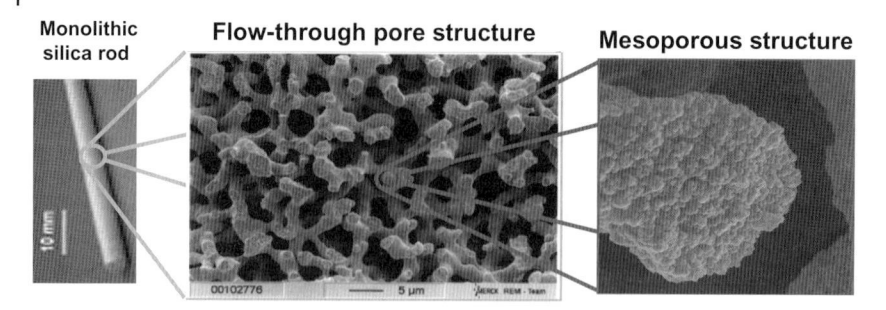

Figure 4.1 The structure of monolithic silica as a rod with the bimodal pore structure: flow-through pore structure, and mesopore structure in the silica skeleton.

monolithic silicas, namely: mercury intrusion-extrusion porosimetry, nitrogen sorption, inverse size-exclusion chromatography, permeation of a liquid, imaging and image analysis, and other alternative techniques and to assess the reproducibility and the validity and of these methods.

4.2
General Aspects Describing Porous Materials

The pore is a small opening or cavity in a solid substance. Based on the diameter of the pore, they are classified in three major groups: micro- <2 nm, meso- 2–50 nm, and macropores >50 nm (see Figure 4.2). Each pore has a certain void space which is defined by the pore volume. In porous oxides micropores exhibit low specific pore volumes (usually smaller then $0.2\,cm^3/g$), mesopores posses moderate specific pore volumes (in the range between $0.5\,cm^3/g$ to $1.5\,cm^3/g$), whereas macropores generate higher specific pore volumes. The sum of the specific pore volumes is expressed as material porosity. It is a measure of the void spaces in a material and is a fraction of the volume of voids in total volume, where the void may contain, for example, air or water. It is important to distinguish between the pores that form a continuous phase within the porous medium (interconnected or effective pore space), and the ones that are isolated or noninterconnected voids, that are dispersed over the medium. The isolated void or pore space does not contribute to transport across the porous medium, while only effective pore space does.

The monolithic silicas exhibit bimodal pores, macropores (flow-through pores) and mesopores (pores in the silica skeleton). Therefore, the macropore volume is referred to as macroporosity or external porosity – the porosity of the flow-through pores, and mesopore volume is referred to as the mesoporosity or internal porosity – the porosity of pores in the silica skeleton. The total silica monolith porosity combines both of the porosities, internal and external. The total porosity is a measure of a fraction of the total volume of all voids over the total volume and as a percentage is between 0–100%. Totally nonporous supports exhibit a 0% total

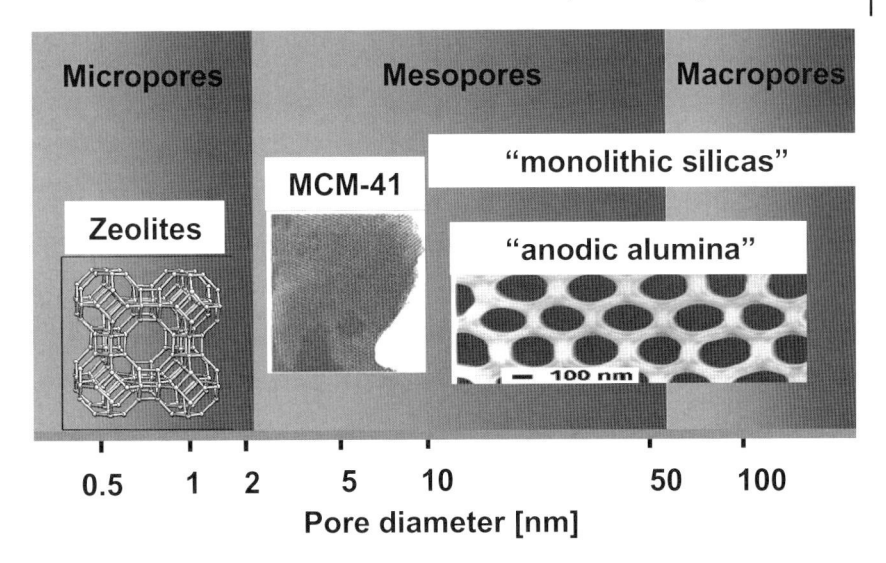

Figure 4.2 Some materials with characteristic pore sizes.

porosity, while simply void or air would be 100% porous. Typically, the total porosity ranges from less than 1% for the solid granite, to more than 50% in clay and even to 95% in monolithic silicas. The specific porosity is related to the specific surface area. Porous oxides with micropores have high specific surface areas of >500 m^2/g, while those with mesopores have moderate specific surface areas between 100 and 500 m^2/g, and macropores oxides have specific surface areas of <50 m^2/g. The monolithic silicas exhibit specific surface areas of 300 m^2/g due to the large amount of mesopores. The specific surface area plays an important role in a variety of applications of porous media. It is a measure of the adsorption capacity and is decisive for the effectiveness of catalysts, ion-exchange columns, filters, etc. It is related to the fluid conductivity or permeability of porous media.

The specific surface area, pore diameter and specific pore volume are interrelated and, therefore, can be varied independently only to a certain extent.

$$p_d = \gamma(v_p / a_s) \tag{4.1}$$

where p_d is the average pore diameter, γ is the obstruction factor for diffusion or external tortuosity, v_p is the volume of the pore system, and a_s is the specific surface area.

The pore structure of an adsorbent is characterized not only by the specific surface area, the pore-size distribution (expressed as number or volume distribution) and specific pore volume, but also by the pore shape and pore connectivity.

Pores are classified as closed pores and opened pores. Since closed pores do not participate in adsorption desorption phenomena, that is discussed here, we will not give any further attention to those. The open pores can be classified into

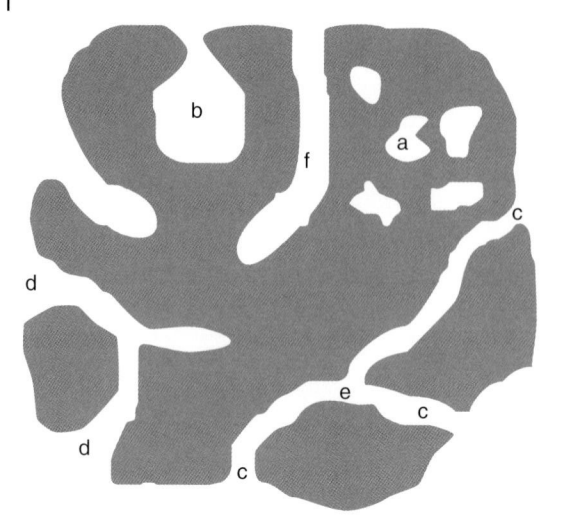

Figure 4.3 The different shape of pores: a–closed pores, b–blind ink-bottle-shaped pores, c–connected, cylindrical pores, d–connected, cone-shaped pores, e–interconnection point of pores, f–blind cylindrical pores.

two main groups: blind pores and connected pores. The shape of pores can be cylindrical, slit-shaped, ink-bottle-shaped, cone-shaped, etc (see Figure 4.3).

The most favorable pores for the mass-transfer kinetics in the porous material are highly connected cylindrical pores. The most unfavorable are blind ink-bottle-shaped pores that usually contribute only negligibly to the mass transport.

Pore network is a term or measure of how the pores are interconnected across a porous adsorbent. The most useful parameter in this context is the (dimensionless) pore connectivity, n_T, which is derived from the pore-network model applied to the experimental isotherm [15, 16]. Pore connectivity describes the number of pore channels meeting at a node at a fixed lattice size. High n_T values indicate a high interconnectivity, which leads to fast mass-transfer kinetics and favorable mass-transfer coefficients [17]. Since the pore texture of monolithic silica is a coherent flow-through pore system with mesopores as primary pores that are connected and accessible through the flow-through pores, it is essential to measure the connectivity of the mesopores.

While high mesopore connectivity values assure fast mass-transfer kinetics in mesopores, the permeability is a measure of external access. The permeability of monolithic silicas is directly dependent on the ratio of the external pore volume and the external surface area. Monolithic silicas have large external porosities (the void volume of flow-through pores) and moderate external surface areas (the surface area of the flow-through pores). Therefore they are highly permeable, which favors the bulk diffusion of the fluid through such a solid structure. The molecules have more space to move around in and less surface area, where their free movement is limited. They lose moving speed the closer they are to the wall.

All named pore characteristic parameters are important in understanding and optimization of any process. For example, a decrease in flow resistance is reached through the increase of the flow-through pore size. There are many suggestions of the characteristic pore parameters that influence the performance of LC [18–22] but a discussion on their impact would be fairly meaningless if the characterization methods to monolithic silicas would provide irrelevant and unreliable data.

4.3
Characterization Methods of the Pore Structure of Monolithic Silicas

There is a number of characterization techniques commonly used to assess the pore structural parameters of monolithic silicas containing flow-through pores (i.e. pores of width >50 nm), *and mesopores* (i.e. pores of width >3 nm). The most common methods for flow-through pore analysis are mercury intrusion-extrusion porosimetry, permeation of a liquid, imaging and image analysis. Mercury intrusion-extrusion porosimetry is one of the liquid-intrusion techniques [23, 24], that provides information about the specific surface area, pore-size distribution (from 40 μm to 4 nm size range), and is used to assess parameters such as the tortuosity, permeability, fractal dimension and compressibility of porous materials.

Liquid permeability enables the assessment of the external porosity and permeability characteristics [25–27], while microscopy is a valuable tool for the investigation of the flow-through pore diameter and the skeleton diameter of monolithic silicas [28–30]. Recent development in the three-dimensional (3D) images of porous structures from magnetic resonance imaging [31, 32] and microscopic images enable the estimation of several very complex parameters, such as the chromatographic separation efficiency [20], domain-size-induced heterogeneity effects [21, 30], optimal external porosity effects [33], and provide a correlation with the column pressure drop [34].

The usual methods of choice for the mesopore characterization are gas adsorption (typically nitrogen, argon, carbon dioxide) and inverse size-exclusion chromatography (ISEC). Mercury porosimetry (MP) can also be applied for the characterization of mesopores of >4 nm.

The nitrogen adsorption–desorption isotherm at 77 K provides information on the specific surface area, the specific pore volume and the mesopore-size distribution in the range between 3–200 nm [35–37]. On the other hand, ISEC [38–41] is a valuable chromatographic technique used for the estimation of mesopore structural parameters based on the size separation of polymers. The porosity and mesopore-size distribution are obtained from a relationship between the silica monolith mesopore size and the molecular size of polymer standards.

4.3.1
Mercury Intrusion-Extrusion

Mercury porosimetry is based on the intrusion/extrusion of a nonwetting liquid, that is, the contact angle is >90 degrees while applying pressure to force mercury into the pores. A progressive increase in hydrostatic pressure is applied to enter the pores in decreasing order of width. Correspondingly, there is an inverse relationship between the applied pressure p and the pore diameter p_d, which in the simplest case of cylindrical pores is given by the Washburn equation

$$p_d = -(4\gamma/p)\cos\theta \qquad (4.2)$$

where γ is the surface tension of mercury, p the pressure and θ the contact angle between the solid sample and mercury.

Generally, γ is assumed to be 484 mN m^{-1}, which is the surface tension of pure mercury at 303 K. The contact angle θ depends on the nature of the solid surface. It is measured for a mercury drop lying on a flat surface of a material of similar composition. The obtained value might be just approximate, since differences may exist in crystalline structure and purity between the flat surface and the pore wall. This uncertainty makes it desirable to distinguish between the main objectives: are the results gained in order to share and compare them (standard calculation procedure and a value of 140° is customarily used [42]), or in the scope of a special reasoning and comparison with other methods. It may be worthwhile to try other values of θ and even adjust it, but this should be clearly stated and justified.

For example, a contact angle θ of 145° was found on recent contact angle measurements of mercury on amorphous silica [43]. This value was used to obtain the pore-size distribution curves of monolithic silica samples and to compare it with the pore-size distribution curves obtained from mercury intrusion by applying the Washburn equation with an assumed standard value contact angle of 140°. As a reference, the pore-size distribution curve obtained by applying the BJH method [44], based on the Kelvin equation using the nitrogen desorption branch is given in Figure 4.4.

Similar results were reported for porous Vycor® and controlled-pore glass [45]. However, it is important to note that both the Washburn equation and the Kelvin equation (which is the basis for the BJH method) are macroscopic, thermodynamic approaches. It is meanwhile well known that these classic methods fail to describe correctly the adsorption and phase behavior of fluids in small mesopores, which leads to a significant underestimation of pore size (for pore widths that are smaller than ca. 20 nm [46–48]), but the given figure is a classical sample for the differences between objectives. For the comparison between different laboratories it is recommended to use an angle of 140°, for the comparison between methods – an angle of 145°.

In mercury porosimetry, the volume of mercury entering the pores is measured as the applied (hydrostatic) pressure is gradually increased. The value $v_i(Hg)$ at the applied pressure p_i apparently gives the cumulative volume of all available pores of radius equal to, or greater than, p_d.

With modern commercial equipment, the pressure can be increased from 0.003 MPa to 400 MPa corresponding to cylindrical pore diameters of 400 μm down to 3.6 nm (by assuming a contact angle of 140°). Allowance must be made for the compressibility and thermal effects of mercury, the hydraulic fluid, the sample, and for thermophysical elastic distortion of the sample cell. Good initial evacuation of the sample, purity of the mercury, and proper choice of equilibra-

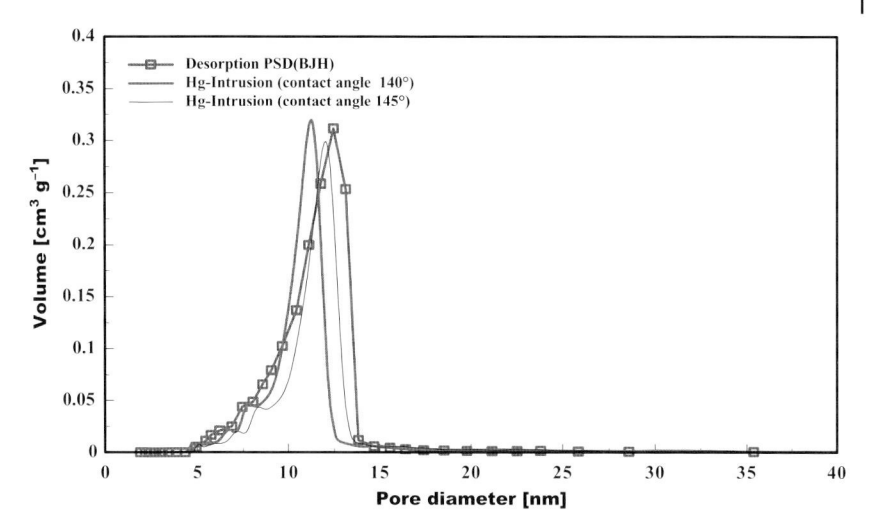

Figure 4.4 Pore-size analysis from the gas adsorption (BJH method applied to the desorption branch of nitrogen) and from the mercury porosimetry (Washburn equation calculated for contact angles of 140° and 145°).

tion times are also mandatory for a successful experiment and consequently a proper analysis of the sample.

Figure 4.5 shows an example of mercury intrusion into the pore system of a controlled-pore glass and the intrusion of mercury into a monolithic silica expressed as the cumulative specific pore volume as a function of the pore diameter. Region A_2 corresponds to a rearrangement of particles within the powder bed of controlled pore class, followed by intrusion of the interparticle voids (B_2), or spaces between the particles. The region A_1 in the monolithic silica sample stays constant until the pressure is reached for pressurizing mercury in the flow-through pores (B_1). After pressurizing mercury in the external pore volume, the pressure is increased to fill the internal pores (region C_1 and C_2), until the constant values are reached (region D_1 and D_2).

The significant improvements in the pore-size characterization with mercury porosimetry was not only supported by the discovery of novel highly ordered micromesoporous model substances such as MCM-41, MCM-48, SBA-15, which exhibit a uniform pore structure and morphology and can therefore be used as model adsorbents to test theories of adsorption, but also the application of methods, such as the nonlocal density functional theory (NLDFT) and computer-simulation methods (e.g., Monte-Carlo and molecular-dynamic simulations). These methods are based on statistical mechanics and allow the configuration of adsorbed molecules in pores to be described on a molecular level (e.g., [47, 48].).

Pore-size analysis data for microporous and mesoporous molecular sieves obtained with these novel methods agree very well with the results obtained from independent methods (based on X-ray diffraction [XRD]; transmission electron microscopy [TEM]), and allow characterization of a sample over the complete

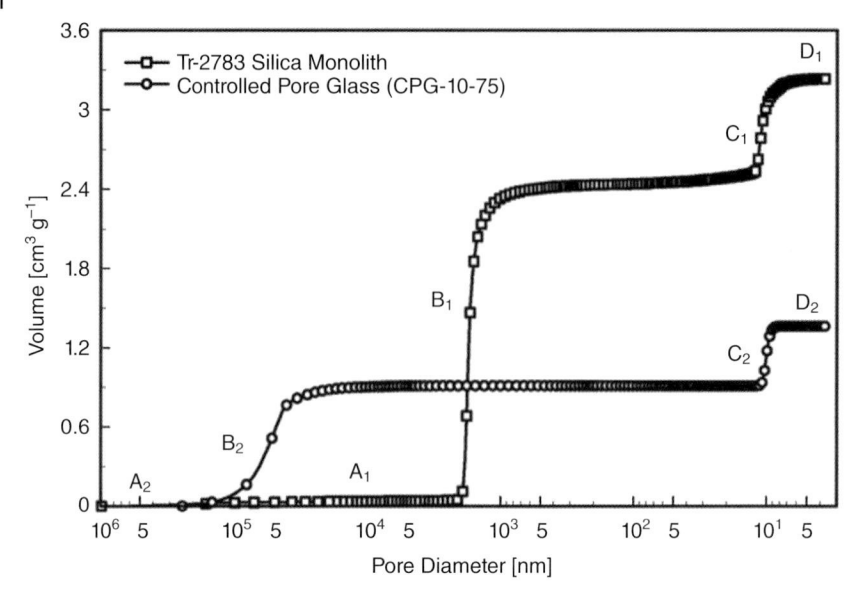

Figure 4.5 The intrusion cycle of mercury into the pore system of controlled-pore glass (CPG-10-75) and a native silica monolith (Tr2783/1). The cumulative specific pore volume is shown as a function of the pore diameter.

micropore/mesopore size range. Appropriate methods for pore-size analysis based on NLDFT and molecular simulation are meanwhile commercially available for many adsorptive/adsorbent systems. If applied to gas adsorption, it allows assessment of pore sizes from the micropore range (pore widths <2 nm) and mesopore range (pores widths between 2 and 50 nm).

If the extrusion of mercury occurs at different pressures from then the intrusion, the hysteresis is observed. On completion of a first intrusion–extrusion cycle, some mercury might be retained by the sample, thereby preventing the loop from closing (see Figure 4.6, controlled-pore glass sample). Intrusion–extrusion curves for the monolithic silica sample show a hysteresis between the intrusion and extrusion branch, but the loop closes, showing that there is no entrapment of mercury by the sample.

Very often, entrapment is observed, that is, mercury remains contained in the porous network, therefore an understanding of the hysteresis and entrapment phenomena is most important. Classically, the entrapment phenomenon is believed to be associated with kinetic effects during mercury extrusion, coupled with the tortuosity of the disordered pore network and the surface chemistry of the material [49]. Experiments with model pore networks and molecular-simulation studies appear to confirm that mercury entrapment is caused by the rupture of mercury bridges in pore constrictions during extrusion, leading to mercury entrapment in ink-bottle pores [50–52].

This is in agreement with recent studies involving grand canonical Monte Carlo simulations using both Glauber dynamics and Kawasaki dynamics [45, 52–64],

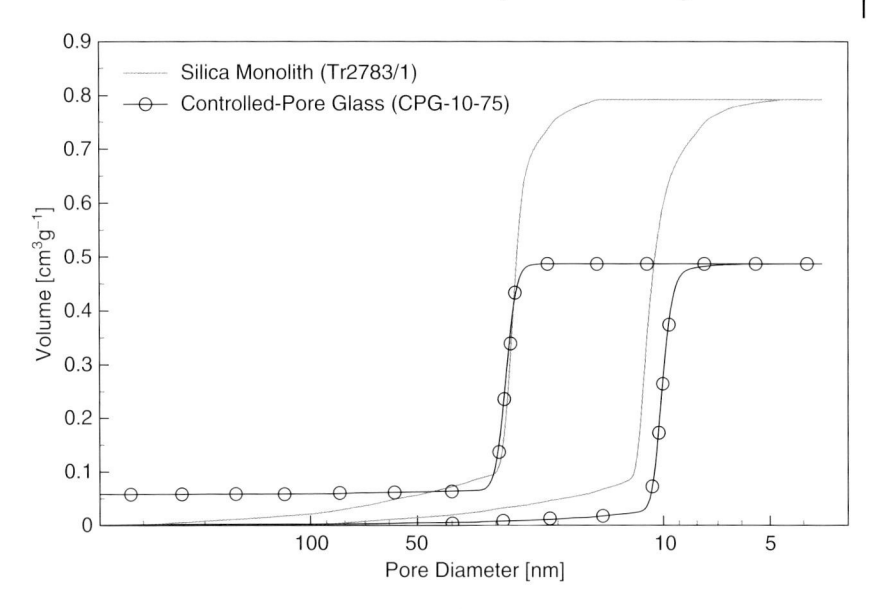

Figure 4.6 Mercury-intrusion–extrusion curves into the mesopore systems of controlled-pore glass (CPG-10-75) and a native silica monolith (Tr2783/1; please note that only the mesopore system is shown here). The specific pore volume is shown as a function of the pore diameter.

which suggest that mercury entrapment and hysteresis are, in principle, of different origin. The entrapment is caused by a decrease in the rate of mass transfer associated with the fragmentation of liquid during extrusion, or in other words due to the differences in advancing and receding contact angles plus the percolation effects involved in the desorption of gases from porous networks. This leads to a configuration where droplets of mercury are surrounded by a vapor phase. The fragmentation slows down the rate of mass transfer of fluid from the porous material. It reflects a mechanism of evaporation of liquid from the entrapped droplets and diffusion of this vapor to the external surface. The pore-blocking/percolation effects are dominant in disordered pore networks, and a reliable pore-size distribution can only be calculated from the intrusion branch by applying complex network models based on percolation theory. The application of such models also allows one to obtain a limited amount of structural information from the intrusion–extrusion hysteresis loop.

The striking feature of the intrusion–extrusion behavior into the sample of a native monolithic silica is the fact that they do not show any appreciable amount of entrapment in the flow-through pores as is given in Figure 4.7.

The amount of entrapped mercury also indicates that the flow-through porous framework of some samples (e.g., TG36/2 in Figure 4.8) that showed moderate amount of entrapment appears to be much more heterogeneous/disordered as compared to the monolithic silica samples that do not show entrapment (Figure 4.7). As displayed in Figure 4.8, no entrapment occurs for monoliths Tr2783/1 and Tr2783/2 that have almost an identical flow-through pore system with regard

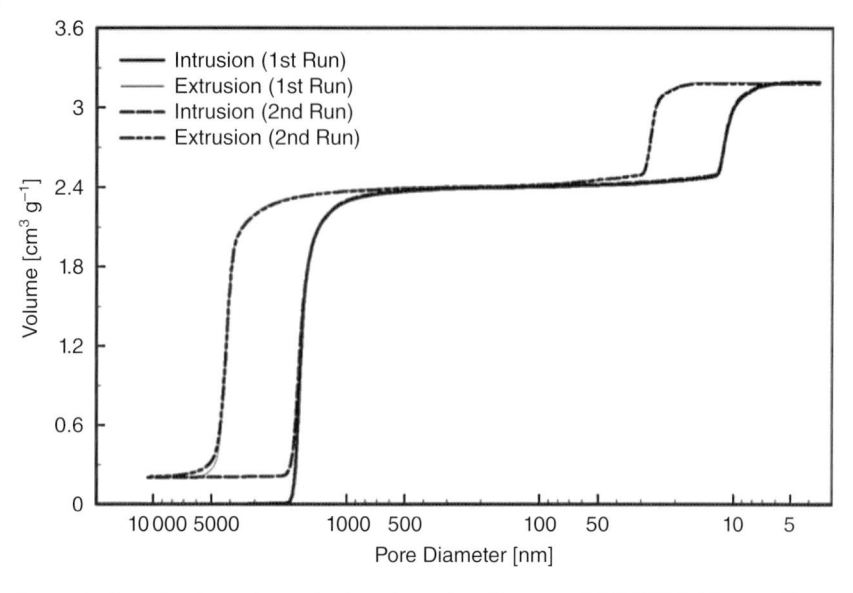

Figure 4.7 Intrusion/extrusion cycles into the native silica monolith Tr2783/1. The specific pore volume is displayed as a function of pore diameter.

to porosity and pore-size distribution, but mercury entrapment is observed in case of the TG36-2 monolith, which exhibits higher flow-through pore sizes and a more disordered framework.

Mercury porosimetry allows one not only to characterize native silica samples, but also to characterize samples with various surface functionalities. An example is given in Figure 4.9 on the cumulative specific pore-volume curves from mercury-porosimetry measurement performed on a native silica monolith (Tr2783/1) and the same sample with grafted n-octadecyl chains on the surface (Fr787).

It clearly reveals that the effective specific pore volume of the functionalized monolithic silica samples is significantly decreased for both the macro- and the mesopore system as a consequence of the grafting. Further, it appears that the shape of the intrusion/extrusion hysteresis has become more of type H2 as compared to the native silica monoliths indicating a wider pore-size distribution. The inspection of the mercury intrusion–extrusion curves reveal that the macropore-size distribution is essentially not affected by the grafting with n-alkyl groups (the small shift to larger values might be due to an effective change in the contact angle of mercury in contact with the grafted surface as compared to a pure silica surface) whereas the mode pore size for the mesopores has been slightly shifted to smaller values (here the reduction in effective pore size due to the surface groups is significant, despite the possibility of a contact-angle change).

The results of a systematic study of the mercury intrusion/extrusion behavior into native and monolithic silicas with bonded n-alkyl groups reveals that the

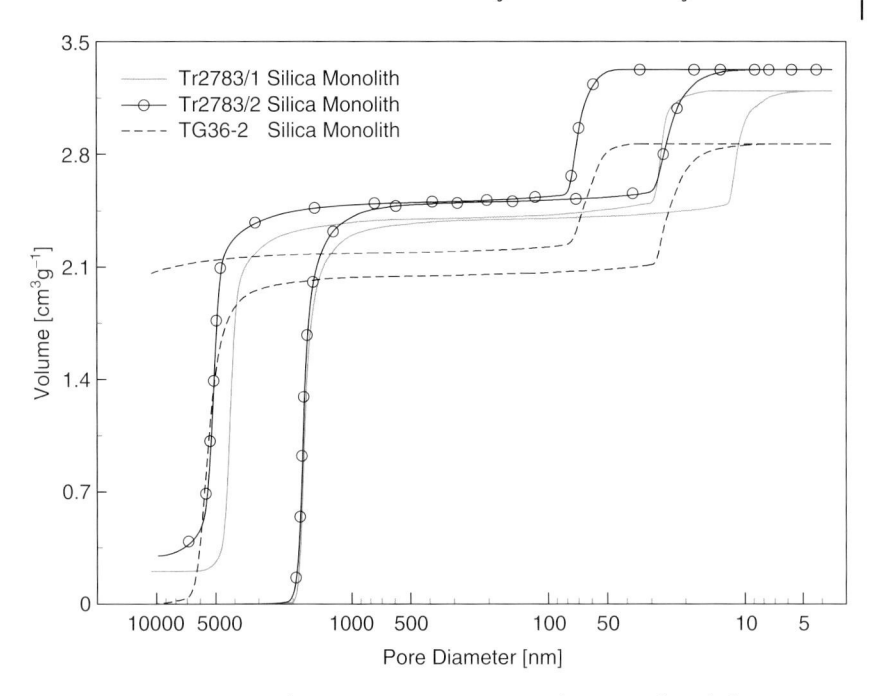

4.3 Characterization Methods of the Pore Structure of Monolithic Silicas | 59

Figure 4.8 Hysteresis curves between mercury intrusion and extrusion branch for some selected native silica monoliths.

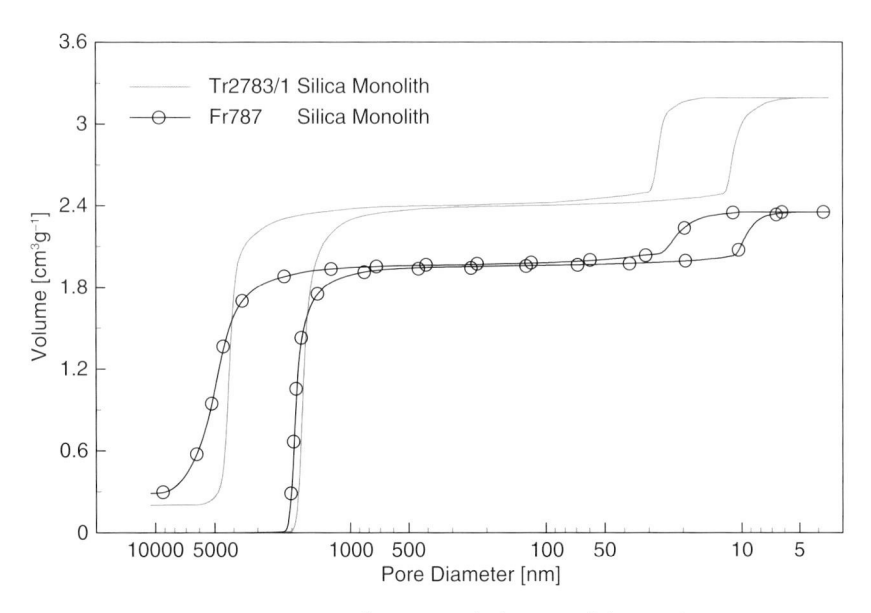

Figure 4.9 Mercury-intrusion curves of investigated silica monoliths: Tr2783/1 – a native silica monolith, Fr787 – silica monolith grafted with n-octadecyl functional groups.

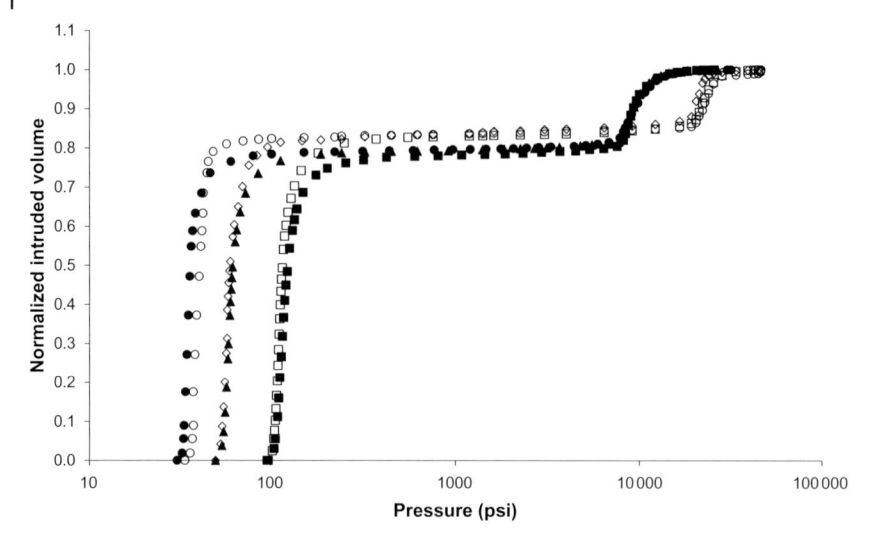

Figure 4.10 Mercury-intrusion curves of investigated silica monoliths: □–"787" monolithic silica rod grafted with n-octadecyl chains, ■–"800" monolithic silica rod grafted with n-octadecyl chains, ◇–"803" monolithic silica rod grafted with n-octadecyl chains, ▲–"811" monolithic silica rod grafted with n-octadecyl chains, ○–"842" monolithic silica rod grafted with n-octadecyl chains, ●–"843" monolithic silica rod grafted with n-octadecyl chains.

flow-through pore (or through pore) structure, which controls the mass transfer to and from the mesopores, mainly controls the entrapment behavior. It appears that entrapment is more likely to occur when the flow-through pore system is heterogeneous and disordered (which would restrict mass transfer) as indicated by a wide pore-size distribution coupled with relatively low porosity. *Vice versa*, the lack of entrapment after extrusion from the monolith mesopore system indicates enhanced transport properties, which is in accord with an ordered, highly porous flow-through pore system.

In conclusion, one can state that the mercury-intrusion curves (Figure 4.10) indicate that the pore structure of monolithic silicas is bimodal, having pores located in the silica skeleton (mesopores) and the flow-through pores. The plateau region at pressures of 1000 psi in the mercury-intrusion curves clearly separates the two pore-size regimes and allows one to calculate the corresponding surface area and porosity values of the two types of pores.

Therefore, MP could be used to assess not only the mesopore size and distribution but also provides a possibility for comprehensive structural characterization of bimodal monolithic silica, namely the flow-through pores as well (see Table 4.1).

4.3.2
Inverse Size-Exclusion Chromatography

A promising method for pore-structural investigations of chromatographic adsorbents is inverse size-exclusion chromatography (ISEC). Unlike nitrogen sorption,

Table 4.1 The pore characteristics of some silica monolith research samples assessed by mercury porosimetry.

Monolithic silica research sample	Surface functionality (native or n-octadecyl bonded and endcapped)	Flow-through pore diameter (μm)	Mesopore diameter (nm)
787	C18e	1.93	10.3
800	C18e	1.86	24.1
803	C18e	3.52	10.2
811	C18e	3.62	23.5
842	C18e	5.74	9.5
843	C18e	6.13	23.9
Tr2783/1	native	1.81	10.9
Tr2783/2c	native	1.76	25.1
Tr2786/1	native	3.52	10.9
Tr2786/2	native	3.40	24.0
TG36/1	native	5.74	10.1
TG36/2	native	5.64	24.4

mercury intrusion or microscopic techniques, ISEC is performed under liquid-phase conditions similar to those used in liquid-chromatographic separations.

The particular advantage of ISEC is that it is a chromatographic method, which provides a comparative statistical representation of the pore space that is accessible to a solute transport. The values derived correspond to the pore volume that is accessible in liquid-based separations, rather than the ones obtained from static characterization methods.

ISEC is a method used to determine the intraparticle pore-size distribution of a column based on measuring the residence times of solutes (polystyrenes, dextrans, etc.) of varying molecular diameter under the conditions where solute adsorption, intraparticle diffusion resistance, longitudinal (hydrodynamic) dispersion and other mass-transfer processes are minimized [38, 65–68] (see Figure 4.11).

ISEC was introduced in 1978 by Halasz et al. [69] who determined the pore-size distribution of porous materials, followed by Knox et al. [70]. Gorbunov et al. [38] provided a thorough review of ISEC and suggested an accurate method for determining the pore-volume distributions for slit-like, cylindrical and spherical pores. Calculated values of the average pore size and width of the distribution are based on the experimental partition coefficient of macromolecules chosen that is explicitly defined so that the volume partitioning of solutes was considered only to occur in the intraparticle pores (mesopores).

As a consequence, the described model is dependent on the morphology of the porous adsorbent, and inapplicable for the characterization of adsorbent where the volume partitioning of solutes is occuring in the intraparticle and interparticle pores. To overcome such a drawback, a general model was developed based on the first moments of the column response to a pulse injection making it independent of the morphology of the porous material [71].

4 Characterization of the Pore Structure of Monolithic Silicas

(a)

(b)

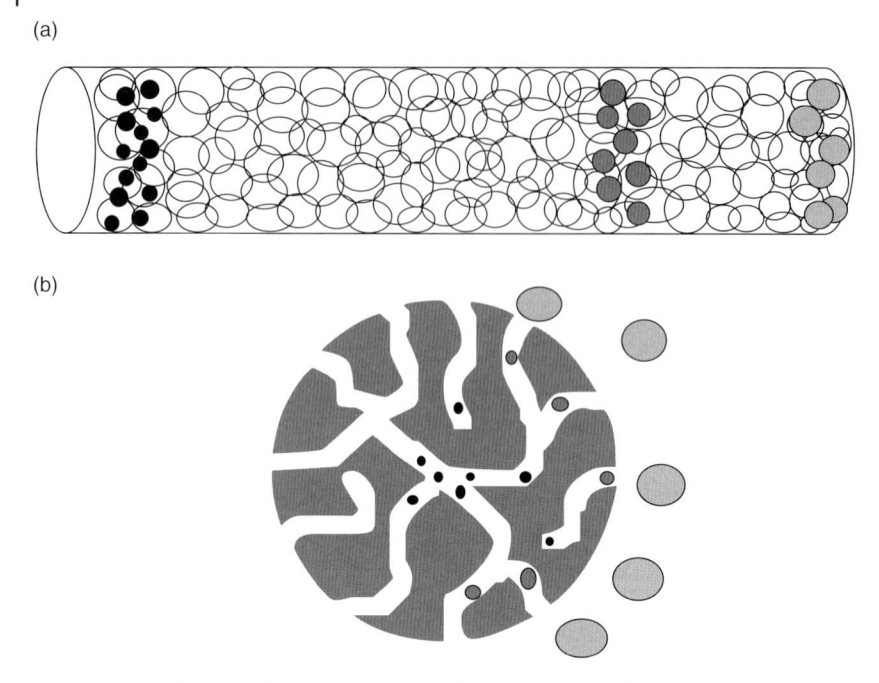

Figure 4.11 Mechanism of size separation in a chromatographic column (a) and illustration of accessible pore volume (b).

A parallel-pore model (PPM) and a pore-network model (PNM) were applied to provide the state-of-the-art methods for the calculation of various pore characteristics from the ISEC experiments [71]. The PPM provides the state-of-the-art method for the calculation of the mesopore and flow-through pore-volume distribution from the experimentally measured partition coefficients of a homologous set of polymer standards based on the first moments of the column response to a pulse injection. The PNM allows the pore connectivity factor, an indirect measure of mass-transfer resistance in the mesopores, and the mesopore number distribution to be obtained. Since the PNM is an idealization that assumes infinite pore connectivity, it does not adequately represent the finite connected porous network occurring in the real porous network. But it is used to provide an initial guess for the nonlinear regression of the experimental data that later is used in the PNM to obtain the actual pore-connectivity values.

In Figure 4.12, the experimentally measured ISEC curves for two monolithic silicas (787 and 800) alongside the theoretical predictions obtained by fitting the PPM and PNM are presented.

The values for the flow-through pore void fraction, ε_b (0.5 for column 787 and 0.48 for column 800) and the average mesopore size based on volume distribution (16.3 nm for column 787 with a distribution of 4.92 nm, and 36.4 nm for column 800 with a distribution of 11.3, respectively) were obtained from regression of the parallel pore model to the experimental data.

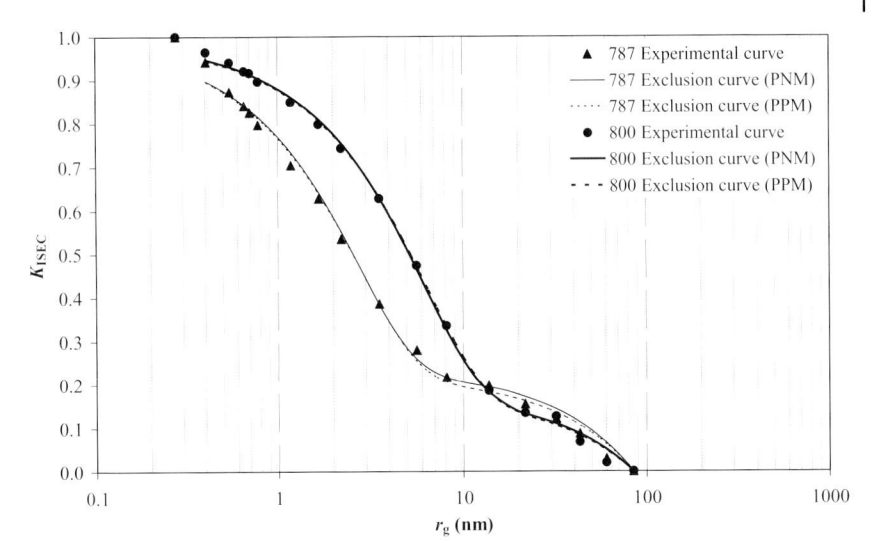

Figure 4.12 Experimentally measured exclusion curves by ISEC for monolithic silicas 787 and 800 along with theoretical exclusion curves obtained from regressing the data with the parallel pore model and the pore-network model.

The values for pore connectivity n_T can be obtained if the pore-network model is applied to the experimental data from ISEC. Values of n_T were greater than 10 for both columns, the parameters of the average pore size and distribution based on the number distribution (average pore size based 11.0 nm for column 787 and 23.3 for column 800, and pore-size distribution was 4.82 nm for column 787 and 11.1 for column 800) and parameters for the average pore size and distribution based on volume distribution (average pore size was 15.8 nm for column 787 and 34.8 for column 800, and pore-size distribution was 4.15 nm for column 787 and 9.14 for column 800) were obtained from the regression of the pore-network model to the experimental data.

Mesopore-structural characterization results obtained from the pore-network model indicated that the values of the pore connectivity, n_T, for all the monoliths studied were 10 or higher. This is an indication that the mesopore topology within the silica skeleton could be considered to be almost infinitely connected with the respect to the volume-partitioning phenomenon and not hindering the mass-transfer kinetics.

Although ISEC proved its high potential in the pore-structural characterization of particulate supports, certain drawbacks were acknowledged when the method was applied to the characterization of monolithic supports. Numerous authors stated that pore-characterization data obtained by ISEC [28, 41, 72] were not in agreement with the data obtained by nitrogen sorption, mercury porosimetry and transmission electron microscopy [28]. This confusion was due to the misinterpretation of the data obtained by ISEC. The exclusion of the solutes from the flow-through pores of monoliths was interpreted as a result of a secondary mesoporosity. Grimes et al. [71] cleared this misunderstanding by applying the pore-network model to obtain the pore-structural data from the ISEC measurement on monolithic silica columns. Grimes stated that an exclusion of high-molecular weight solutes is due to the exclusion mechanism in the flow-through pores (separation by flow). Another potential of ISEC was recognized by Kuga [73], who stated that the most important feature of ISEC is its applicability to polymeric gels in the swollen state to which the conventional porosimetry cannot be applied. This feature was extensively studied in numerous publications as the only method for obtaining reliable pore-characterization data on polymer-based supports [74–76].

4.3.3
Nitrogen Sorption

Adsorption is the adhesion process that forms a layer of adsorbate on the adsorbent surface. The reverse process of adsorption is desorption. The adsorption process might be defined as physisorption (characteristic of weak van der Waals forces) or chemisorption (characteristic of covalent bonding). In physisorption (physical adsorption), there is a weak van der Waals attraction of the adsorbate to the surface when the chemical identity of the adsorbate remains intact, that is, no breakage of the covalent structure. In *chemisorption* (chemical adsorption), the adsorbate sticks to the solid by the formation of a chemical bond with the surface, which can result in the dissociation of the adsorbate upon adsorption (dissociative adsorption).

The amount of adsorbate on the adsorbent as a function of its pressure (if gas) or concentration (if liquid) at constant temperature describes the adsorption isotherm (see Figure 4.13).

Once the isotherm curve is accurately expressed as a series of pressure *vs.* quantity adsorbed data pairs, a number of different theories or models can be applied to determine the pore-size distribution.

Langmuir's equation is usually used to describe the chemisorption isotherms through: $A_{(g)} + S \rightleftharpoons AS$, where A is a gas molecule and S is an adsorption site. The BET isotherm [77] describes the physisorption (see Figure 4.14).

Depending on the properties of the characterized material, the adsorption isotherms might have specific shape, which reflects the pore structure of the tested porous support [78] (see Figure 4.14).

The adsorption process is generally taken as completely reversible, but, under some conditions, the isotherm may exhibit a different shape upon desorption as compared to absorption. This is called *hysteresis*. Sometimes, hysteresis data can be used to determine the structure and size of pores in the absorbent. We will therefore need to generate an isotherm for both absorption and desorption. Depending on the hysteresis form one can differentiate between opened cylindrical pores with narrow pore-size distribution or wide distribution in pore size and shape, or slit-like pores, or microporous, slit-porous materials.

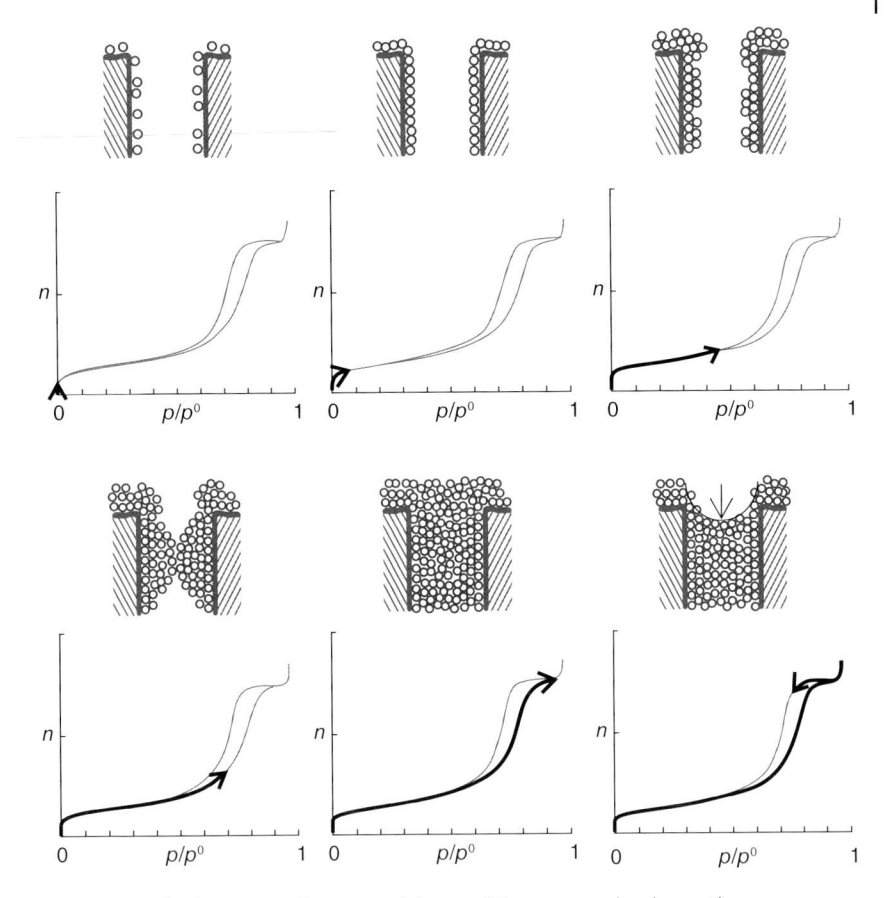

Figure 4.13 The formation of mono- and then multilayer gas molecules on the porous substance surface, followed by the capillary condensation and desorption.

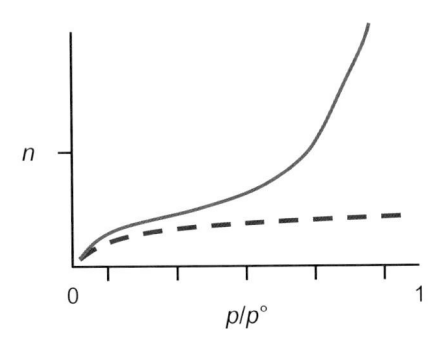

Figure 4.14 Langmuir isotherm (– – –) and BET isotherm (–).

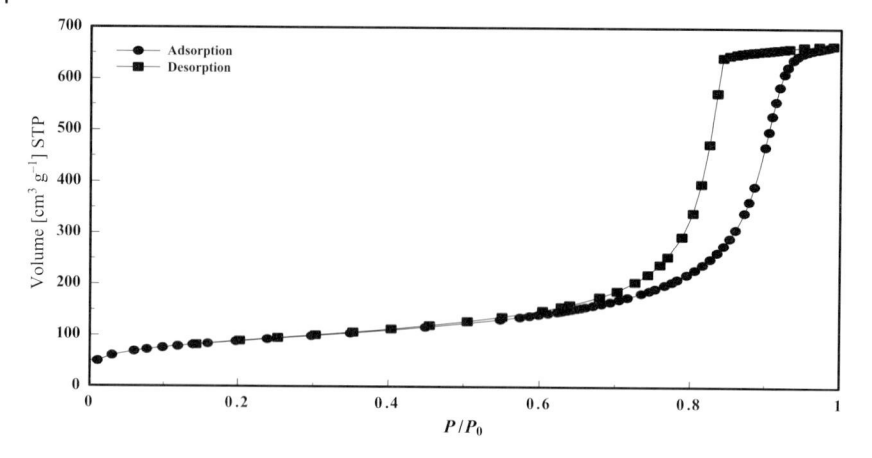

Figure 4.15 Nitrogen sorption at 77.4 K on the native silica monolith Tr2783/1.

Nitrogen sorption is commonly used for the characterization of porous materials measuring the nitrogen adsorption and desorption isotherms at 77 K. The volume of adsorbed nitrogen as a function of the partial equilibrium pressure represents the adsorption branch of the isotherm, and the volume of desorbed nitrogen as a function of the partial equilibrium pressure represents the desorption branch of the isotherm.

From the isotherm data the specific surface area according to the BET method, the specific pore volume according to the Gutwitsch (G) method [79] and the pore-volume distribution according to the method of Barrett–Joyner–Halenda (BJH) [77] are calculated.

Nitrogen adsorption/desorption data obtained at 77.4 K on the native silica monolith Tr2783/1 are displayed in Figure 4.15. The type-IV isotherm [80] reveals a hysteresis loop indicative of pore condensation. The hysteresis loop can be classified as to between type H1 and H2. This would indicate that in addition to the intrinsic reasons for the hysteresis, (i.e. the delay in condensation is caused by the metastable pore fluid) and the pore-blocking/percolation effects are present that lead to the delay in the position of the desorption branch. The calculated specific surface area (BET method) was 298 m^2/g for the monolithic silica sample Tr2783/1, the specific pore volume (G method) – 0.88 cm^3/g, and the average pore size according to the volume distribution (BJH method) was 10.9 nm.

Due to the fact that the BJH method fails to describe correctly the adsorption and the phase behavior of fluids in small mesopores, it leads to a significant underestimation of the pore size (for pore widths that are smaller than. 20 nm). Theoretical approaches such as the nonlocal density functional theory (NLDFT) [46, 47] are able to describe the configuration of the adsorbed phase on a molecular level, and therefore allow an accurate pore-size distribution to be obtained. In addition, the application of the NLDFT correctly predicts that the adsorption branch of a hysteretic adsorption isotherm is not at the thermodynamic equilibrium, that is, the pore condensation occurs with a delay due to the metastable-pore fluid. Hence, when hysteresis is only caused by the metastable-pore fluid and no networking effects are present, the desorption branch of the hysteresis loop reflects the thermodynamic equilibrium transition and the pore-size distribution calculated from the desorption branch by applying the NLDFT equilibrium method and from the adsorption branch by applying the so-called NLDFT metastable adsorption branch kernel (which takes into account the delay in condensation) should agree. Data comparison of applying these two kernels on the adsorption and desorption branches of the nitrogen isotherm are in Chapter 4.4 alongside the results of the results of MP and ISEC.

4.3.4
Liquid Permeation

Liquid permeability (LP) has been applied to a lesser extent than mercury porosimetry although it bears a high potential of information. The liquid permeability (the flow resistance) is dependent on the external surface area to volume ratio of the porous material. For the laminar flow regime, permeation through a porous material is described by Darcy's law:

$$u_{sf} = \frac{\kappa}{\eta} \frac{\Delta p}{L} \tag{4.3}$$

where u_{sf} is the superficial fluid velocity, η–fluid viscosity (Pa s), Δp–pressure drop (Pa) over a layer of thickness L (m) of material, and the permeability κ (m²). This equation holds for the laminar or viscous flow regimes.

To link the permeability to the pore size or other characteristics of the material, a model is needed for structure or texture. When the structure is not exactly known or difficult to describe, the flow of a liquid through a porous material can still be described by the Hagen–Poiseuille (the dynamic invasion of liquid into capillaries) or by the Kozeny–Carman equation (the concept of hydraulic radius to define equilibrium positions of fluid-fluid interfaces in tubes of different cross section).

The *Hagen–Poiseuille* equation is based on the assumption of parallel uniform cylindrical pores:

$$\frac{\Delta p}{L} = \frac{32 \eta u_{sf}}{\varepsilon d_H^2} \tag{4.4}$$

where ε is the porosity accessible to flow (e.g., external porosity of packed bed). For noncylindrical pores, the hydraulic diameter can be used (flow area divided by wetted perimeter). One can correct this equation for tortuosity of the pores (if known) by correcting the length L by the tortuosity factor τ (longer flow path). Formulas similar to, but somewhat deviating from Equation 1.2 are used to extract pore size, for example, for monolithic chromatography packings [81].

$$d_H = 2 \left(\frac{5\kappa}{\varepsilon} \right)^{1/2} \tag{4.5}$$

The *Kozeny–Carman* equation [82] assumes the pores to be voids between closely packed spheres of equal size, using the concept of hydraulic radius to obtain the equivalent cylindrical pore diameter:

$$\frac{\Delta P}{L} = \frac{180\eta(1-\varepsilon)^2}{\varepsilon^3}\frac{u_{sf}}{d_p^2} \qquad (4.6)$$

which is valid in the laminar regime or for small particle Reynolds numbers below 1:

$$\text{Re}_p = \frac{u_{sf}d_p\rho}{\eta(1-\varepsilon)} \qquad (4.7)$$

with ρ the fluid density (kg/m^3)

The Kozeny–Carman (approach assuming pores as voids between closely packed spheres of equal size) as well as the Hagen–Poiseuille equation (approach based on the assumption of cylindrical pores) do not fit the experimental data, and hence do not give a relevant particle or pore size when the texture of the porous material exhibits at least one of the following characteristics:

1. high porosity, significantly deviating from the 0.4 for packed spheres, say above 0.7;
2. particles very far from spherical shape (flaky materials, fibers);
3. consolidated porous media (compressed deformable particles);
4. multimode or very large grain or pore-size distribution.

But the major problem with the liquid-permeability method is that when a material is composed of pores with different pore sizes, and considering that the flow rate through the larger pores will be more proportionally larger than the flow through the smaller pores (flow rate ~ [pore diameter]^4) it is obvious to expect that the observed flow resistance in the case of a distribution of pore sizes will always be smaller than the flow resistance calculated on the basis of the average pore size. Hence, the estimated pore size will always be large compared to other methods. The method is good at predicting flow or pressure drop, certainly in comparing particle beds or monoliths in chromatography, but may not yield the correct pore measure for other phenomena such as diffusion [83]. Recently, attempts have been made to find the appropriate length scale in a porous structure to insert in the Kozeny–Carman equation [33, 34, 84].

Although the measurement principle of permeation is fairly straightforward, it is not a trivial task to perform a reproducible and accurate measurement, and no standard equipment is readily available with broad applicability. Basically, one has to either generate a flow and measure the resulting pressure drop, or apply a differential pressure and measure the resulting flow. Both gas and liquid can be used, but in the case of a gas one has to assure flow is by viscous forces and not by diffusion that might dominate for very small pores [85]. Some of the issues involved are mainly linked to sample preparation and finding an adequate flow and pressure-measurement window. A sample through which one wants to measure flow should be cut (often cylindrical form) from the material and fitted in a tube avoiding at all cost leaks between sample and wall. Usually, the sample will be glued in place with epoxy glue or an inorganic very fine cement that should not penetrate appreciably in the material and disturb the flow pattern. The sample diameter should be large enough across the flow direction to be representative and avoid wall effects due to porosity changes close to the wall, usually extending more than 5 particle diameters from the wall (preferably 20 to 100 times the largest flow pore or channel). Sample length can be adjusted to adapt to available flow and pressure measurement. Before starting a measurement the material should be evacuated to allow full penetration of the liquid and avoid dead zones where air is trapped. Obviously, a wetting inert liquid should be used properly degassed. Liquid syringe pumps and liquid chromatography pumps are often suitable for such measurements as they can cover a wide pressure range up to several 100 bar. Provided one can measure low flow rates (e.g., collecting the liquid and measuring gravimetrically), differential pressure can be reduced to avoid material damage or compression, especially a problem with soft compressible or fragile materials. Sample size can range from a fraction of a gram or ml to large blocks of the material. Smaller samples are obviously more difficult to machine or form in the right shape and mount for the flow experiment. Small pores then also lead to very small flow rates for reasonable pressures.

The experiment can be arranged so as to have either horizontal or vertical flow through the sample. Great care must be taken in every case to prevent bypassing of the sample by the liquid (see Figure 4.16). It should be taken into account that some liquids might change the pore structure and therefore the permeability, due to the rearrangement of some particles, swelling of certain materials, chemical reactions, etc.

The measurement of permeability is a simple and fast method for the estimation of the pore hydraulic radius, but simplicity in this case is viewed with certain precautions already mentioned. Among them, the correct theoretical approach should be chosen, together with a nonwetting liquid, and the influence of the

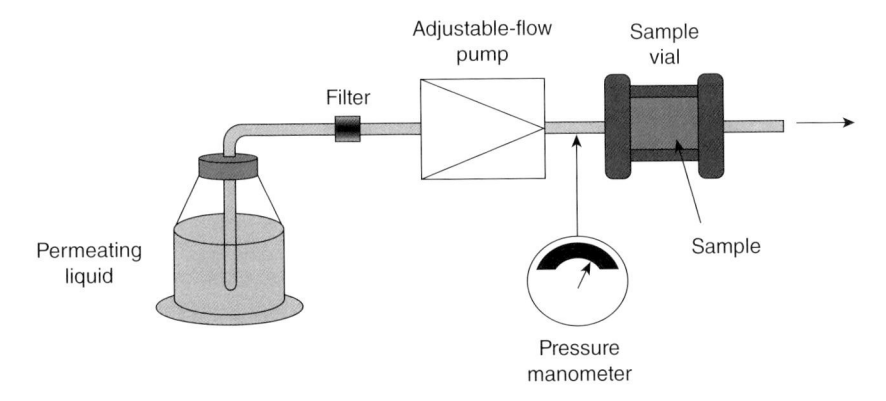

Figure 4.16 Principle of liquid-flow permeametry.

4 Characterization of the Pore Structure of Monolithic Silicas

Table 4.2 Flow-through pore characteristics of silica monoliths assessed by the permeability of a liquid.

Monolithic silica research sample	Average flow-through pore diameter (μm)	
	Hagen–Poiseuille method	Kozeny–Carman method
787	1.931	5.563
800	1.879	4.771
803	3.490	5.652
811	3.821	9.851
842	6.051	13.662
843	6.542	13.691

pore geometry and tortuosity on the interpretation of the results should be considered [86]. This method is applicable in the 0.1 to 1000 μm range.

Various papers have been published dealing with the Kozeny–Carman approach at the characterization of silica-based monoliths. The most fundamental work was done by Minakushi [29], Leinweber [13, 83], and Vervoort [30, 34] resulting in various interpretations of the flow-resistance values. The same relative particle diameters were obtained, but certain specific approximations were applied for the calculation of the flow-through pore sizes. A notable progress was made by introducing the term of domain size, being the sum of the skeleton diameter and the diameter of the flow-through pore [33]. Though this assumption did not reflect the real silica-based monolithic support, it enabled the use of liquid-permeability data to characterize the monolith regardless of its format and chemistry.

Alternatively, an approach was developed for monolithic silica flow-through pore characterization [87] *via* the liquid-permeability method, which was based not on the Kozeny–Carman approach but on the Hagen–Poiseuille equation. The assumption of cylindrical pores fits best to the flow-through pore morphology of the silica monolith. No approximations are necessary to link the particle size with the flow-through pore diameter (see Table 4.2).

The results in Table 4.2 first clearly reveal major differences between the Kozeny–Carman method and the Hagen–Poiseuille approach.

The permeability method could be used not only to obtain flow-through pore values but also skeleton diameters [87]. Moreover, the surface area of the skeletons and volume of the skeleton ratios could be calculated as well.

4.3.5
Microscopy and Image Analysis

Scanning electron microscopy is a method of imaging the sample surface by scanning it with a high-energy beam of electrons in a raster scan pattern. The electrons interact with the atoms that make up the sample producing signals that contain information about the sample's surface topography, composition and other properties such as electrical conductivity. Resolution of a few nanometers is possible but it is limited by sample size (e.g., <~500 nm for ~1 nm resolution) and the angle through which the sample stage may be turned. A particular problem with SEM is the possibility of image artifacts due to sample ablation from the electron beam.

The first SEM image was obtained by Knoll, who in 1935 obtained an image of silicon steel showing electron channeling contrast [88]. Further pioneering work on the physical principles of the SEM and beam–specimen interactions was performed by von Ardenne in 1937 [89, 90] who produced a British patent [91] but never made a practical instrument. The SEM was further developed by Prof. C. Oatley and his postgraduate student Stewart and was the first marketed in 1965 by the Cambridge Instrument Company as the "Stereoscan".

SEM was used to give a direct image of the flow-through pore system of monolithic silica samples (Figure 4.17). Different methods can be applied to assess the average flow-through pore diameter and skeleton diameter: for example – the direct scale measurement or the "Pixcavator" program. The direct measurement is performed *via* segmenting each image into relative areas and measuring the characteristic parameters *via* given scale. The values estimated by the "Pixcavator" program is based on the area estimation *via* integrating the number of pixels in that area. The area is defined by the color. Since the flow-through pores are in most cases black, the estimation of all the pixels in the area is performed by calculation of their number and multiplying by the area of one pixel. The obtained relative area of a single flow-through pore is then assumed to be equal to the area of a circle (meaning a round-shaped pore) and the diameter of this circle is taken as the diameter of the flow-through pore.

The detailed image analysis reveals not only cylindrical silica structure skeletons, but also various forms by which these skeletons are connected. This causes a wide distribution of skeleton diameter values as well [92]. As a consequence, the external surface area of such monolithic structure is smaller if compared to the ideal cylindrical morphology when the volume and the average structural diameters are equal. This leads to smaller flow-resistance and enhanced mass-transfer values.

Through SEM is the most popular technique, there exists a number of alternative imaging techniques:

Magnetic resonance imaging (MRI) – is based on aligning of water-molecule hydrogen proton magnetic moments with the direction of the applied field. This method is used for the estimation of the saturated void space. The resolution limit for direct visualization is ~10 µm. Computerized X-ray tomography – is based on passing X-rays through an object *via* different paths in different directions and measuring the intensity of the beam. The image is created *via* summarizing the gathered data. The possible resolution is approximately 0.1–1 µm.

Confocal laser scanning microscopy – a technique that is used for obtaining high-resolution images with depth selectivity.

The given techniques could be combined to increase the method resolution or amount of characteristic information gained in a single measurement, for

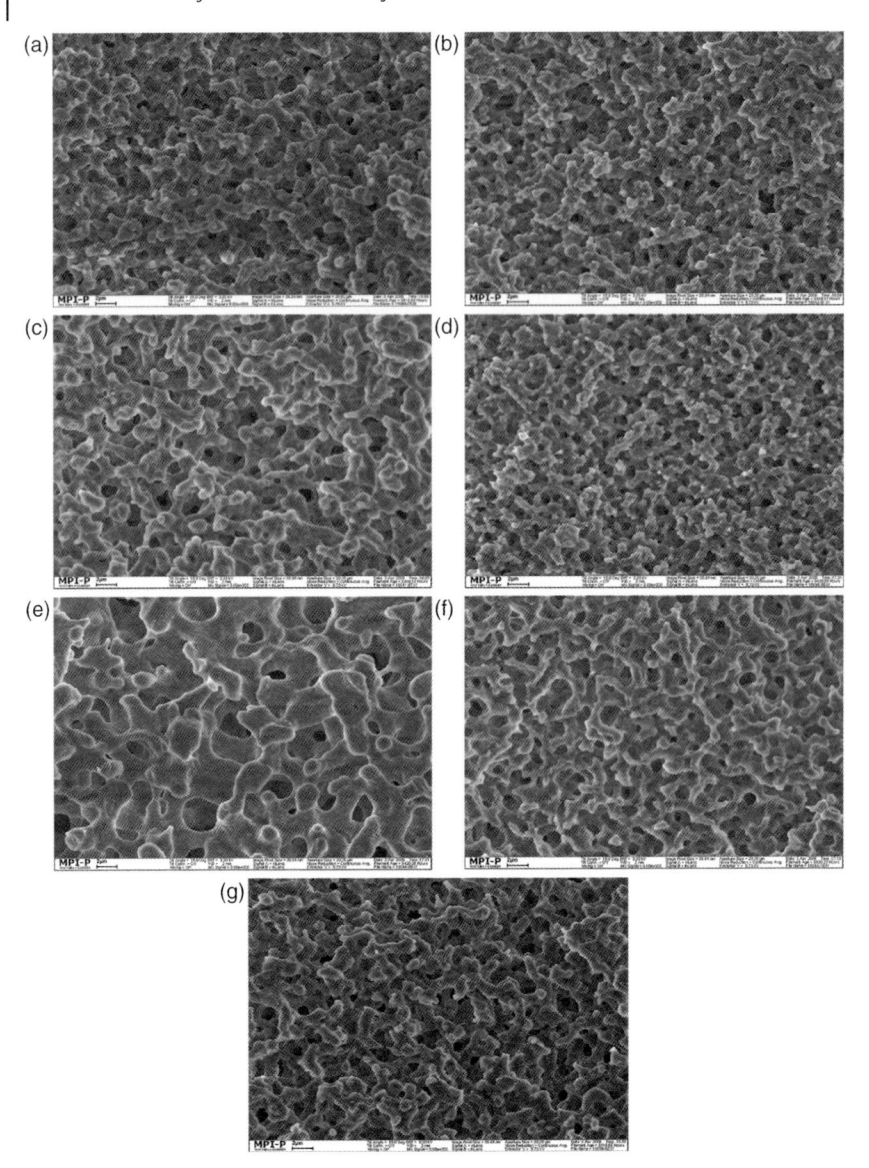

Figure 4.17 The electron scanning micrographs of monolithic research silica samples: (a) KN 253, (b) KN 341, (c) KN 255, (d) KN 349, (e) KN 345, (f) KN 344, (g) KN 252.

example combination of computerized X-ray tomography with mercury porosimetry allows more information to be obtained on the pore structure of macroporous materials with pores too small to be imaged directly; The combination of MRI with mercury porosimetry removes some of the freely adjustable model parameters giving more reliable data on pore connectivity, etc.

The image analysis is based on simple representations of the porous medium, with a single pore shape: cylindrical, slit-shaped, formed between close-packed solid spheres. A more refined use of the experimental data necessarily raises the challenge of describing the geometry of the pore network.

For disordered materials, it becomes evident that porous media must be characterized in statistical terms [92]. Here, direct 2D and 3D observations can be used through an imaging protocol followed by an image analysis. In parallel, mathematical pore models or statistical reconstruction stay interesting as "toy models", allowing elaboration of a possible classification of these media and a better understanding of the relation between geometry, transport and thermodynamics processes and mechanical comportment [93].

Moreover, statistical modeling of a porous medium may conveniently lend itself to the further modeling of the phenomena involved in those methods of characterization (such as adsorption, capillary condensation, permeation, molecular diffusion, excitation relaxation, or phase transitions…). It should therefore be of great help for their understanding and comparison.

4.4
Comparison of the Silica Monolith Mesopore-Characterization Data

The mesopore-size distribution obtained from MP, ISEC, and nitrogen sorption are compared in Figure 4.18.

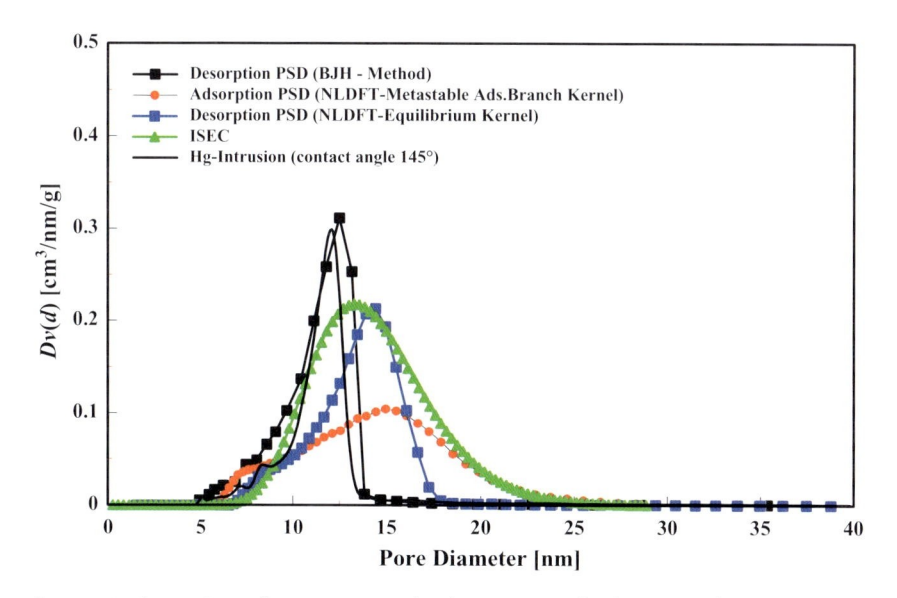

Figure 4.18 Comparison of mesopore-size distribution curves for the native silica monolith Tr2783/1 obtained from nitrogen sorption by applying NLDFT and the BJH method, ISEC and MP.

A good agreement is obtained between the pore-size distribution (psd) curves by applying the BJH method using the desorption branch of the nitrogen sorption method with the mesopore size curves obtained from MP by applying the Washburn equation and a contact angle of 145°. Applying the NLDFT-metastable adsorption branch kernel and the NLDFT-equilibrium kernel methods on the nitrogen isotherm of sample Tr2783/1 reveals that the mode pore diameters (most frequent pore diameters, or maximum of the mesopore-size distribution [psd] curve) agree well, but the mode pore diameter obtained from the desorption branch is slightly shifted to the smaller values. Furthermore, the psd curve is much narrower as compared to the psd obtained from the adsorption branch of the nitrogen isotherm indicating the presence of some network/pore blocking effects.

The data presented in Figure 4.18 also indicates that (as is expected) the BJH method (nitrogen sorption) and mercury intrusion method applying 145° contact angle significantly underestimate the pore size compared to the NLDFT pore-size method (nitrogen sorption). Interestingly, the width of the pore-size distribution curve obtained by ISEC agrees reasonably well with the psd curve obtained from the NLDFT adsorption branch.

4.5
Comparison of the Silica Monolith Flow-Through Pore-Characterization Data

For the comparison of the flow-through pore size values of monolithic silicas the Pearson correlation coefficient, r_p, was calculated (see Table 4.3). The best correlation coefficient values were obtained for mercury intrusion and image analysis using the "Pixcavator" program, $r_p = 0.998$; mercury intrusion and the permeability Hagen model gave $r_p = 0.998$.

Table 4.3 Pearson correlation coefficient (r_p), calculated for the flow-through pore diameters obtained by mercury intrusion, image analysis and liquid permeability for monolithic research silica columns.

Characterization methods		Mercury porosimetry	Imaging and image analysis		Permeability	
			Direct analysis	Indirect analysis	Kozeny	Hagen
Mercury porosimetry			0.991	0.998	0.958	0.998
Imaging and image analysis	Direct analysis	0.991		0.986	0.971	0.992
	Indirect analysis	0.998	0.986		0.957	0.997
Permeability	Kozeny	0.958	0.971	0.957		0.968
	Hagen	0.998	0.992	0.997	0.968	

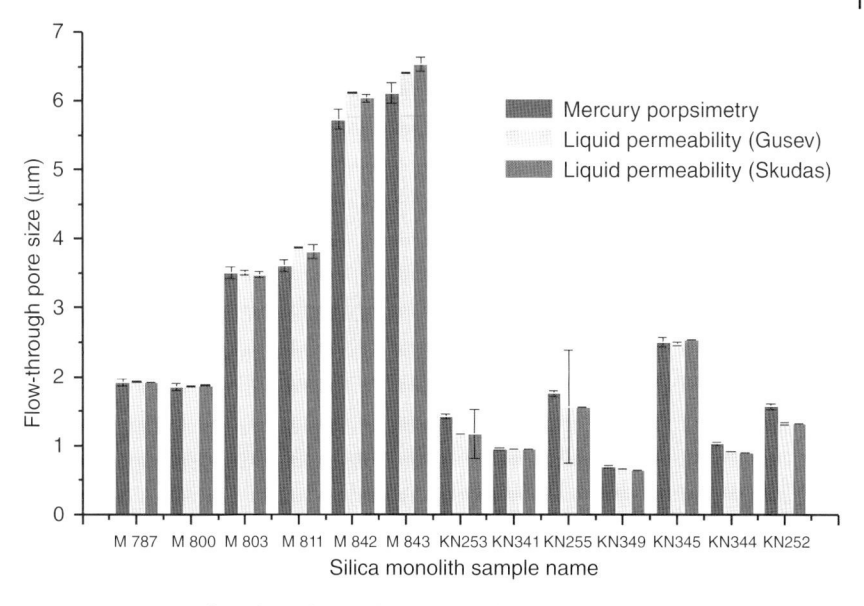

Figure 4.19 Average flow-through pore diameters and standard deviation of silica monoliths obtained by mercury porosimetry, liquid-permeability method according to Kozeny and Hagen methodology.

As mentioned before, the Kozeny–Carman theory does not fit to the calculation of pore characteristic values of silica monoliths. The obtained Pearson's correlation coefficients obtained were <0.95.

When one compares the flow-through pore diameters by liquid permeability approaches, and mercury intrusion, one notices slight differences (Figure 4.19).

These are due to the fact that not all the pores were permeable at the same extent. The larger pores (explicitly columns 842 and 843) have a higher permeability than the others since the flow rate is proportional to the pore diameter to the fourth. Therefore, the estimated flow-through pore diameter values are larger as well. It could be concluded that obviously not all of the surface area has access to the liquid at the same ratio.

The study of other characterization techniques is ongoing but the main guidelines could be found in the latest UIPAC recommendation for characterization methods applied on macroporous supports.

References

1 Nakanishi, K., and Soga, N. (1991) Phase gelation in gelling silica-organic polymer solution: systems containing poly(sodium styrenesulfonate). *J. Am. Ceram. Soc.*, **74**, 2518–2530.

2 Nakanishi, K., and Soga, N. (1992) Phase separation in silica sol-gel system containing polyacrylic acid I. Gel formation behaviour and effect of solvent composition. *J. Non-Cryst. Solids*, **139**, 1–13.

3 Nakanishi, K., and Soga, N. (1992) Phase separation in silica sol-gel system containing polyacrylic acid II. Effects of molecular weight and temperature. *J. Non-Cryst. Solids*, **139**, 14–24.

4 Minakuchi, H., Nakanishi, K., Soga, N., Ishizuka, N., and Tanaka, N. (1996) Octadecylsilylated porous silica rods as separation media for reversed-phase liquid chromatography. *Anal. Chem.*, **68**, 3498–3501.

5 Nakanishi, K., Minakuchi, H., Soga, N., and Tanaka, N. (1997) Double pore silica gel monolith applied to liquid chromatography. *J. Sol-Gel. Sci. Technol.*, **8**, 547–552.

6 Minakuchi, H., Nakanishi, K., Soga, N., Ishizuka, N., and Tanaka, N. (1997) Effect of skeleton size on the performance of octadecylsilylated continuous porous silica columns in reversed-phase liquid chromatography. *J. Chromatogr. A*, **762**, 135–146.

7 Cabrera, K., Wieland, G., Lubda, D., Nakanishi, K., Soga, N., Minakuchi, H., and Unger, K.K. (1998) SilicaRod™-A new challenge in fast high-performance liquid chromatography separations. *TrAC*, **17**, 50–53.

8 Cabrera, K., Lubda, D., Minakuchi, H., and Nakanishi, K. (2000) A new monolithic-type HPLC column for fast separations. *J. High Resolut. Chromatogr. Commun.*, **23**, 93–99.

9 Lubda, D., Cabrera, K., Minakuchi, H., and Nakanishi, K. (2002) Monolithic HPLC silica columns. *J. Sol-Gel Sci. Technol.*, **23**, 185–189.

10 Motokawa, M., Kobayashi, H., Ishizuka, N., Minakuchi, H., Nakanishi, K., Jinnai, H., Hosoya, K., Ikegami, T., and Tanaka, N. (2002) Monolithic silica columns with various skeleton sizes and through-pore sizes for capillary liquid chromatography. *J. Chromatogr. A*, **961**, 53–63.

11 Tanaka, N., Kobayashi, H., Ishizuka, N., Minakuchi, H., Nakanishi, K., Hosoya, K., and Ikegami, T. (2002) Monolithic silica columns for high-efficiency chromatographic separations. *J. Chromatogr. A*, **965**, 35–49.

12 Siouffi, A.-M. (2003) Silica gel-based monoliths prepared by the sol–gel method: facts and figures. *J. Chromatogr. A*, **1000**, 801–818.

13 Leinweber, F.C., Lubda, D., Cabrera, K., and Tallarek, U. (2002) Characterization of silica-based monoliths with bimodal pore-size distribution. *Anal. Chem.*, **74**, 2470–2477.

14 Skudas, R., Grimes, B.A., Machtejevas, E., Kudirkaite, V., Kornysova, O., Hennessy, T.P., Lubda, D., and Unger, K.K. (2007) Impact of pore structure parameters on column performance and resolution of reversed-phase monolithic silica columns for peptides and proteins. *J. Chromatogr. A*, **1144**, 72–84.

15 Meyers, J.J., and Liapis, A.I. (1998) Network modelling of the intraparticle convective flow and diffusion of molecules adsorbing in monoliths and in porous particles packed in a chromatographic column. *J. Chromatogr. A*, **827**, 197–213.

16 Meyers, J.J., and Liapis, A.I. (1999) Network modelling of the convective flow and diffusion of molecules in porous particles packed in a chromatographic column. *J. Chromatogr. A*, **852**, 3–23.

17 Unger, K.K., Bidlingmaier, B., du Fresne von Hohenesche, C., and Lubda, D. (2002) Evaluation and comparison of the pore structure and related properties of particulate and monolithic silicas for liquid phase separation processes. *COPS VI Proc. Stud. Surf. Sci. Catal.*, **144**, 115–122.

18 Unger, K.K., Skudas, R., and Schulte, M.M. (2008) Particle packed columns and monolithic columns in high-performance liquid chromatography– comparison and critical appraisal. *J. Chromatogr. A*, **1184**, 393–415.

19 Guiochon, G. (2007) Monolithic columns in high-performance liquid chromatography. *J. Chromatogr. A*, **1168**, 101–168.

20 Gzil, P., Smet, J.D., and Desmet, G. (2006) A discussion of the possible ways to improve the performance of silica monoliths using a kinetic plot analysis of experimental and computational plate height data. *J. Sep. Sci.*, **29**, 1675–1685.

21 Billen, J., Gzil, P., and Desmet, G. (2006) Domain size-induced heterogeneity as performance limitation of small-domain monolithic columns and other LC support types. *Anal. Chem.*, **78**, 6191–6201.

22 Liapis, A.I., Meyers, J.J., and Crosser, O.K. (1999) Modeling and simulation of the dynamic behavior of monoliths Effects of pore structure from pore-network model analysis and comparison with columns packed with porous spherical particles. *J. Chromatogr. A*, **865**, 13–25.

23 Ritter, H.L., and Drake, L.C. (1945) Pore-size distribution in porous. Materials. *Ind. Eng. Chem. Anal. Chem.*, **17**, 782–786.

24 Drake, L.C., and Ritter, H.L. (1945) Macropore-size distributions in some. Typical porous substances. *Ind. Eng. Chem. Anal. Ed.*, **17**, 787–789.

25 Washburn, W. (1921) The dynamic of capillary flow. *Phys. Rev.*, **17**, 273–283.

26 Carman, P.C. (1941) Capillary rise and capillary movement of moisture in fine sand. *Soil Sci.*, **52**, 1–14.

27 Carman, P.C. (1956) *Flow of Gases through Porous Media*, Butterworths, London, Ch 1.

28 Lubda, D., Lindner, W., Quaglia, M., du Fresne von Hoheneschne, C., and Unger, K.K. (2005) Comprehensive pore structure characterization of silica monoliths with controlled mesopore size and macropore size by nitrogen sorption, mercury porosimetry, transmission electron microscopy and inverse size-exclusion chromatography. *J. Chromatogr. A*, **1083**, 14–22.

29 Minakuchi, H., Nakanishi, K., Soga, N., Ishizuka, N., and Tanaka, N. (1998) Effect of domain size on the performance of octadecylsilylated continuous porous silica columns in reversed-phase liquid chromatography. *J. Chromatogr. A*, **797**, 121–131.

30 Vervoort, N., Saito, H., Nakanishi, K., and Desmet, G. (2005) Experimental validation of the tetrahedral skeleton model pressure drop correlation for silica monoliths and the influence of column heterogeneity. *Anal. Chem.*, **77**, 3986–3992.

31 Montminy, M.D., Tannenbaum, A.R., and Macosko, C.W. (2004) The 3D structure of real polymer foams. *J. Colloid Interf. Sci.*, **280**, 202–211.

32 Baldwin, C.A., Sederman, A.J., Mantle, M.D., Alexander, P., and Gladden, L.F. (1996) Determination and Characterization of the Structure of a Pore Space from 3D Volume Images. *J. Colloid Interf. Sci.*, **181**, 79–92.

33 Gzil, P., Vervoort, N., Baron, G.V., and Desmet, G. (2004) General rules for the optimal external porosity of LC supports. *Anal. Chem.*, **76**, 6707–6718.

34 Vervoort, N., Gzil, P., Baron, G.V., and Desmet, G. (2003) A correlation for the pressure drop in monolithic silica columns. *Anal. Chem.*, **75**, 843–850.

35 Gregg, S.J., and Sing, K.S.W. (1982) *Adsorption Surface Area and Porosity*, 2nd edn, Academic Press, London, p. 303.

36 Rouquerol, F., Rouquerol, J., and Sing, K.S.W. (1991) *Adsorption by Powders and Porous Solids*, Academic Press, London, p. 467.

37 Lowell, S., and Shields, J.E. (1991) *Powder, Surface Area and Porosity*, 3rd edn, Chapman & Hall, London, p. 250.

38 Gorbunov, A.A., Solovyova, L.Y., and Pasechnik, V.A. (1998) Fundaments of the theory and practice of polymer gel permeation chromatography as a method of chromatographic porosimetry. *J. Chromatogr.*, **448**, 307–332.

39 Hagel, L., Östberg, M., and Andersson, T. (1996) Apparent pore-size distributions of chromatographic media. *J. Chromatogr. A*, **743**, 33–42.

40 Goto, M., and McCoy, B.J. (2000) Inverse size-exclusion chromatography for distributed pore and solute sizes. *Chem. Eng. Sci.*, **55**, 723–732.

41 Al-Bokari, M., Cherrak, D., and Guiochon, G. (2002) Determination of the porosities of monolithic columns by inverse size-exclusion chromatography. *J. Chromatogr. A*, **975**, 275–284.

42 ISO (2005) 15901-1. Pore-size distribution and Porosity of Solid Materials by Mercury Porosimetry and Gas Adsorption. Part 2. Analysis of Macropores by Mercury Porosimetry.

43 Simon, J., Saffer, S., and Kim, C.J. (1997) A liquid-filled microrelay with a moving mercury micro-drop. *J. Microelectromech. Syst.*, **6**, 208–216.

44 Barrett, E.P., Joyner, L.G., and Halenda, P.P. (1951) The determination of pore volume and area distributions in porous substances. I. Computations from nitrogen isotherms. *J. Am. Chem. Soc.*, **73**, 373–380.

45 Porcheron, F., Thommes, M., Ahmad, R., and Monson, P.A. (2007) Mercury porosimetry in mesoporous glasses: a comparison of experiments with results from a molecular model. *Langmuir*, **23**, 3372–3380.

46 Thommes, M. (2004) Physical adsorption characterization of ordered and amorphous mesoporous materials, in *Nanoporous Materials, Science & Engineering* (eds G.Q. Lu and X.S. Zhao), Imperial College Press, London, Ch 11, pp. 317–364.

47 Neimark, V.A., and Ravikovitch, P.I. (2001) Capillary condensation in MMS and pore structure characterization. *Micropor. Mesopor. Mater.*, **44**, 697–707.

48 Ravikovitch, P.I., and Neimark, V.A. (2001) Characterization of nanoporous materials from adsorption and desorption isotherms. *Colloids Surf. A*, **187**, 11–21.

49 Moscou, L., and Lub, S. (1981) Practical use of mercury porosimetry in the study of porous solids. *Powder Technol.*, **29**, 45–52.

50 Rigby, S. (2002) *Characterization of Porous Solids VI. Studies in Surface Science and Catalysis*, vol. 144 (eds F. Rodriguez-Reinoso, B. Mac Enaney, J. Rouquerol, and K.K. Unger), Elsevier, Amsterdam, p. 185.

51 Rigby, S.P., Evbuoumwan, I.O., Watt-Smith, M.J., Edler, K., and Fletcher, R.S. (2006) Using nano-cast model porous media and integrated gas sorption to improve fundamental understanding and data interpretation in mercury porosimetry. *Part. Part. Syst. Charact.*, **23**, 82–93.

52 Felipe, C., Rojas, F., Kornhauser, I., Thommes, M., and Zgrablich, G. (2006) Mechanistic and experimental aspects of the structural characterization of some model and real systems by nitrogen sorption and mercury porosimetry. *Adsorpt. Sci. Technol.*, **24**, 623–643.

53 Giesche, H. (2006) Mercury porosimetry: a general (practical) overview. *Part. Part. Syst. Charact.*, **23**, 9–19.

54 Giesche, H. (1996) *Mater. Res. Soc. Symp. Proc.*, **431**, 151.

55 Lowell, S., and Shields, J.E. (1981) Influence of contact angle on hysteresis in mercury porosimetry. *J. Colloid Interf. Sci.*, **80**, 192–196.

56 Lowell, S., and Shields, J.E. (1981) Hysteresis, entrapment and wetting angle in mercury porosimetry. *J. Colloid Interf. Sci.*, **83**, 273–278.

57 Salmas, G., and Androutsopolous, G.J. (2001) Mercury porosimetry: contact angle hysteresis of materials with controlled pore structure. *J. Colloid Interf. Sci.*, **239**, 178–189.

58 Day, M., Parker, I.B., Bell, J., Thomas, M., Fletcher, R., and Duffie, J. (1991) *Characterization of Porous Solids II* (eds J. Rouquerol, F. Rodriguez-Reinoso, K.S.W. Sing, and K.K. Unger), Elsevier, Amsterdam, p. 75.

59 Day, M., Parker, I.B., Bell, J., Fletcher, R., Duffe, J., Sing, K.S.W., and Nicholson, D. (1994) *Characterization of Porous Solids III. Studies in Surface Science and Catalysis*, vol. 87 (eds J. Rouquerol, F. Rodriguez-Reinoso, K.S.W. Sing, and K.K. Unger), Elsevier, Amsterdam, p. 225.

60 Zgrablich, G., Mendioroz, S., Daza, L., Pajares, J., Mayagotia, V., Rojas, F., and Conner, W.C. (1991) Effect of porous structure on the determination of pore-size distribution by mercury porosimetry and nitrogen sorption. *Langmuir*, **7**, 779–785.

61 Felipe, C., Cordero, S., Kornhauser, I., Zgrablich, G., Lopez, R., and Rojas, F. (2006) Domain complexion diagrams related to mercury intrusion-extrusion in Monte Carlo-simulated porous networks. *Part. Part. Syst. Charact.*, **23**, 48–60.

62 Murray, K.L., Seaton, N.A., and Day, M.A. (1999) Use of mercury intrusion data, combined with nitrogen adsorption measurements, as a probe of pore network connectivity. *Langmuir*, **15**, 8155–8160.

63 Porcheron, F., Monson, P.A., and Thommes, M. (2004) Modeling mercury porosimetry using density functional theory. *Langmuir,* **20**, 6482–6489.

64 Porcheron, F., and Monson, P.A. (2005) Dynamic aspects of mercury porosimetry: a lattice model study. *Langmuir,* **21**, 3179–3186.

65 Parcher, J.F. (1997) *Anal. Chem.,* **69**, 229A–234A.

66 Casassa, E.F. (1967) Equilibrium distribution of flexible polymer chains between macroscopic solution phase and small voids. *J. Polym. Sci. Part B,* **5**, 773–778.

67 Casassa, E.F., and Tagami, Y. (1969) An equilibrium theory for exclusion chromatography of branched and linear polymer chains. *Macromolecules,* **2**, 14–19.

68 Casassa, E.F. (1976) Comments on exclusion of polymer chains from small pores and its relation to gel permeation chromatography. *Macromolecules,* **9**, 182–185.

69 Halasz, I. (1978) *Angew. Chem. Int. Ed. Engl.,* **17**, 901–908.

70 Knox, J., and Scott, H.P. (1984) Theoretical models for size-exclusion chromatography and calculation of pore-size distribution from size-exclusion chromatography data. *J. Chromatogr.,* **316**, 311–332.

71 Grimes, B.A., Skudas, R., Unger, K.K., and Lubda, D. (2007) Pore-structural characterization of monolithic silica columns by inverse size-exclusion chromatography. *J. Chromatogr. A,* **1144**, 14–29.

72 Ishizuka, N., Minakuchi, H., Nakanishi, K., Soga, N., Nagayama, H., Hosoya, K., and Tanaka, N. (2000) Performance of a monolithic silica column in a capillary under pressure-driven and electrodriven conditions. *Anal. Chem.,* **72**, 1275–1280.

73 Kuga, S. (1988) *Aqueous Size-Exclusion Chromatography* (ed. P.L. Dublin), Elsevier, Amsterdam, p. 157.

74 Hradil, J., Horak, D., Pelzbauer, Z., Votavova, E., Svec, F., and Kalal, J. (1983) *J. Chromatogr.,* **259**, 269.

75 Rao, T.P., Praveen, R.S., and Daniel, S. (2004) *Crit. Rev. Anal. Chem.,* **34**, 177.

76 Ousalem, M., Zhu, X.X., and Hradil, J. (2000) *J. Chromatogr. A,* **903**, 13.

77 Brunauer, S., Emmett, P.H., and Teller, E. (1938) *J. Am. Chem. Soc.,* **60**, 309.

78 Sing, K.S.W., Everett, D.H., Haul, R.A.W., Moscou, L., Pierotti, R.A., Rouquerol, J., and Siemieniewska, T. (1985) *Pure Appl. Chem.,* **57**, 603.

79 Gurwitsch, L.G. (1915) *J. Phys. Chem. Soc. Russia,* **47**, 805.

80 Jinnai, H. (2001) *Langmuir,* **17**, 619.

81 Gusev, I. (1999) Capillary columns with in situ formed porous monolithic packing for micro high-performance liquid chromatography and capillary electrochromatography. *J. Chromatogr. A,* **855**, 273–290.

82 Carman, P.C. (1937) *Trans. Inst. Chem. Engrs.,* **15**, 150.

83 Leinweber, F.C. (2003) Chromatographic performance of monolithic and particulate stationary phases Hydrodynamics and adsorption capacity. *J. Chromatogr. A.,* **1006**, 207–228.

84 Leinweber, F.C. (2002) *Chem. Eng. Technol.,* **11**, 1177–1181.

85 Miguel, F.A., and Serrenho, A. (2007) On the experimental evaluation of permeability in porous media using a gas flow method. *J. Phys. D Appl. Phys.,* **40**, 6824–6828.

86 Skudas, R., Grimes, B.A., Thommes, M., and Unger, K.K. (2009) Flow-through pore characteristics of monolithic silicas and their impact on column performance in high-performance liquid chromatography. *J. Chromatogr. A,* **1216**, 2635–2636.

87 Thommes, M., Skudas, R., Unger, K.K., and Lubda, D. (2008) Textural characterization of native and n-alkyl-bonded silica monoliths by mercury intrusion/extrusion, inverse size-exclusion chromatography and nitrogen sorption. *J. Chromatogr. A,* **1191**, 57–66.

88 Knoll, M. (1935) *Zeitschrift für technische Physik,* **16**, 467–475.

89 von Ardenne, M. (1937) Das Elektronen-Rastermikroskop. Theoretische Grundlagen. *Zietschrift für Physik,* **108**, 553–572.

90 von Ardenne, M. (1937) Das Elektronen-Rastermikroskop. Praktische Ausführung. *Zietschrift für technische Physik,* **19**, 407–416.

91 von Ardenne, M. (1937) Improvements in electron microscopes. GB patent 511204.

92 Wang, Y., De Carlo, F., Mancini, D.C., McNulty, I., Tieman, B., Bresnahan, J., Foster, I., Insley, J., Lane, P., von Laszewski, G., Kesselman, C., Su, M.H., and Thiebaux, M. (2001) A high-throughput X-ray microtomography system at the Advanced Photon Source. *Rev. Sci. Instrum.*, **72**, 2062–2068.

93 Holzer, L., Indutnyi, F., Gasser, P.H., Munch, B., and Wegmann, M. (2004) Three-dimensional analysis of porous $BaTiO_3$ ceramics using FIB nanotomography. *J. Microsc.*, **216**, 84–95.

5
Microscopic Characterizations

Haruko Saito, Kazuyoshi Kanamori, and Kazuki Nakanishi

5.1
Introduction

The cocontinuous macropores of silica monolith provide excellent liquid transport, and have been successfully applied to separation media in the field of high-performance liquid chromatography (HPLC) for over a decade [1, 2]. The characteristic macroporous morphology brings a significant influence on the liquid transport and separation efficiencies [3–5], and the detailed examination on the macropore structure is required for the further optimization of macroporous silica as separation media, however, the structural properties of cocontinuous macropores are still poorly understood due to its complexity.

Conventional methods to characterize macropore structure have been scanning electron microscopy (SEM) and the mercury-intrusion method. One of the characteristics of SEM is a large focal depth. It provides an excellent picture of macropore morphology; however, a quantitative analysis of macropore interfaces on the SEM image is not easy since geometrical information from the thick focal depth is projected onto a single 2-dimensional (2D) image. The mercury-intrusion method, on the other hand, provides quantitative information on macropore size and volume. However, the accuracy in geometrical information remains ambiguous, since the cylindrical macropore shape and a specific contact angle have to be assumed in the conversion of intrusion pressure into macropore diameter using the Washburn equation [6].

We have proposed a novel method to characterize macropore structure using laser scanning confocal microscopy (LSCM) [7–9]. The LSCM supplies a thin (ca. 150 nm) focal plane and 3D images can be reconstructed via marching cube algorithm (MCA) using a computer by stacking the series of 2D LSCM images in the axial direction. A vector analysis calculation on the reconstructed images provides surface curvatures and gel skeleton thicknesses and other important parameters to characterize macropore structure [10–13].

In this section, the observation of cocontinuous macropores of silica monolith using LSCM, the structural parameters that can be obtained, and the application of LSCM method for the further improvement of monolith are discussed. An

Monolithic Silicas in Separation Science. Edited by K.K. Unger, N. Tanaka, and E. Machtejevas
© 2011 WILEY-VCH Verlag GmbH & Co. KGaA, Weinheim
ISBN: 978-3-527-32575-7

overview of the fundamental techniques is followed by the characterization of monoliths prepared in a macroscopic scale. Deformation of macroporous structure due to the presence of a microsurface of a mold, which becomes more important when preparing miniaturized separation media, is also characterized by LSCM.

5.2
Preparation of Macroporous Silica Monolith

First, poly(ethylene glycol) (M_w = 100 000, denoted as PEO10) and D-sorbitol were homogeneously dissolved in 1 M nitric acid. Then, 6.5 g of tetraethoxysilane (TEOS) was added under vigorous stirring under ice-cooled conditions. The weight ratios of the starting compositions were TEOS:1 M HNO_3:PEO10:D-sorbitol = 1 : 1.23–1.54 : 0.12–0.16 : 0–0.08. The starting compositions of the samples are listed in Table 5.1. After 30 min of stirring, the resultant homogeneous solution was transferred into a polypropylene tube (6 mm inner diameter, 20 cm long) and allowed to gel at 40 °C in a closed tube. After aging at the same temperature for 10 h, the resultant gel was immersed in water for 3 h in order to remove the residual acid, and solvent-exchanged with 1 M aqueous urea for 3 h. The following hydrothermal treatment took 5 h at 110 °C in 1.5 M urea aq. After 2 h of solvent exchange with water, obtained gels were dried and heat treated at 600 °C for 5 h in order to remove residual organics.

Table 5.1 Starting compositions of monoliths.

	1MNH$_3$/g	PEO10/g	D-Sorbitol/g	TEOS/g
A	7.0	0.90	0.0	6.4
B	7.0	0.90	0.8	6.4
C	8.0	0.90	0.4	6.4
D	8.0	0.90	0.0	6.4
E	8.0	0.90	0.6	6.4
F	9.0	0.90	0.0	6.4
G	9.0	0.90	0.8	6.4
H	9.0	0.90	0.2	6.4
I	10.0	0.90	0.2	6.4
J	10.0	0.97	0.0	6.4
K	9.0	0.90	0.0	6.4
L	10.0	0.95	0.0	6.4

5.3
Laser Scanning Confocal Microscope Observation

For the LSCM observation, silica monoliths are sliced into pieces of $2.5 \times 5 \times 5\,mm^3$ in size, and immersed in a mixture of formamide, benzyl alcohol and fluorescein as a marker dye. The typical ratio is formamide:benzyl alcohol:fluorescein = 1 : 3 : 0.05 by weight, where the mixed-solvent solution has the same refractive index as silica within the range of ±0.01. This process allows the laser light to transmit the sample with suppressed scattering and provides a better contrast in the obtained LSCM images.

A laser with 488 nm wavelength is used to excite fluorescein. The wavelength is determined depending on the marker dye. A long-pass filter (LP505) was installed in front of a photomultiplier in order to detect only the fluorescent light (approximately 519 nm) and an oil-immersed 100×/NA = 1.40 (Plan-Apochromat, Carl Zeiss, Germany) objective lens is employed. The typical setup of LSCM is shown in Figure 5.1. The laser is scanned in the lateral plane, measuring the fluorescent intensity in a 2D optically sliced image composed of N^2 ($N = 512$) pixels [2], where N is the number of pixels along the edge of the 2D image. The observation is performed in planes with varied depth by changing the distance of the objective and the sample, with the increment along the optical axis, Δz. The increment is determined according to the structural dimension, ranging

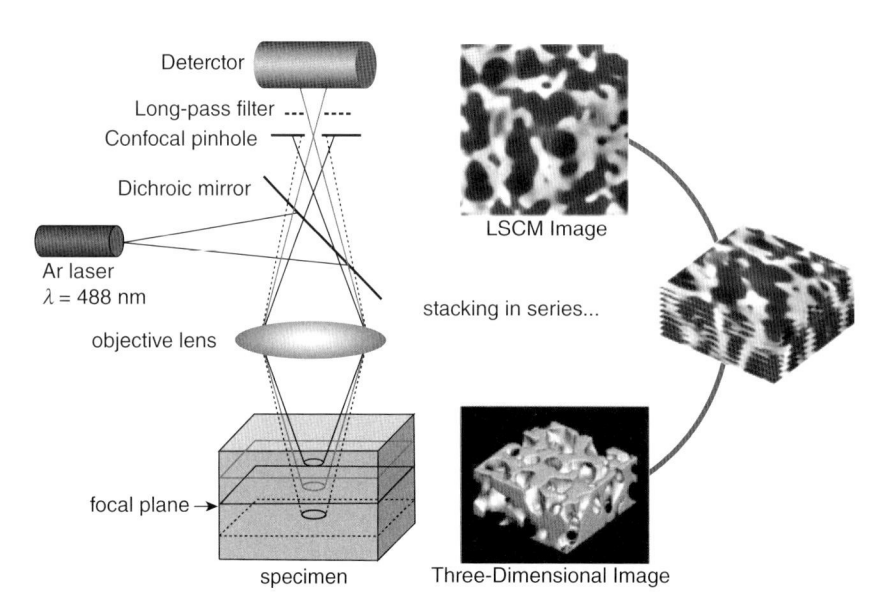

Figure 5.1 Schematic principle of LSCM and image processings. A long-pass filter (LP505) is installed in front of the photomultiplier in order to detect only the fluorescent light (approximately 519 nm) and oil-immersed 63× or 100×/NA = 1.40, or 40×/NA = 1.30 objective is employed.

from 0.15 to 0.45 µm. Since the image of an ideal point object always spreads, or is elongated in the direction of incidence, the obtained LSCM images further have to be taken through a deconvolution process to eliminate the "point spread", and then are provided for image processings.

5.4
Image Processing

The LSCM images have to be binarized before reconstructions into 3D images and a proper threshold of brightness is needed for the binarization process. The raw LSCM images sometimes suffer from noises and spatial intensity variations (i.e. inhomogeneous illuminations and high-frequency noises) due to aberrations of the objective, which complicate the threshold determination. In order to remove the intensity variation, the LSCM images are processed by "contrast-variance enhancement (CVE)" method [12]. The CVE is a technique that makes all local regions of an output image have an equal variance in terms of their image intensities. The noises are removed by a median filter, a class of nonlinear spatial filter that was designed to remove outlier pixels with completely inconsistent intensities with their surrounding values. The thresholds are calculated on the processed images according to Otsu's method [14]. Otsu's method depends on statistical calculations and determines the threshold so as to minimize the intraclass variance and to maximize the interclass variance. The series of binarized images are reconstructed with marching cubes algorithm into a 3D image as shown in Figure 5.2. Each of the reconstructed images is composed of $512 \times 512 \times$ (the number of optical slices) voxels, which have black or white in color (0 or 1 in digital information), and the voxel size is typically around $0.05 \times 0.05 \times 0.010 \, \text{mm}^3$.

5.5
Fundamental Parameters

Fundamental parameters are calculated for the prepared monoliths (A)–(L), with various combinations of macropore size and porosity. The fundamental parameters include porosity, specific surface area, characteristic wavelength, skeleton thickness, macropore size, and chord length. Since the size of mesopores is below the resolution of LSCM, ca. 0.2 µm, only the macropore surfaces are examined. In Figure 5.2, a polymer blend system consisting of poly(styrene) ($M_w = 200\,000$) and poly(methyl methacrylate) ($M_w = 60\,000$) is exhibited as a representative of the isotropic spinodal decomposition (SD). Besides, a gyroid model, which can be treated as 3D ordered cocontinuous morphology with cubic crystalline symmetry, is considered. A single pixel is assumed to be 1 µm for the gyroid structure. The calculation was performed on the 3D images reconstructed from a series of optical sliced images obtained by LSCM observation.

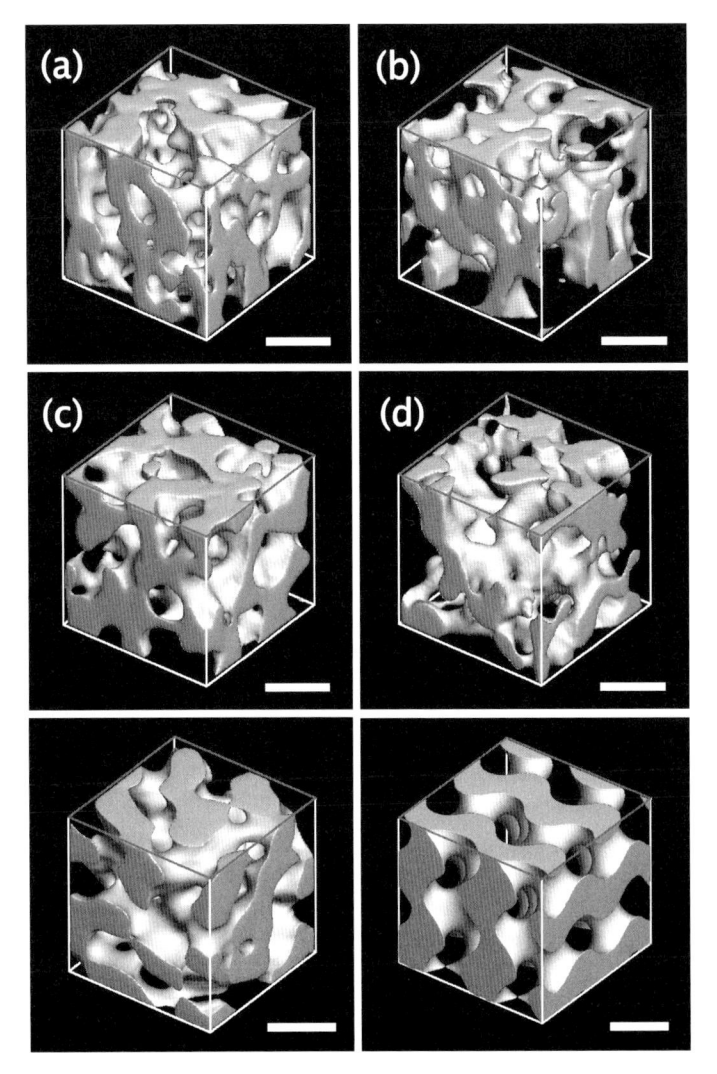

Figure 5.2 Overview of the 3D structures reconstructed by the image processings of the samples (a) **A**, (b) **C**, (c) **F**, (d) **K**, (e) polymer blend, and (f) the gyroid model. The lengths of the bars are 5 μm for (a)–(d), and 100 μm for (e) and (f).

5.5.1
Porosity

Porosity is an important parameter to characterize the porous structure and is defined as a volume fraction of pore space in a porous material. In the LSCM method, porosity ε was simply calculated as "voxel number in pore region/voxel number of entire region". Porosity has significant influences upon physical properties of porous materials such as thermal conductivity, elasticity, mechanical strength, and liquid transport, the latter being a major concern in this study. As described above, there are several methods to measure porosity: the traditional mercury-intrusion method and image processing using LSCM. The mercury-intrusion method can measure larger volumes of a sample at one time and the obtained data is less affected by inhomogeneity of the macropore structure. Although LSCM has a small observation volume, it can measure the surfaces of the macropore structure point by point in 3D, and the obtained data is based on the actual macropore structure. Porosity estimation by mercury intrusion and LSCM methods each contain several drawbacks and advantages, and LSCM has supremacy because the characterization is based on a real macroporous structure.

5.5.2
Surface Area

Traditionally, structural dimension of cocontinuous structure has been evaluated by the macropore size. However, macropore size is not always a reasonable representation for structural size. In the case of comparing macropore structures with the same macropore sizes and different porosities, structural dimensions are observed to be smaller with increasing porosity. One of the reasonable ways to compare structural dimension between nonhomothetic structures is to compare specific surface areas, a_s.

Interface areas are calculated by measuring the total area of the triangles that model interface polygons generated by MCA. Volumetric specific surface areas are then obtained by dividing the interface area by the total volume of the reconstructed 3D image, and are listed in Table 5.2. These values are related to macropore sizes and porosities. A close investigation of these relationships is left for the following discussion.

5.5.3
Characteristic Wavelength

Characteristic wavelength, Λ_m has been the principal structural parameter of the cocontinuous structure, which derives from SD [15–17]. It is well known that any pattern can be interpreted as a superposition of sinusoidal curves with various wavelengths, and the contribution of each wavelength (wavelength distribution) is calculated by the fast Fourier transformation (FFT) of the structure.

In the case of the real monoliths, the wave-number distribution of the cocontinuous pattern has a single peak at the characteristic wavelength, which supports the periodicity of the macropore structure of silica monoliths. The characteristic wavelength is close to the summation of average thicknesses of the two phases for the polymer blend with isotropic SD structure and the gyroid model structure, as exhibited in Table 5.2. However, the characteristic wavelength shows a significant discrepancy from the summation of skeletons and macropores in the case of the real monoliths.

5.5 Fundamental Parameters

Table 5.2 An overview of structural parameters of real monoliths.

	ε	$a_s/\mu m^{-1}$	$\Lambda_m/\mu m$	$d_s/\mu m$	σ_s	$d_p/\mu m$	σ_p	$d_c/\mu m$	H^*	K^*
A	0.515	0.836	9.2	2.73	0.296	3.16	0.39	5.87	0.175	−1.241
B	0.534	0.762	11.8	3.45	0.312	4.17	0.41	6.80	0.183	−1.267
C	0.551	0.750	15.4	2.88	0.324	3.67	0.43	7.05	0.060	−1.569
D	0.555	0.841	10.5	2.55	0.311	3.20	0.43	6.27	0.038	−1.500
E	0.565	0.668	17.5	3.21	0.321	4.02	0.44	7.84	0.022	−1.624
F	0.575	0.668	14.6	3.13	0.335	4.03	0.47	7.76	−0.012	−1.672
G	0.584	0.280	40.9	7.41	0.304	9.37	0.50	16.99	0.006	−1.355
H	0.587	0.618	19.4	3.36	0.335	4.40	0.47	8.72	−0.064	−1.762
I	0.613	0.411	25.5	4.70	0.313	6.17	0.50	11.39	−0.162	−1.820
J	0.634	0.794	13.8	2.39	0.309	3.26	0.46	6.13	−0.174	−1.880
K	0.664	0.632	16.4	2.91	0.327	4.36	0.51	6.99	−0.314	−1.754
L	0.576	0.623	18.3	3.13	0.329	5.26	0.50	8.13	−0.129	−1.872
Polymer blend	0.494	0.161	76.7	36.5	0.07	36.5	0.07	73.5	0.007	−0.002
Gyroid model	0.495	0.025	110.6	55.4	0.01	55.4	0.00	109.0	0.000	−0.001

* One pixel assumed to be one µm for gyroid structure.

The real monoliths studied in the present measurement are synthesized using PEO as a phase-separation inducer. In the process of phase separation, PEO, a water-soluble polymer, exhibits strong specific attractive interaction with silanol groups on polymerizing silica oligomers, and forms a PEO–silica complex. The hydrogen-bonded PEO–silica complex becomes less-soluble in the solvent due to the consumption of hydrophilic sites, and the system eventually phase separates into two phases: one phase that is rich in the PEO–silica complex and the other rich in the solvent mixture. Phase separation in PEO-derived systems occurs at the relatively early stage of polymerization with a low degree of cross-linking, and the polymerization still continues even after the onset of phase separation. The PEO–silica complex thus formed exhibits high viscosity, and at the same time, contains a considerable amount of silanols loosely blocked by the PEO chains from free condensation with other silanols. Both inter- and intramolecular condensation is hence slowed down. The separated phase domains remain viscous and deformable before being firmly cross-linked by prolonged condensations. The difference in viscoelasticity between the two separating phases in monolith systems elongates the gel skeletons and thickens the junctions that connect the gel skeletons. It accordingly deforms the phase-separating structure from isotropic SD, and the characteristic wavelength determined by FFT is largely overestimated by the increased separation of interjunction distances.

The process enhances the heterogeneity of the macropore structure of PEO-derived silica monolith. The normalized wave-number distribution, which is known as structure function [12], $F(q,t)$, in Figure 5.3, exhibits the broader dis-

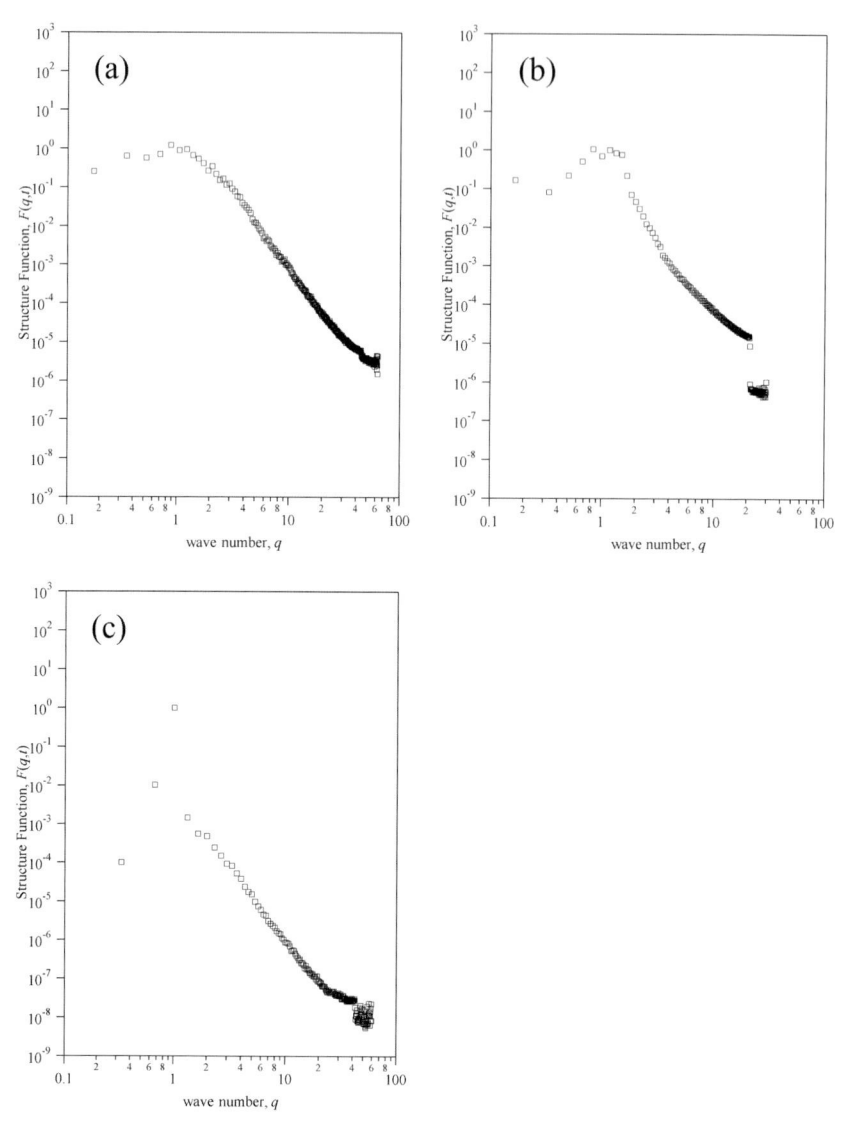

Figure 5.3 Structure functions of (a) the real monolith (**A**), (b) polymer blend, and (c) the gyroid model.

tribution with increasing heterogeneity; a broad distribution for the real monolith, intermediate for the polymer blend underwent isotropic SD, and a sharp distribution for the gyroid model. The relation between heterogeneity and the broadness of structure function can be confirmed by the standard deviations of skeleton thickness and macropore size. The FFT of one structure can thus be a measure of the periodicity and periodic length, or characteristic wavelength, of the structure.

5.5.4
Macropore Size and Skeleton Thickness

The distributions of macropore size and skeleton thickness are obtained from the skeletonization process [18] and the results are exhibited in Figure 5.4. The informal definition of the skeletonization is a line representation of an object, that is; (1) one-pixel thick, (2) through the middle of the object, and (3) preserves the topology of the object. By skeletonizing macropores and gel skeletons, the thin core of the macropores and gel skeletons can be determined, and by calculating the distances between the thin core and the interface, the distributions of macropore size and skeleton thickness can be obtained. The skeleton thickness and macropore size are determined further by using the Gaussian fitting procedure on the distributions of those obtained from the skeletonization.

Macropore size has a broader distribution compared to skeleton thickness, as shown in Figure 5.4. The scaled standard deviations of skeleton and macropores, σ_s, σ_p, calculated from the distributions are listed in Table 5.2. Each of the standard deviations is normalized with the mean value in order to eliminate the influence of dimension.

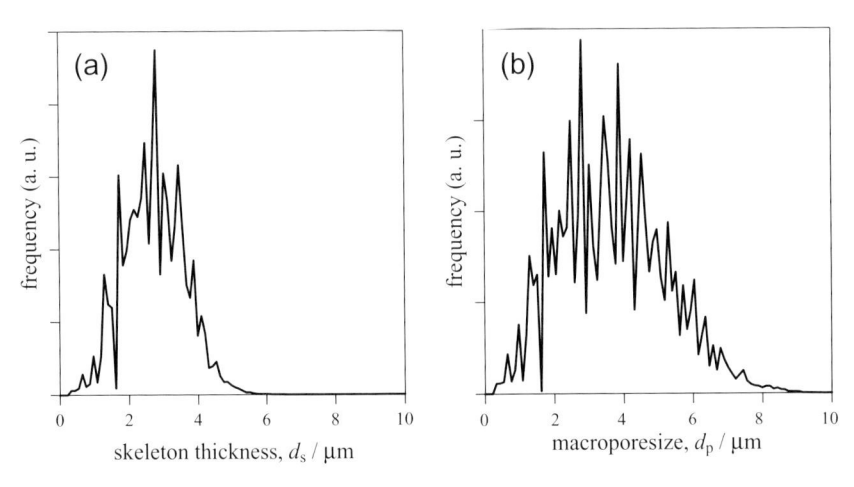

Figure 5.4 Distributions of (a) skeleton thickness d_s and (b) macropore size d_p are depicted. Macropore size has a broader distribution compared to the skeleton thickness.

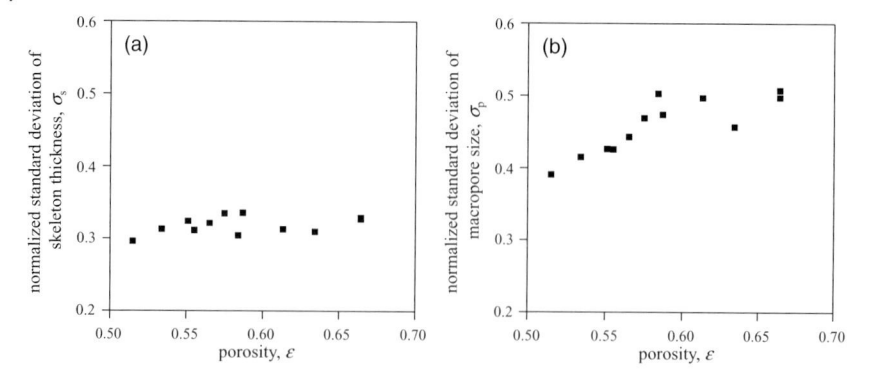

Figure 5.5 Normalized standard deviations of (a) skeleton thickness σ_s and (b) macropore size σ_p are depicted. Macropore size has the larger standard deviations compared to the skeleton thickness.

The normalized standard deviation is plotted against porosity in Figure 5.5. The heterogeneity in skeleton thickness increased slightly with porosity; however, the change is negligible. The heterogeneity in macropore size, on the other hand, took on a steeper increase with increasing porosity, suggesting that the structural deformation during phase separation is influenced by the volume fraction of siloxane phase. That is, the viscosity of the siloxane phase is much higher than that of the solvent phase and the decreased ratio of siloxane phase enhances the fluidity of the overall phase-separating liquid. The macropores therefore would be disturbed by a hydrodynamic flow induced by the development of phase-separating structure and/or an inhomogeneous temperature distribution in the sol. However, the connectivity of the siloxane phase is conserved by its viscoelasticity and hence the heterogeneity in the skeleton thickness hardly changes.

5.5.5
Chord Length

Chord length, d_c is also one of the measures for the structural dimension in addition to the specific surface area. The original definition of chord length is the distance between two intersections of a line and an arc; however, the interpretation of chord length is extended to the distance between intersections of a line and a curved surface in the study of the phase-separating structure. The distance between two neighboring intersections however is not a reasonable measure for the structural dimension since the value is likely influenced by the porosity difference. With higher porosity, the skeleton thickness becomes smaller compared to the case of lower porosity with comparable structural dimension.

The most reasonable measure is to calculate the distance between second-neighboring intersections. The influence of porosity is canceled out by adding the chord lengths of macropore and gel skeleton parts.

Chord lengths are calculated by drawing 341 lines in different direction in the obtained 3D image, counting the number of voxel for the pore and skeleton sec-

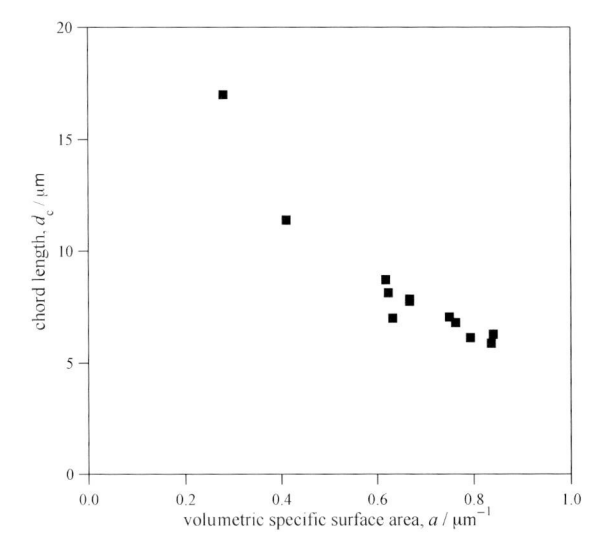

Figure 5.6 Chord lengths d_c are calculated from reconstructed images and plotted against specific surface area a. The chord length increases proportionally to the specific surface area. This implies that the chord length can be a reasonable measure for the structural dimension.

tions on every drawn line, and calculating most probable value by the Gaussian fitting. The calculated values are listed in Table 5.2 and plotted against specific surface area in Figure 5.6. The chord length changes proportionally with the specific surface area. This implies that the chord length can be a reasonable measure for the structural dimension; however, it is difficult to comprehend what the chord length represents among the structural parameters of a cocontinuous structure. It is not macropore size, skeleton thickness, nor characteristic wavelength, Λ_m, which is approximated by the summation of macropore size and skeleton thickness.

5.5.6
Mean Curvature and Gaussian Curvature

The curvature of a curved surface is the measure for a deviation from a flat surface. The curvature of a site on an arc is equal to the reciprocal of the radius of a circle that fits best to the curve at the site (osculating circle). Similarly, local principal curvatures, κ_1 and κ_2, on a curved surface are defined as reciprocals of the largest and the smallest osculating circles at a specific point. Mean curvature is defined as the mean of the local principal curvatures and Gaussian curvature is the product of the local principal curvatures.

A parallel surface to the interface is formed by translating the original interface along its normal by an equal distance at any point on the surface. The relation between the area of the infinitesimal patch at a point p, $da(0, p)$, and that of the parallel patch, $da(t, p)$, is

$$da(t,p) = da(0,p)(1 + 2H(p)t + K(p)t^2) \tag{5.1}$$

where $H(p)$ and $K(p)$ are, respectively, the mean and Gaussian curvatures at the point p and t is a signed displacement of the parallel surface from the interface; it is positive if the direction of the displacement points to one side of the surface, and negative otherwise. Summing Equation 5.1 over the whole area of the interface by changing the position of the point of interest, p, gives

$$A(t) = A(0)(1 + 2\langle H \rangle t + \langle K \rangle t^2) \tag{5.2}$$

where $A(0)$ is the total area of the original surface and $A(t)$ is that of the parallel surface with displacement t. Thus, area-averaged mean and Gaussian curvatures are calculated using Equation 5.2.

In order to eliminate the influence of the structural dimension, area-averaged mean and Gaussian curvatures calculated by this parallel surface method (PSM) [10, 19] have been further scaled by inverse of specific surface area, S^{-1} ($S = S'/V$, where S' and V corresponds to the total gel skeleton surface area and the total sample volume, respectively). The inverse of specific surface area is employed to eliminate the influence of the structural dimension.

Scaled area averaged mean and Gaussian curvatures, H^* and K^*, respectively, are plotted against porosity, ε. The values of H^* have positive value at lower porosity, and then decrease with increasing porosity and finally take negative values, as show in Figure 5.7a. The values of K^*, on the other hand, are located in $K^* < 0$ region for all samples as shown in Figure 5.7b, and decrease with increasing porosity.

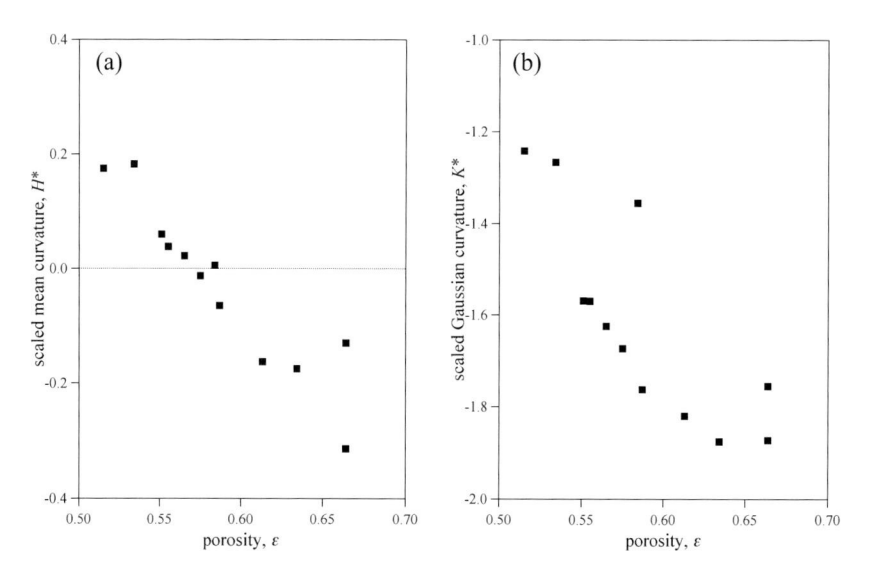

Figure 5.7 (a) Scaled mean curvatures, H^*, and (b) scaled Gaussian curvatures, K^*, are plotted against porosity ε.

At $\varepsilon < 0.6$, H^* takes positive values and K^* takes negative values, which suggests that the macropore surface on an average is curved inward relative to the macropore side. Such kind of the surface forms convexly curved macropores, while keeping the continuity according to the SEM and reconstructed 3D images. Such macropores that meet both of the requirements should have the convexly curved domains for the most part and bottleneck domains that connect the convexly curved parts.

At a porosity around $\varepsilon = 0.6$, H^* takes values around zero and K^* takes negative values, which suggests that the macropore surface on an average is curved evenly to both of the macropore and gel skeleton phases. The expectation from this kind of surface is symmetry between macropore and gel skeleton structures. Macropore surface is also expected to be evenly curved since the driving force of the structural evolution during SD is a decrease in the interfacial energy, or interface area between the polymerizing siloxane and solvent phases, which later on forms gel skeletons and macropores, respectively. The size distribution of macropores with porosity around 0.6 is hence expected to be sharper when compared to those with porosity less than 0.6.

At porosity over $\varepsilon > 0.6$, both H^* and K^* take negative values, which suggests that the macropore surface on average is curved inward relative to the gel skeleton side, or that the gel skeleton surface is curved outward relative to the macropore side. The more negative values of K^* represent the fact that the macropore surface is concave relative to the macropore side, and hence the macropores are expected to have a kind of structure generated by fiber-like gel skeletons.

5.5.7
Curvature Distributions

As shown in Figure 5.8a, the mean and Gaussian curvatures exhibit the principal shape of surfaces, for example, $K > 0$ corresponds to an elliptic surface (such as a sphere), $K = 0$ to a parabolic surface (such as a cylinder) and $K < 0$ to a hyperbolic surface (like saddle shape) and H is related to the symmetrical property of two coexisting phases. By plotting the probability density distribution of the combination of the mean and Gaussian curvatures, $P(H, K)$, the overall characteristic of the macropore surfaces can be examined. The curvature distribution is calculated by a "sectioning and fitting method (SFM)" [20].

5.5.7.1 Comparison between Different Porosities
For monoliths **B**, **H** and **K**, probability density distributions of the local curvature are shown in Figures 5.8b, d and c, respectively. The distribution area increases with increasing porosity, suggesting that the surface geometry becomes more manifold with increasing porosity. The peak position at the lower porosity locates in the $H > 0$ region and is eventually shifted to $H < 0$ region with porosity.

At the same time, the tail on the $H < 0$ side grows following the $K = H^2$ curve. Mean and Gaussian curvatures related by $K = H^2$ represent a spherical surface of the gel skeletons. This suggests that the macropore structure at higher porosity

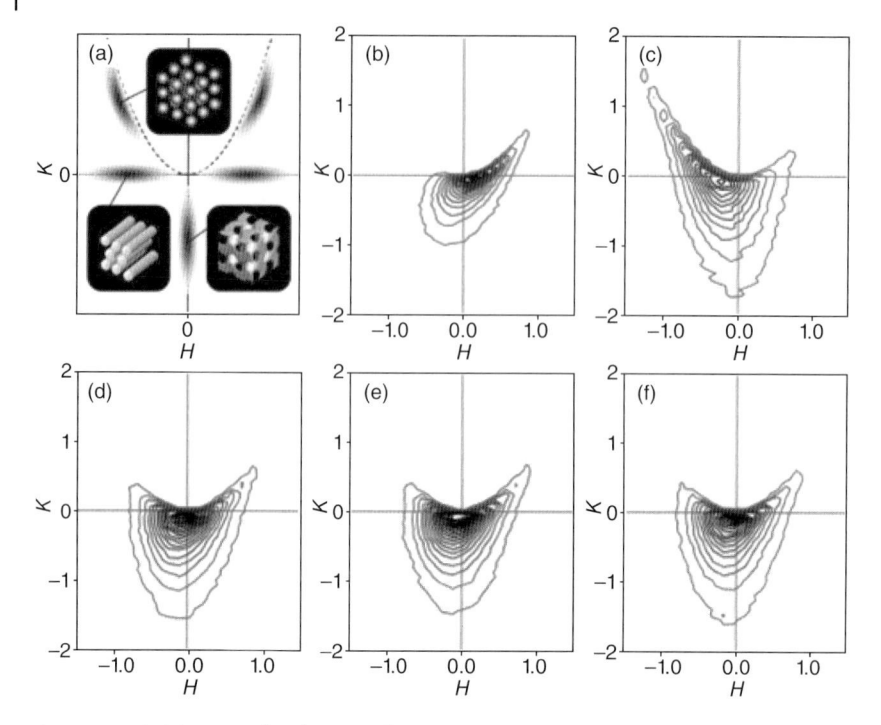

Figure 5.8 (a) Schematic distributions of local curvatures and the corresponding structures, where H is the scaled mean curvature and G is the scaled Gaussian curvature. Curvature distributions of the samples (b) **B** ($d_p = 4.17\,\mu m$ and $\varepsilon = 0.534$), (c) **K** ($d_p = 4.36\,\mu m$ and $\varepsilon = 0.664$), (d) **H** ($d_p = 4.40\,\mu m$ and $\varepsilon = 0.587$), (e) **L** ($d_p = 5.26\,\mu m$ and $\varepsilon = 0.576$), and (f) **G** ($d_p = 9.37\,\mu m$, $\varepsilon = 0.584$).

is not simply constructed only by fiber-like gel skeletons, but also by sphere-like surfaces caused by the Rayleigh instability of the cylindrical thread of fluids [21]. A cylindrical thread of fluid is known to become unstable to surface wave fluctuations whose length is larger than the circumference of the cylinder. The capillary wave fluctuations, which are brought about in the thread, build up pressure gradients that drive the thread-forming fluid from a thinner part to the thicker part. The thread is eventually broken up into a number of spherical droplets aligned along the thread. Once the breakups of the bridge occur, it is expected to split into a pair of elliptic surfaces, that is, corns facing each other.

5.5.7.2 Comparison between Different Macropore Sizes

For the monoliths **H**, **L** and **G**, probability density distributions of local curvatures are also shown in Figures 5.8d, e and f, respectively. The peak position and the distribution range are almost the same for the monoliths **H** and **L**. This suggests that the self-similar growth of phase-separating structure during SD. The peak position for the monolith **G** is also located closely to those for the monoliths **H** and **L**, but the distribution range has shifted to the smaller H direction following $K = 0$. This suggests the existence of fibrous gel skeletons.

5.6
Three-Dimensional Observation of Deformations in Confined Geometry

In the preceding sections, we have shown the methodology and advantage of 3D observation utilizing LSCM and of image analyses of 3D interfacial data. In the following sections, we will present the observation of deformations of macroporous structures in confined geometry by LSCM. In the applications of macroporous monoliths to miniaturized formats such as capillary and microfluidic channels, it is known that different porous structures from the corresponding bulk emerge [22–28]. Here, "bulk" means the isotropic porous monoliths prepared in a free volume, that is, without effects of confinement. The fact that polyphasic structure of polymer blends and solutions is influenced by the presence of surfaces has been investigated by several groups in polymer science [29–31]. Above all, Tanaka *et al.* has revealed that hydrodynamic flow in liquid polymer blends/solutions plays an important role in the phase-formation process in spinodal decomposition [32]. Preferential wetting of one component toward the surfaces is induced by the hydrodynamic flow in the intermediate and late stages of spinodal decomposition, which results in deformation of the polyphasic structure. The similar conclusion has been drawn by Wang *et al.* who reported that wetting is driven by the hydrodynamic flow in a critical mixture of PMMA/dPMMA [33]. Later, Jinnai *et al.* observed these phenomena by LSCM and vividly showed that wetting severely influences the overall structure of a polymer blend confined in a thick gap between plates [34]. Since silica and hybrid monoliths are derived from spinodal decomposition, a similar effect would occur in a gap between plates, in a space in a capillary and microchannels, and even inside a macroporous monolith. These kinds of confined geometry are utilized for the preparation of films for thin-layer chromatography, capillary column [35–37] and a lab-on-a-chip system, and a surface coating [38] of a preformed monolith.

5.6.1
Synthesis and Characters of Organic–Inorganic Hybrid Monoliths

In order to obtain functionality and/or improving mechanical properties, organotrialkoxysilanes, instead of tetraalkoxysilanes, are widely used as the siloxane sources. Methyltrimethoxysilane (MTMS) is one of the typical trifunctional alkoxysilanes in which a nonreactive methyl group is directly attached to the silicon atom. Whereas trifunctional alkoxysilanes possess such functional and mechanical advantages, they are difficult to form a homogeneous gel due to the formation of 4-membered cyclic species (T_4), 8-membered cubic crystals (T_8) and other incomplete species based on closed rings [39–41]. Moreover, since the organic group attached to each silicon atom exhibits hydrophobicity, the condensed species tend to phase separate from the mother solution in the course of the polycondensation, and then form hydrophobic resins and precipitates in most cases. To overcome this difficulty, we prepared MTMS-derived gels using limited amounts of water and polar solvent such as methanol and formamide [37]. In such a condition, condensates form a relatively homogeneous branched network

Table 5.3 Phase separation time t_{ps} and gelation time t_{gel} of the typical compositions in MF and MM systems.

	t_{ps}/min	t_{gel}/min	$t_{gel} - t_{ps}$/min
MTMS:methanol:H_2O = 1:1.0:2.0	120	184	64
MTMS:formamide:H_2O = 1:2.3:2.5	<<3039[a]	3148	>>109

a) It was difficult to precisely determine t_{ps} due to the close refractive indices between MTMS-derived gel-rich and formamide-rich phases.

due to the high concentration of MTMS and phase separation is moderately suppressed by the solvent phase that is less polar than water. Homogeneous gelation occurs prior to macroscopic phase separation (precipitation) as a result.

To prepare MTMS-derived macroporous monoliths in a confined geometry, mainly two systems have been investigated; MTMS-formamide system (denoted as the MF system) and MTMS-methanol system (denoted as the MM system). The phase-separation time (t_{ps}) and the gelation time (t_{gel}) are listed in Table 5.3 for the typical compositions in both systems. The sol-gel transition and the relevant structural formation by spinodal decomposition are significantly more sluggish in the MF system because the polymerization reaction proceeds under the pH at around the isoelectric point (IEP) of MTMS [42]. The pH increase is caused by the hydrolysis of formamide into formic acid and ammonia in the strong acid solution. Higher viscosity in MF system is another reason for the slow polycondensation and spinodal decomposition. This means that the MF system remains fluidic during the structural formation by spinodal decomposition for the longer duration. In contrast, since the reactions in MM system proceeds under low pH and viscosity is lower, gelation time and phase-separation time are considerably shorter than in MF system.

5.6.2
Deformed Macroporous Structures between Plates in MF System

The overall procedure and some important definitions in determining the deformation by LSCM are exhibited in Figure 5.9. The upper half part describes the preparation procedure in which bulk gel (prepared without surface effects) and confined gel in a mold are simultaneously obtained from the identical sol in which the mold is immersed. The periodic length of the bulk gel Λ_m can be determined as the sum of skeleton thickness and pore diameter. The LSCM observation of the mold is also shown, in which a thick gel film (thickness = D) between plates is observed in along the direction of the thickness. Other detailed experimental conditions and setups are available in our previous reports [22, 37].

Figure 5.10 depicts 3D reconstructed LSCM images of confined gels prepared with MF system. Here, we determine the value D/Λ_m, which gives an intuitive

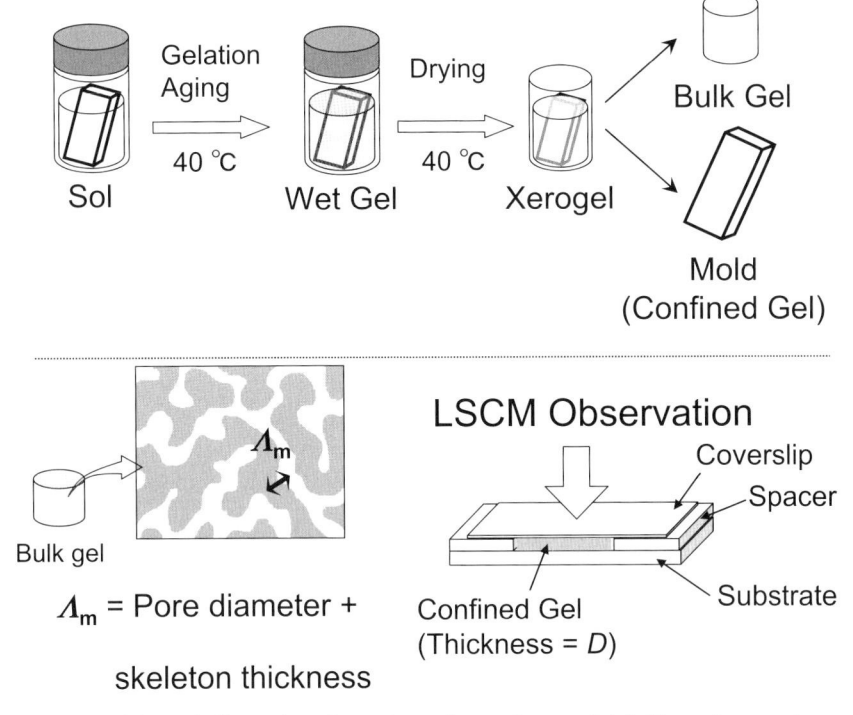

5.6 Three-Dimensional Observation of Deformations in Confined Geometry | 97

Figure 5.9 Schematic illustration of experimental procedures and definitions of parameters for the confinement experiments.

Λ_m = Pore diameter + skeleton thickness

Confined Gel (Thickness = D)

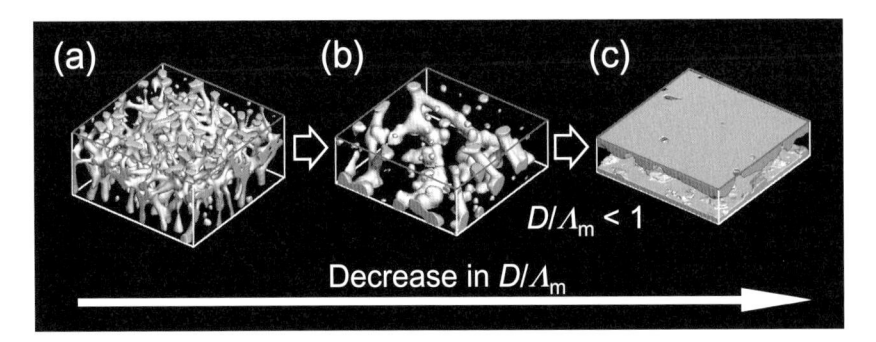

Figure 5.10 Deformation of phase-separated structure of gels prepared with MF system confined between parallel plates. Deformation develops with decreasing D/Λ_m and wetting transition occurs when $D/\Lambda_m < 1$. (a) $D/\Lambda_m = 2.9$, (b) $D/\Lambda_m = 2.1$, and (c) $D/\Lambda_m = 0.67$.

expression of "how large is the confining space with respect to the characteristic length of spinodal decomposition in bulk". In the present case, the values of D/Λ_m are (a) 2.9, (b) 2.1 and (c) 0.67. That is, the thickness of confining spaces of (a) and (b) are respectively 2.9 and 2.1 times longer than the characteristic length of spinodal decomposition, whereas that of (c) is shorter than the characteristic length. The uppermost and the lowermost planes in the reconstructed images correspond to the planes that had been touched with soda-lime silica glass slide and coverslip surfaces during gelation. The elongated gel skeletons near the surfaces can be already confirmed in part (a) where the confinement effect is not very strong. The wetting effect becomes stronger in part (b), which mainly consists of columns connecting two parallel surfaces. In part (c), it is shown that all the gel phase completely wets the surfaces when the confining space becomes smaller than the characteristic length D, which is termed a "wetting transition".

To closely investigate the structural differences, the sample shown in Figure 5.10a is cut into three regions according to the volume fraction of gel phase and shown in Figure 5.11. The volume fractions at a certain depth can be simply

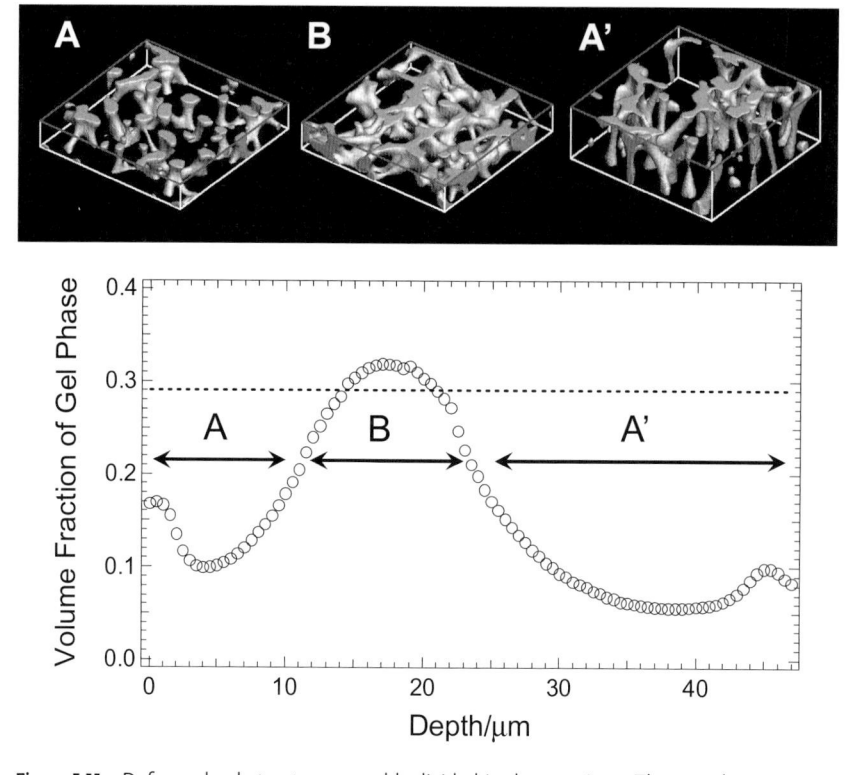

Figure 5.11 Deformed gel structures roughly divided in three regions. The sample is identical to that shown in Figure 5.2a. The dotted line in the depth-volume fraction profile shows the volume fraction of gel phase in bulk.

obtained by counting the numbers of black or white pixels in a digitized image at a desired depth. The profiles such as those shown in Figure 5.11 actually indicate the variation of porosity in the depth direction if we take the vertical axis in "volume fraction of pores". In the upper and lower parts (parts A and A'), the columnar gel skeletons elongated perpendicular to the mold surfaces are more clearly seen and the volume fraction of gel phase in these regions is smaller than that of bulk (= 0.29, dotted line). Meanwhile, the middle region labeled with "B" consists of the bicontinuous structure, which is similar to the bulk. However, as confirmed in Figure 5.11, the volume fraction of gel phase is oscillating and the characteristic length in this region is 1.5 times longer than that of bulk. Scaled mean curvature H^* and Gaussian curvature K^* obtained by the PSM revealed that both curvatures in the region B are similar to the bulk, that is, $H^* \sim 0$ and $K^* < 0$. By contrast, those for regions A and A' show the larger absolute value of H that results from the elongated columnar shape. This indicates, according to the Laplace equation describing the relationship between H and pressure, that the pressure in the columns during the fluidic-phase formation process is larger than the pressure in the wetting layer adsorbed on the mold surfaces. This is good evidence for the hydrodynamic pumping mechanism proposed by Tanaka *et al.* [32, 43], in which the hydrodynamic preferential wetting is accelerated by the pressure difference between columnar "channels" and wetting layers. This mechanism finally leads to the complete wetting morphology as shown in Figure 5.10c.

5.6.3
Deformed Macroporous Structures between Plates in MM System

Compared to MF system that allows considerably longer duration of fluidic-phase separation, the confined gels prepared with MM system show less-deformed structures, as shown in Figure 5.12a. Although D/Λ_m (=2.6) is smaller than the sample in Figure 5.10a, elongated columns cannot be seen in this sample. The sliced digitized image displayed inset shows the one at a depth of a local minimum of volume fraction of gel phase, in which many lateral connections of gel phase (shown in black) can be confirmed. The reduced deformation in the MM system is attributed to the fact that the deformation due to wetting develops only during the fluidic-phase separation. After the gel obtains a certain elasticity when the system comes near the gel point, the deformation is no longer possible. In this viewpoint, the MM system is more suitable to tailor macroporous monoliths inside narrow spaces without being significantly influenced by the surfaces. However, in a thick film ($D/\Lambda_m \sim 18$), deformation near the surface becomes significant, as shown in Figure 5.12b. This is due to the gravity effect, which becomes dominant in a system with relatively small surface tension between phases and with large asymmetry in densities of the phases [44]. The siloxane gel-rich phase is much higher in density (1.35 g cm^{-3} by He pycnometry) compared to methanol (0.79 g cm^{-3}), which is a main constituent of the solvent-rich phase. The elongated part near the upper surface is more rigorously stretched by the large volume of the confined gel. In other words, a rather large tensile stress

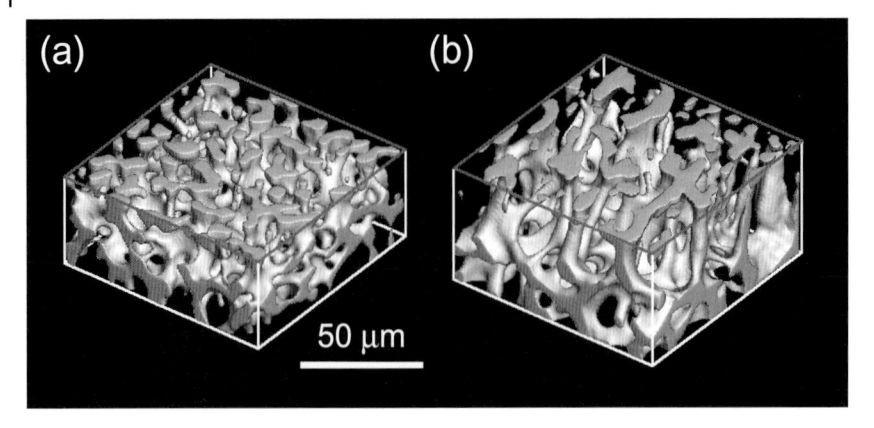

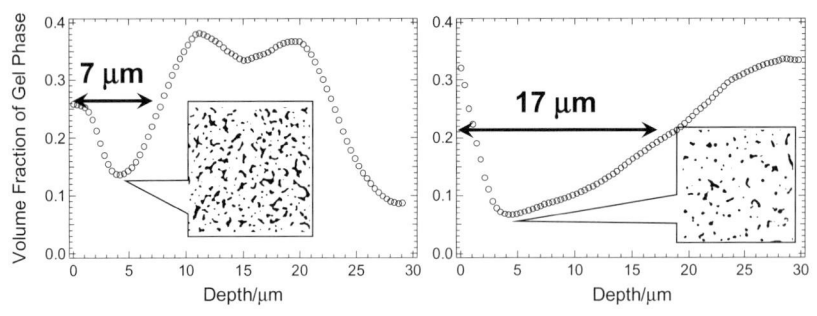

Figure 5.12 Phase-separated structure of gels prepared with MM system confined between parallel plates. (a) $D/\Lambda_m = 2.6$ and (b) $D/\Lambda_m \sim 18$ (only the near-surface region is shown).

is imposed on the columnar skeletons, which are the weakest portion in the whole film. The more deformed structure is also confirmed by the larger absolute values of H^* in this region.

By positively utilizing the gravity effect, it is possible to design a pillar structure inside a capillary with rectangular cross-sections [25]. Figure 5.13 demonstrates the pillar formation in a modified system containing N,N-dimethylacetamide (DMA) as solvent, which acts as cosolvent of MTMS-derived condensates and water. Increasing amount of DMA thus suppresses the phase separation tendency. In a 30-μm gap between parallel plates of the capillary, it can be confirmed that the pillar structure forms as a result of wetting and stretching by the gravity. This tendency also becomes stronger on decreasing D/Λ_m, that is, decreasing D and/or increasing Λ_m. In the present case, since the decreasing amount of DMA leads to the higher phase-separation tendency, that is, increasing Λ_m, the lateral continuity of the gel phase becomes less on decreasing the DMA/MTMS ratio, as can be seen in the digitized sliced LSCM image. The porosity of the pillar structure becomes as high as >0.80 at the middle of the capillary thickness, and becomes higher with decreasing DMA/MTMS ratio.

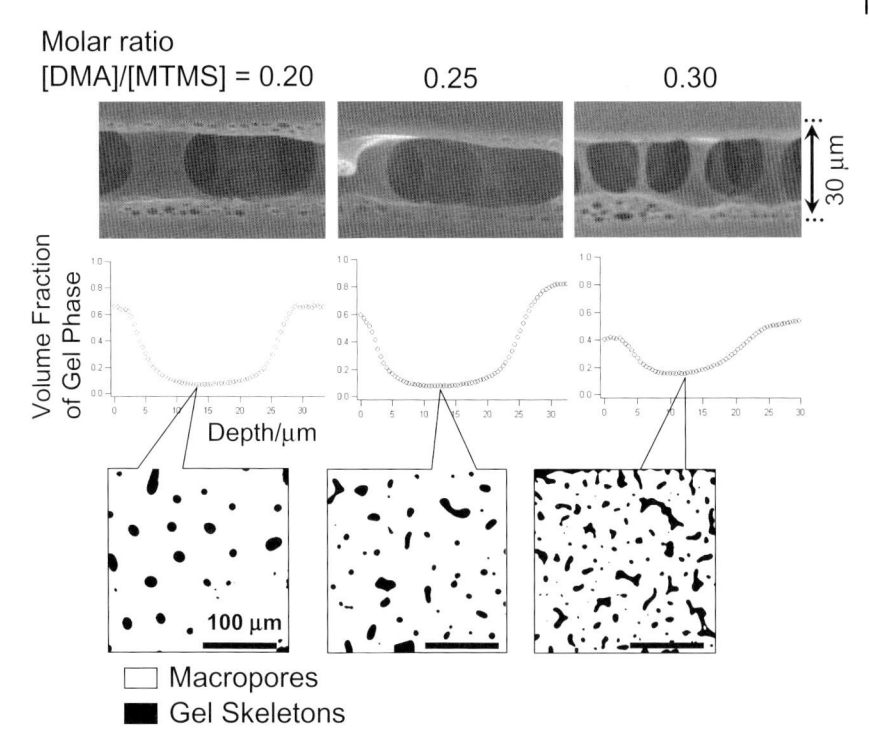

5.6 Three-Dimensional Observation of Deformations in Confined Geometry | 101

Figure 5.13 Pillar structure developed in a capillary with a rectangular cross-section (thickness = 30 μm). The volume fraction profile data and digitized images obtained by LSCM are also shown.

The structural feature and high porosity of the pillar structure allows faster liquid transport, as demonstrated in Figure 5.14. In the case where the MTMS-derived gel forms bicontinuous-type structure (at molar ratio [DMA]/[MTMS] = 0.3 and 0.4), the pressure drop of the rectangular capillary monolith (thickness = 50 μm) is as large as a regular capillary monolith with pore diameter of 3 μm. The pressure drop of the commercialized regular capillary monolith MonoCap with pore diameter of 2 μm is even larger due to the small pore size. The rectangular capillary monolith with the pillar structure obtained at [DMA]/[MTMS] = 0.2 shows a considerably small pressure drop, which indicates the possibility of increased linear velocity while maintaining a small pressure drop. The small pressure drop can be attributed to the pillar structure, which does not possess the connections perpendicular to the pillars, and large through-pore size (distance in-between pillars). However, since the through-pore size is more effective among these two, a further improvement of the gel structure such as adjusting through-pore size and increasing specific surface area will be required in order to utilize the pillared capillary monolith as a separation medium.

Molar ratio
[DMA]/[MTMS] = 0.20   0.30   0.40

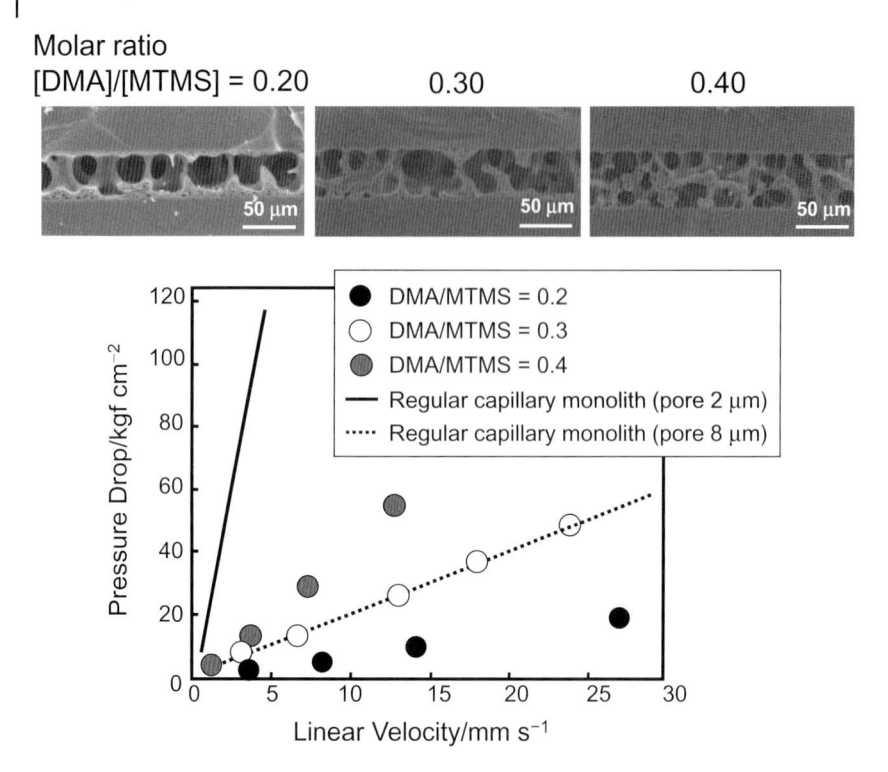

Figure 5.14 Phase-separated gel structures in a capillary with thickness of 50 μm and the relationship between pressure drop and linear velocity of capillary columns at a length of 50 mm. Data of regular capillary monoliths with pore diameter 2 and 8 μm are also shown for comparison.

References

1. Minakuchi, H., Nakanishi, K., Soga, N., Ishizuka, N., and Tanaka, N. (1996) *Anal. Chem.*, **68**, 3498–3501.
2. Ishizuka, N., Minakuchi, H., Nakanishi, K., Soga, N., Nagayama, H., Hosoya, K., and Tanaka, N. (2002) *Anal. Chem.*, **75**, 1275–1280.
3. Vervoort, N., Gzil, P., Baron, G.V., and Desmet, G. (2003) *Anal. Chem.*, **75**, 843–850.
4. Vervoort, N., Gzil, P., Baron, G.V., and Desmet, G. (2004) *J. Chromatogr. A*, **1030**, 177–186.
5. Gzil, P., Vervoort, N., Baron, G.V., and Desmet, G. (2004) *J. Sep. Sci.*, **27**, 887–896.
6. Washburn, E.W. (1912) *Phys. Rev.*, **17**, 273–283.
7. Saito, H., Nakanishi, K., Hirao, K., and Jinnai, H. (2006) *J. Chromatogr. A*, **1119**, 95–104.
8. Saito, H., Kanamori, K., Nakanishi, K., Hirao, K., Nishikawa, Y., and Jinnai, H. (2007) *Colloids Surf. A*, **300**, 245–252.
9. Saito, H., Kanamori, K., Nakanishi, K., and Hirao, K. (2007) *J. Sep. Sci.*, **17**, 2881–2887.
10. Jinnai, H., Koga, T., Nishikawa, Y., Hashimoto, T., and Hyde, S.T. (1997) *Phys. Rev. Lett.*, **78**, 2248–2251.
11. Nishikawa, Y., Jinnai, H., Koga, T., Hashimoto, T., and Hyde, S.T. (1998) *Langmuir*, **14**, 1242–1249.
12. Jinnai, H., Nishikawa, Y., Morimoto, H., Koga, T., and Hashimoto, T. (2000) *Langmuir*, **16**, 4380–4393.

13 Jinnai, H., Nakanishi, K., Nishikawa, Y., Yamanaka, J., and Hashimoto, T. (2001) *Langmuir*, **17**, 619–625.
14 Otsu, N. (1979) *IEEE Trans. Syst. Man Cybern.*, **9**, 62–66.
15 Gunton, J.D., Minguel, M.S., and Sahni, P.S. (1983) *Phase Transition and Critical Phenomena*, Academic Press, London, pp. 269–275.
16 Hashimoto, T. (1988) *Phase Trans.*, **12**, 47–119.
17 Cahn, J.W. (1965) *J. Chem. Phys.*, **42**, 93–99.
18 Jinnai, H., Watashiba, H., Kajihara, T., and Takahashi, M. (2003) *J. Chem. Phys.*, **119**, 7554–7559.
19 Hyde, S.T. (1989) *J. Chem. Phys.*, **93**, 1458–1464.
20 Jinnai, H., Nishikawa, Y., Spontak, R.J., Smith, S.D., Agard, D.A., and Hashimoto, T. (2000) *Phys. Rev. Lett.*, **84**, 518–521.
21 Rayleigh, L. (1878) *Proc. Lond. Math. Soc.*, **s1-10**, 4–13.
22 Kanamori, K., Nakanishi, K., Hirao, K., and Jinnai, H. (2003) *Langmuir*, **19**, 5581–5585.
23 Kanamori, K., Nakanishi, K., Hirao, K., and Jinnai, H. (2003) *Langmuir*, **19**, 9101–9103.
24 Kanamori, K., Nakanishi, K., Hirao, K., and Jinnai, H. (2004) *Colloids Surf. A*, **241**, 215–224.
25 Suzumura, Y., Kanamori, K., Nakanishi, K., Hirao, K., and Yamamichi, J. (2006) *J. Chromatogr. A*, **1119**, 88–94.
26 He, M., Zeng, Y., Sun, X., and Harrison, D.J. (2008) *Electrophoresis*, **29**, 2980–2986.
27 Kanamori, K., Nakanishi, K., and Hanada, T. (2009) *Soft Matter*, **5**, 3106–3113.
28 Detobel, F., Eghbali, H., De Bruyne, S., Terryn, H., Gardeniers, H., and Desmet, G. (2009) *J. Chromatogr. A*. **1216**, 7360–7367.
29 Jones, R.A., Norton, L.J., Kramer, E.J., Bates, F.S., and Wiltzius, P. (1991) *Phys. Rev. Lett.*, **66**, 1326–1329.
30 Geoghegan, M., and Krausch, G. (2003) *Prog. Polym. Sci.*, **28**, 261–302.
31 Puri, S. (2007) *Physica A*, **384**, 100–107.
32 Tanaka, H. (2001) *J. Phys. Condens. Matter*, **13**, 4637–4674.
33 Wang, H., and Composto, R.J. (2000) *Phys. Rev. E*, **61**, 1659–1663.
34 Jinnai, H., Kitagishi, H., Hamano, K., Nishikawa, Y., and Takahashi, M. (2003) *Phys. Rev. E*, **67**, 021801.
35 Peters, E.C., Petro, M., Svec, F., and Fréchet, J.M.J. (1997) *Anal. Chem.*, **69**, 3646–3649.
36 Ishizuka, N., Minakuchi, H., Nakanishi, K., Soga, N., Nagayama, H., Hosoya, K., and Tanaka, N. (2000) *Anal. Chem.*, **72**, 1275–1280.
37 Kanamori, K., Yonezawa, H., Nakanishi, K., Hirao, K., and Jinnai, H. (2004) *J. Sep. Sci.*, **27**, 874–886.
38 Kanamori, K., Nakanishi, K., and Hanada, T. (2006) *J. Sep. Sci.*, **29**, 2463–2470.
39 Roy, D.A., Baugher, B.M., Baugher, C.R., Schneider, D.A., and Rahimian, K. (2000) *Chem. Mater.*, **12**, 3624–3632.
40 Matjka, L., Dukh, O., Hlavatá, D., Meissner, B., and Brus, J. (2001) *Macromolecules*, **34**, 6904–6914.
41 Dong, H., Lee, M., Thomas, R.D., Zhang, Z., Feidy, R.F., and Mueller, D.W. (2003) *J. Sol-Gel Sci. Technol.*, **28**, 5–14.
42 Brinker, C.J., and Scherer, G.W. (1990) *Sol-Gel Science: The Physics and Chemistry of Sol-Gel Processing*, Academic Press, New York, pp. 140–141.
43 Tanaka, H. (1996) *Phys. Rev. E*, **54**, 1709–1714.
44 Ma, W.-J., Maritan, A., Banavar, J.R., and Koplik, J. (1992) *Phys. Rev. A*, **45**, R5347–R5350.

6
Modeling Chromatographic Band Broadening in Monolithic Columns

Frederik Detobel and Gert Desmet

6.1
Introduction

Since their introduction in the early and mid-1990s [1–3], monolithic columns have revolutionized the way of thinking in liquid chromatography (LC) and have led to a number of groundbreaking performances and applications. To guide the current research on their further improvement, the availability of mathematical models that describe the band broadening as a function of the geometrical characteristics and the physicochemical parameters of the analytes and mobile phase (diffusion coefficients, retention factors) is crucial [4]. The simplest and most frequently used band broadening models in LC are the traditional "A,B,C-term"-models [5, 6]. However, the use of these models often does not go beyond the stage wherein the A,B,C-term constants are considered as black-box variables, whereas the ultimate chromatographic model would provide mathematical expressions for A, B and C leading to an accurate estimate for the plate-height curve just by filling in the geometrical characteristics of the bed, and the diffusion coefficients and retention coefficients of the analytes.

The geometrical characteristics of a packing structure are difficult to quantify, especially for the case of monolithic columns, because of the higher degree of geometrical freedom compared to a packed-bed column. Monolithic columns have also been less extensively studied so that only a limited number of empirical and semiempirical correlations are available to express the relation between the A, B and C-parameters and their morphology. Important work on the establishment of such relationships has been done by Meyers and Liapis [7], Tanaka and coworkers [8–11], Miyabe and coworkers [12, 13] and Tallarek and coworkers [14–16], involving the introduction of geometrical concepts such as random capillary networks [7], through-pore and domain sizes [9–11], equivalent sphere diameters [14–16] and cylindrical skeleton equivalents [14–16].

Correlations that describe the dependency of the A, B and C terms on the retention factor and the (intra- and extraskeleton) diffusion coefficients of the analytes are generally obtained by working out the general diffusion and convection mass balance. This mass balance has for the first time been solved by Lapidus and

Monolithic Silicas in Separation Science. Edited by K.K. Unger, N. Tanaka, and E. Machtejevas
© 2011 WILEY-VCH Verlag GmbH & Co. KGaA, Weinheim
ISBN: 978-3-527-32575-7

Amundson [17], using the method of moments. The same approach has been used by van Deemter et al. [18] to establish their famous expressions, which already contained the same explicit retention factor dependency as the models that have been proposed later. Giddings [19, 20] arrived at the same expressions as van Deemter et al., but obtained them in an alternative way, using the so-called non-equilibrium theory. For the specific case of the silica monolith columns developed by Nakanishi and Tanaka [9–11], Miyabe and Guiochon [12, 13, 21] approximated the monolith geometry as a cylindrical skeleton concentrically surrounded by a cylindrical pore and used the techniques of Lapidus and Amundson (so-called method of moments). The obtained expressions for the mobile phase and the stationary phase mass-transfer contribution are identical to those obtained by Giddings in 1961 [19]. Furthermore, it has recently been shown [22] that, although the different analytical LC plate-height expressions introduced by Giddings, van Deemter, Huber, Horvath and all those who followed in their footsteps [12, 18, 20, 21, 23–32] often appear very differently and employ totally different symbols, they can all be reduced to one single expression, further referred to by "the general plate-height model". Even the most recent work on the band-broadening theory [31, 33], yields expressions that are identical to the solutions obtained by van Deemter and Giddings, who in turn followed the 1952 paper of Lapidus and Amundson [17]. The only (occasional) difference among the different literature plate-height expressions is due to differences in the expressions used for eddy dispersion and for the mass-transfer coefficients. These expressions are mainly empirical correlations that are most certainly only valid for a limited number of geometries and support shapes.

Although these general rate models appear sound, they are still subject to a number of hypotheses and approximations. For example, the flow in the pores is approximated by a perfect plug flow, whereas the flow has a parabolic profile. A second assumption is that the mass transfer in the mobile and stationary zone occurs independent of each other and that their contribution to the observed plate height is additive. Another problem is that the mass-transfer rate in the mobile zone still has to be estimated from semiempirical correlations, like the Wilson–Geankoplis [34] or the penetration model [12], that have not been truly validated for the specific geometry of silica monoliths [12, 13, 21].

In this chapter, the different parameters appearing in the general plate-height model are discussed and some modeling guidelines are formulated for the case specific to monolithic columns. As an illustration, these guidelines are then applied to the numerical band-broadening data of a tetrahedral skeleton model (TSM), a simplified mimic of the sponge-like morphology of silica monoliths, as these data give the clearest view on the accuracy of the general plate-height model.

6.2
The General Plate-Height Model

According to Ref. [22] and with a notation that has been adapted to the presently considered case of monolithic columns, the general plate-height expression can be written as:

$$H = H_{ax} + H_{Cm} + H_{Cs} \qquad (6.1)$$

where H_{ax} is the band broadening originating from the eddy-dispersion and the longitudinal diffusion, and H_{Cm} and H_{Cs}, respectively, represent the contributions from the mass transfer resistance in the through-pore and the skeleton region, given by:

$$H_{Cm} = 2 \cdot \frac{k''^2}{(1+k'')^2} \cdot u_i \cdot \frac{V_{tp}}{\Omega} \cdot \frac{1}{k_{f,tp}} \qquad (6.2)$$

$$H_{Cs} = 2 \cdot \frac{k''^2}{(1+k'')^2} \cdot u_i \cdot \frac{V_{tp}}{\Omega} \cdot \frac{1}{k_{f,skel} \cdot K_P} \qquad (6.3)$$

6.2.1
Meaning of k'', u_i and K_P

A first important feature of Equations 6.2 and 6.3 is that they do not depend on the phase retention factor k' (defined with respect to the linear chromatographic velocity u_0) but on the zone-retention factor k'' (defined with respect to the linear average through-pore velocity u_i):

$$k'' = \frac{u_i t_R}{L} - 1 \qquad (6.4)$$

The linear interstitial or through-pore velocity u_i is generally defined on the basis of the flow rate F, the cross-sectional area S of the column and the external porosity ε:

$$u_i = \frac{F}{S \cdot \varepsilon} \qquad (6.5)$$

whereas the linear chromatographic velocity u_0 is generally defined on the basis of t_0 the time of a nonretained marker:

$$u_0 = \frac{L}{t_0} \qquad (6.6)$$

Both velocities become identical for a nonretained component that is excluded from the mesopores of the skeleton.

Whereas k' is much more frequently used (simply because it can be directly assessed from the velocity of a nonretained marker), k'' is the intrinsic parameter determining the rate of band broadening, because the band-broadening process only depends on how long a species element is retained in the stagnant zone (monolithic skeleton in present case) and not upon whether this time is spent there in the unretained state (i.e. in the mobile phase liquid filling the mesopores of the skeleton) or in the retained state (i.e. in or on the stationary phase covering the mesopore surface).

If the volumetric fractions of the column that are accessible to the species outside and inside the skeleton (respectively, represented by the external and the internal porosity ε and ε_{int}) are known, k'' can easily be calculated from the chromatographically measurable k', using [22]:

$$k'' = k'(1+k_0'') + k_0'' \tag{6.7}$$

with

$$k_0'' = \frac{(1-\varepsilon)}{\varepsilon}\varepsilon_{\mathrm{int}} = \frac{\varepsilon_T}{\varepsilon} - 1 \tag{6.8}$$

where k_0'' is the zone-retention factor of a nonretained component and ε_T the total porosity. The latter also allows establishing the following relationship between u_i and u_0 [22]:

$$u_i = (1+k_0'')u_0 = \frac{\varepsilon_T}{\varepsilon} u_0 \tag{6.9}$$

Another parameter appearing in Equation 6.3 is the skeleton-based equilibrium constant K_p, defined as:

$$K_p = \frac{m_{\mathrm{eq,skel}}/V_{\mathrm{skel}}}{C_{m,\mathrm{eq}}} \tag{6.10}$$

wherein $m_{\mathrm{eq,skel}}$ is the amount (mass or mole) of species occupying the skeleton volume in equilibrium with the mobile phase. K_p is directly related to k'' via [22]:

$$k'' = \frac{1-\varepsilon}{\varepsilon} \cdot K_p \tag{6.11}$$

6.2.2
Expressions for H_{ax}

Basically, H_{ax} represents the dispersion caused by the erratic flow field (eddy-dispersion) and by the longitudinal component of the molecular diffusion through the skeleton and the through-pores. Usually, the latter is represented by the so-called B-term constant. Because the eddy-dispersion is related to the distance between the adjacent parallel flow paths, the characteristic length for this dispersion effect should be a combination of the skeleton size d_{skel} and the through-pore size d_{tp}. An often used characteristic length for monolithic columns is the so-called domain size, customarily defined as the sum of the through-pore and the skeleton diameter [8, 10, 11, 35]:

$$d_{\mathrm{dom}} = d_{\mathrm{tp}} + d_{\mathrm{skel}} \tag{6.12}$$

For the eddy-dispersion, many different expressions have been proposed in the literature. Combining them with the B-term dispersion and writing the eddy-diffusion in terms of the domain size, the set of expressions below covers most literature cases:

$$H_{ax} = d_{\mathrm{dom}} \cdot \left[A + \frac{B}{v_{i,\mathrm{dom}}} \right] \tag{6.13}$$

$$H_{ax} = d_{\mathrm{dom}} \cdot \left[A v_{i,\mathrm{dom}}^{1/3} + \frac{B}{v_{i,\mathrm{dom}}} \right] \tag{6.14}$$

$$H_{ax} = d_{dom} \cdot \left[\sum_j \left(\frac{1}{A_j} + \frac{1}{D_j \cdot v_{i,dom}} \right)^{-1} + \frac{B}{v_{i,dom}} \right] \quad (6.15)$$

In Equations 6.13–6.15, $v_{i,dom}$ is the u_i-based reduced velocity generally defined as:

$$v_{i,ref} = \frac{u_i d_{ref}}{D_{mol}}, \quad (6.16)$$

with the domains size as the reference length. As shown in [36], this implies that B should be defined as:

$$B = 2 \frac{D_{eff}}{D_{mol}} (1 + k'') \text{ or as } B = \frac{H_B \cdot u_i}{D_{mol}}, \quad (6.17)$$

with D_{eff} the apparent mobile phase diffusion coefficient and H_B the contribution of the observed plate height value for the limiting case of a zero velocity or flow rate. Several expressions are available in the literature to determine the value of D_{eff}. These expressions are generally of the form:

$$D_{eff,par} = \frac{\gamma D_{mol} + k'' D_{skel}}{1 + k''} \quad (6.18)$$

or, in terms of k':

$$D_{eff,par} = \frac{\gamma_m D_{mol} + k' \gamma_s D_s}{1 + k'} \quad (6.19)$$

where γ_s is a geometrical constant and γ and γ_m are the so-called obstruction factors of the monolithic structure It was, however, recently shown that the error involved when using these expressions can be very large, up to 100% and more [36, 37]. Therefore, the value of D_{eff} should be determined experimentally by measuring the band broadening at zero flow rate (i.e. peak-parking measurements).

6.2.3
Estimation of D_{mol} and D_{skel}

When modeling the band-broadening data of monoliths, the exact value of the molecular diffusion coefficient D_{mol} and D_{skel} are often unknown and need to be estimated. There are several correlations that allow the calculation of D_{mol}, like the Wilke-Chang correlation [38], the Scheibel equation [39] and others, but one should always be aware that there is a certain error involved with the use of these equations (up to 20%) [40]. D_{mol} can also be determined experimentally by conducting peak-parking measurements in an unobstructed (open tubular) medium. The value of intraskeleton diffusion coefficient is more difficult to achieve. According to Broeckhoven et al. [37], the only accurate calculation of D_{skel} would be based on numerical methods on an exact replica of the studied packing structure. To a first approximation, one could, however, estimate the value of D_{skel} from the measured D_{eff} values using for example the Landauer–Davis (LD) expression wherein the z value is a geometrical factor that need to be fitted to the D_{eff} data [36]:

$$D_{\text{eff,LD}} = D_{\text{skel}} \frac{k''}{(1+k'')}\frac{1}{(1-\varepsilon)}\left[A + \sqrt{A^2 + \frac{2}{(z-2)}\frac{(1-\varepsilon)}{\varepsilon k''}\frac{D_{\text{mol}}}{D_{\text{skel}}}}\right] \quad (6.20)$$

with

$$A = \frac{[(z/2)(1-\varepsilon)-1]\varepsilon k'' D_{\text{skel}} + [(z/2)\varepsilon+1](1-\varepsilon)D_{\text{mol}} - 1}{(z-2)\varepsilon k'' D_{\text{skel}}} \quad (6.21)$$

6.2.4
Expressions for $k_{f,tp}$ and $k_{f,skel}$

The mass-transfer coefficients $k_{f,tp}$ and $k_{f,skel}$ needed in Equations 6.2 and 6.3 are usually written in dimensionless terms using the so-called Sherwood number (Sh), generally defined using a reference length d_{ref} and a diffusion coefficient D_{diff} as:

$$\text{Sh} = \frac{k_f \cdot d_{\text{ref}}}{D_{\text{diff}}} \quad (6.22)$$

For the mass transfer inside a mesoporous region, where the convection is zero, this Sh number is simply a constant geometry-dependent shape factor. Using either the method of moments (see, e.g., Equations 6.22–6.24 of Miyabe et al. [32]) or the procedure described by Giddings (see e.g., the Appendix of De Malsche et al. [41]), it can be calculated that Sh_{part} would be equal to 8 in the case of an infinitely long cylinder (given that $\alpha = 4$, see Section 6.2.5).

Noting that the relevant characteristic size for the intraskeleton mass transfer in a monolithic structure should be the skeleton diameter d_{skel}, while the relevant diffusion coefficient is D_{skel}, it follows from Equation 6.22 that the mass-transfer coefficient $k_{f,skel}$ needed in Equation 6.3 should be expressed as

$$k_{f,\text{skel}} = \text{Sh}_{\text{skel}} \frac{D_{\text{skel}}}{d_{\text{skel}}} \quad (6.23)$$

wherein Sh_{skel} is a shape factor that in the case of monoliths needs to be determined via model fitting, given their complex structure.

To calculate the mass transfer inside the through-pore region ($k_{f,tp}$), convective effects need to be taken into account as well. Given the laminar nature of the flow, implying that the shape of the velocity field is independent of the absolute value of the velocity, these convective effects can all be grouped into a dimensionless velocity of the form given by Equation 6.16 [42].

A very broad and general expression that can be used to calculate the Sh number for the mass transfer in a region where both convection and diffusion play a role is given by:

$$\text{Sh}_{tp} = \frac{k_f \cdot d_{\text{ref}}}{D_{\text{diff}}} = \gamma_d + \gamma_c v_{i,\text{ref}}^n \quad (6.24)$$

wherein γ_d and γ_c are geometrical constants and n is a constant generally lying between 0 and 1. Equation 6.24 also covers the frequently used Wilson–Geankoplis correlation ($d_{ref} = d_{skel}$, $\gamma_d = 0$, $\gamma_c = 1.09/\varepsilon^{2/3}$ and $n = 1/3$) [34]:

$$Sh_{tp} = \frac{1.09}{\varepsilon^{2/3}} v_{i,skel}^{1/3} \tag{6.25}$$

the Kataoka model ($d_{ref} = d_{skel}$, $\gamma_d = 0$, $\gamma_c = 1.85(1-\varepsilon)^{1/3}$ and $n = 1/3$) [43]:

$$Sh_{tp} = 1.85(1-\varepsilon)^{1/3} v_{i,skel}^{1/3} \tag{6.26}$$

and the penetration model ($d_{ref} = d_{skel}$, $\gamma_d = 0$, $\gamma_c = 2/\pi^{1/2}$ and $n = 1/2$) [44]:

$$Sh_{tp} = \frac{2}{\sqrt{\pi}} v_{i,skel}^{1/2} \tag{6.27}$$

All these correlation are based on the skeleton size d_{skel}. It makes, however, physically more sense to relate $k_{f,tp}$ to the through pore size (d_{tp}). With the mobile phase diffusion coefficient (D_{mol}) as the relevant diffusion coefficient, this yields:

$$k_{f,tp} = Sh_{tp} \frac{D_{mol}}{d_{tp}} = (\gamma_d + \gamma_c v_{i,tp}^n) \frac{D_{mol}}{d_{tp}} \tag{6.28}$$

It should be noted that the use of a different characteristic size will also result in different values for γ_d and γ_c.

6.2.5
Selection of the Characteristic Reference Lengths

Although Equations 6.1–6.3 do not explicitly depend upon a specified reference length, the expressions that are needed to calculate H_{ax}, $k_{f,tp}$ and $k_{f,skel}$ do so. Unfortunately, each of these parameters depends on a different characteristic length (d_{dom}, d_{tp} and d_{skel}, respectively). The fact that three different sizes are needed obviously complicates the modeling, but this is only a minor issue compared to the fact that there are no straightforward rules to define and measure these quantities.

It seems, for example, straightforward to define d_{skel} as the diameter at the position where the skeleton branch is purely cylindrical as for example indicated in Figure 6.1, but this does not take into account that the regions where the different branches intersect have a noncylindrical geometry. Furthermore, and more importantly, the volume in the intersection nodes has almost no outer surface, so that the d_{skel} parameter as defined in Figure 6.1, is certainly not the correct characteristic size describing the mass transfer throughout the entire skeleton. An appropriate definition for d_{skel} should also allow the direct calculation of the skeleton-volume-to-outer-surface-ratio (V_{skel}/Ω), which is in turn needed to calculate the V_{tp}/Ω ratio, another crucial factor appearing in Equations 6.2 and 6.3, and given by:

$$\frac{V_{tp}}{\Omega} = \frac{V_{tp}}{V_{skel}} \frac{V_{skel}}{\Omega} = \frac{\varepsilon}{1-\varepsilon} \frac{d_{skel}}{\alpha} \tag{6.29}$$

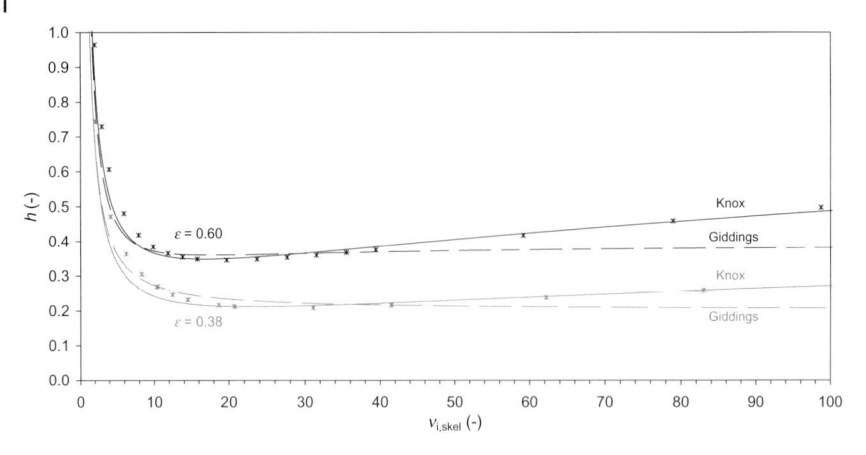

Figure 6.1 Representation of the 3D tetrahedral skeleton model (TSM) with an external porosity of (a) $\varepsilon = 0.60$ and (b) $\varepsilon = 0.38$. The complete structure is defined by the length of the cylindrical branches l_{skel} and the diameter d_{skel}. The arrow indicates the direction of the fluid flow (reproduced from ref. [46]).

wherein α is another shape factor, with $\alpha = 4$ for the case of infinitely long cylindrical pillars, while $\alpha = 6$ for spherical particles.

Finding an appropriate definition for the through-pore size (d_{tp}) is, given the complex structure of the through-pore space, even more difficult. One approach would be to use the hydraulic diameter, a typical engineering parameter representing the diameter one would obtain if each through-pore were replaced by a cylindrical tube having the same liquid volume to solid surface ratio as the skeleton. The general definition [45] of the hydraulic through-pore diameter ($d_{tp,H}$) is closely related to the V_{tp}/Ω ratio already defined in Equation 6.29:

$$d_{tp,H} = \frac{4V_{tp}}{\Omega} = \frac{4\varepsilon}{1-\varepsilon}\frac{d_{skel}}{\alpha} \qquad (6.30)$$

Another obvious measure for the through-pore diameter is the diameter based on the flow resistance of the bed. In this case, each through-pore is replaced by a cylindrical tube with a diameter selected such that it has the same permeability as the through-pore network of the monolith. Using the equivalence [45] between Poiseuille's pressure-drop law for cylindrical tubes and the law of Kozeny–Carman for porous media, and introducing the cylinder shape factor K_0 (=2 for a cylindrical tube) and the tortuosity τ of the through-pores, an equivalent cylinder diameter can be obtained according to:

$$d_{tp,\varphi} = 4\sqrt{2\tau^2 K_{vi}} \qquad (6.31)$$

wherein K_{vi} is the interstitial permeability, defined as:

$$K_{vi} = \frac{F}{S \cdot \varepsilon} \frac{\eta L}{\Delta P} \qquad (6.32)$$

There are, however, no sound arguments why any of these characteristic sizes would represent the true average mass-transfer distance and given that both the skeleton size and through-pore size are ill-defined, it goes without saying that the domain size d_{dom} appearing in the H_{ax} expressions (6.13, 6.14 and 6.15) is ill-defined as well.

Because there are as many different possible reduced-velocity definitions as there are possible characteristic lengths (and none can be preferred over the others on theoretical grounds), one should hence always be very cautious about the exact definition of the reduced velocity. From Equation 6.16 it follows readily that, upon replacing d_{ref} by d_{tp}, d_{dom} and d_{skel}, respectively:

$$v_{i,tp} = \frac{u_i d_{tp}}{D_{mol}} = \frac{d_{tp}}{d_{dom}} v_{i,dom} = \frac{d_{tp}}{d_{skel}} v_{i,skel}, \qquad (6.33)$$

6.2.6
Complete Plate-Height Equation

With Equations 6.23 and 6.28, Equation 6.1 can be expressed in full, and written as a function of the interstitial or through-pore velocity u_i and the diverse characteristic mass-transfer distances:

$$H = H_{ax} + 2\frac{k''^2}{(1+k'')^2}\frac{\varepsilon}{(1-\varepsilon)}\frac{1}{\alpha}\frac{d_{skel}}{d_{tp}}\frac{1}{\left(\gamma_d + \gamma_c\left(\frac{u_i \cdot d_{tp}}{D_{mol}}\right)^n\right)}\frac{d_{tp}^2}{D_{mol}}u_i$$

$$+ 2\frac{k''}{(1+k'')^2}\frac{1}{\alpha}\frac{1}{Sh_{skel}}\frac{d_{skel}^2}{D_{skel}}u_i \qquad (6.34)$$

A dimensionless variant of Equation 6.34 can be established by choosing an appropriate characteristic size and writing all terms of Equation 6.34 as a function of the same reference length. Basically, the two characteristic lengths appearing in Equation 6.34 are equally well suited for this purpose. In the remainder of this chapter, everything is written in terms of the skeleton size d_{skel}, because this size can be most easily measured experimentally. Doing so, and putting $h = H/d_{skel}$, and introducing the ratio of d_{tp}/d_{skel} as an extra shape factor, this yields:

$$h = \frac{H}{d_{skel}} = h_{ax} + 2\frac{k''^2}{(1+k'')^2}\frac{\varepsilon}{(1-\varepsilon)}\frac{1}{\alpha}\frac{d_{tp}}{d_{skel}}\frac{v_{i,skel}}{\left(\gamma_d + \gamma_c\left(\frac{d_{tp}}{d_{skel}}\right)^n v_{i,skel}^n\right)}$$

$$+ 2\frac{k''}{(1+k'')^2}\frac{1}{\alpha}\frac{1}{Sh_{skel}}\frac{D_{mol}}{D_{skel}}v_{i,skel} \qquad (6.35)$$

Given the complex geometry of a monolith, mathematical expressions that allow the prediction of the model shape factors are still lacking and the values of γ_d and

γ_c need to be fitted to band-broadening data. Accordingly, the problem of the ill-defined characteristic sizes can be masked by incorporating the d_{tp}/d_{skel} ratio into the values of these two shape factors (that need to be fitted anyway), resulting in a reduction of the parameter space of the general plate-height model:

$$h = h_{ax} + 2\frac{k''^2}{(1+k'')^2}\frac{\varepsilon}{(1-\varepsilon)}\frac{1}{\alpha}\frac{v_{i,skel}}{(\gamma'_d + \gamma'_c v''_{i,skel})} + 2\frac{k''}{(1+k'')^2}\frac{1}{\alpha}\frac{1}{\text{Sh}_{skel}}\frac{D_{mol}}{D_{skel}}v_{i,skel} \qquad (6.36)$$

with

$$\gamma'_d = \gamma_d \frac{d_{skel}}{d_{tp}} \qquad (6.37)$$

and

$$\gamma'_c = \gamma_c \left(\frac{d_{skel}}{d_{tp}}\right)^{n-1} \qquad (6.38)$$

It can easily be verified that the parameters γ'_d and γ'_c are equal to γ_d and γ_c that would have been obtained if $k_{f,tp}$ in Equation 6.28 was written in relation to the skeleton size d_{skel} instead of the through-pore size d_{tp}.

6.3
Use of the General Plate-Height Model to Predict Band Broadening in TSM Structures

Examining the validity of the general plate-height expression and its ability to predict the band broadening in silica monoliths is a difficult task, as most of the required geometrical constants and diffusion coefficients are not *a priori* known and need to be estimated, thereby introducing an inherent uncertainty. Hence, the most reliable strategy to illustrate the use of the general plate-height expression to model band broadening in monolithic columns is one based on data obtained from computer simulations of the band broadening in a simplified structure with exactly known geometry and diffusion coefficients. Detobel *et al.* [46] used the general plate-height equation to model band-broadening data obtained from tetrahedral skeleton models (TSM). These TSM structures, first presented by Vervoort *et al.* [47, 48], are based on the crystal lattice structure of diamond and consists of cylindrical branches with thickness d_{skel} and length l_{skel}, intersecting in node points connecting four neighboring branches under the tetrahedral angle of 109.47° (Figure 6.1). The TSM has the important advantage that all its structural properties (porosity, flow-through pore size, specific skeleton surface, branch connectivity) are implicitly known from the value of only two parameters: d_{skel} and l_{skel}. In this section, the strategy required to model plate-height data with the general plate-height expression is applied on band-broadening data obtained from two TSM structures ($\varepsilon = 0.38$ and $\varepsilon = 0.60$, Figure 6.1) at different degrees of retention (k'' varies from 0 to 10) and at different values for the intra-skeleton diffusivity ($D_{skel} = 1 \times 10^{-10}\,\text{m}^2\,\text{s}^{-1}$ and $D_{skel} = 5 \times 10^{-10}\,\text{m}^2\,\text{s}^{-1}$).

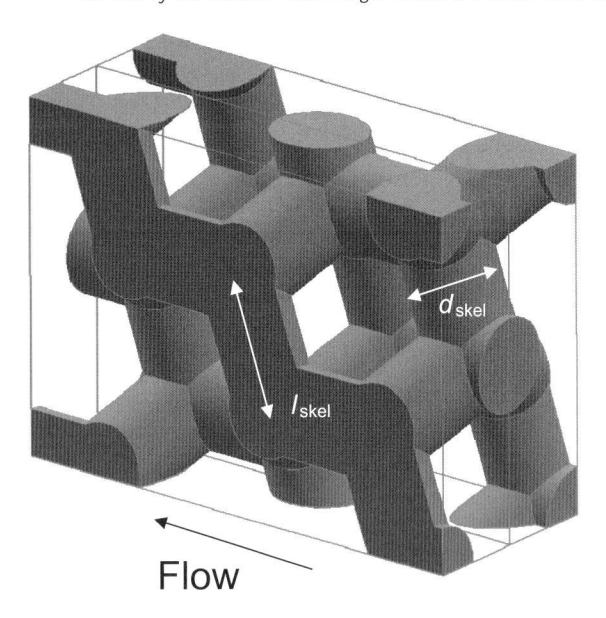

Figure 6.2 d_{skel}-based reduced van Deemter plot for a TSM-structure with $\varepsilon = 0.38$ (gray curves) and $\varepsilon = 0.60$ (black curves) in case of a nonporous skeleton ($k'' = 0$). The numerically simulated data points were fitted with the h_{ax} model of Knox (6.14) (full line) and Giddings (6.40) (dashed line) using the least-squares method. The D_{eff} values were obtained from peak parking simulations. The corresponding values of the fitting parameters A, A' and D' can be found in Tables 6.1 and 6.2 (reproduced from ref. [46]).

6.3.1
Nonporous Skeleton Case

Figure 6.2 shows the plate-height values obtained for the TSM structures in the case of a nonporous skeleton. In this case, the zone-retention factor k'' is zero, so that only the h_{ax} term remains in Equation 6.36. Any of the equations provided for the h_{ax} term (13–15) can be used to model the plate-height data. For the Giddings Equation 6.15, short-range, long-range and transcolumn coupling effects are absent in a perfectly ordered TSM structure, so that only the intrachannel coupling term has to be considered. In this case, the through-pore diameter is a more relevant characteristic distance than the domain size, so that Equation 6.15 can best be rewritten in dimensionless terms as:

$$h_{ax} = 2\frac{D_{eff}}{D_{mol}}(1+k'')\frac{1}{v_{i,skel}} + \frac{D\left(\dfrac{d_{tp}}{d_{skel}}\right)^2 v_{i,skel}}{1+\dfrac{D}{A}\dfrac{d_{tp}}{d_{skel}}v_{i,skel}} \tag{6.39}$$

However, as the values of A and D are geometry dependent and have to be fitted to the plate-height data, the ratio of d_{tp}/d_{skel} can be incorporated into their values, yielding an new set of parameters A' and D':

$$h_{ax} = 2\frac{D_{eff}}{D_{mol}}(1+k'')\frac{1}{v_{i,skel}} + \frac{D'v_{i,skel}}{1+\frac{D'}{A'}v_{i,skel}} \quad (6.40)$$

with

$$A' = A\frac{d_{tp}}{d_{skel}} \quad (6.41)$$

and $$D' = D\left(\frac{d_{tp}}{d_{skel}}\right)^2 \quad (6.42)$$

wherein A' and D' are equal to the values of A and D when Equation 6.15 is based on the skeleton size as the characteristic length instead of the through pore size. Note that for real silica monoliths, the other coupling terms appearing in the Giddings equation [20] should be taken into account as well.

Figure 6.2 displays two fitting results, one based on the Knox expression (6.14) and one based on the Giddings coupling theory (6.40). Both the Knox and the Giddings model do not perfectly fit the nonporous TSM band-broadening data, although the Knox model is slightly better, despite the lack of physical meaning. As described by Detobel et al. [46], the difference between the Knox and the Giddings eddy-dispersion models, however, almost completely vanishes when the plate-height curves are fitted for the porous TSM-cases, where the band broadening is to a very large extent determined by the h_{Cs} and the h_{Cm} term. Yet, the differences between both models will be much more pronounced in real monoliths because of the larger eddy-dispersion contribution to h.

6.3.2
Porous Skeleton Case

Traditionally, the modeling of band-broadening data of real monoliths is mainly performed by making (rough) approximations on the values of the different parameters appearing in the general plate-height model. The parallel connection of resistances expression (see 6.18) is for example used to estimate D_{eff}, while the Sh_i value appearing in the h_{Cm} term is calculated using either the Wilson–Geankoplis, the penetration model or the Kataoka model expression [12, 32]. The values for Sh_{part} and α are generally obtained by assuming that the geometry of a silica monolith can be approximated as an infinitely long cylinder [12, 49–51]. As described in the previous section, it can be calculated that in this case, Sh_{part} would be equal to 8 and the α value is given by $\alpha = 4$.

By applying this traditional modeling approach ($D_{eff} = D_{eff,par}$, $Sh_i = 1.09/\varepsilon.(\varepsilon.v_i)^{1/3}$ (Wilson–Geankoplis), $Sh_{part} = 8$ and $\alpha = 4$) to the TSM band-broadening data obtained for different degrees of retention (k'' varies from 1 to 10) and

6.3 Use of the General Plate-Height Model to Predict Band Broadening in TSM Structures

different intra-skeleton diffusion coefficients ($D_{skel} = 1 \times 10^{-10}\,m^2\,s^{-1}$ and $5 \times 10^{-10}\,m^2\,s^{-1}$) and by using the Giddings model (6.40) to describe the h_{ax} term (as this model theoretically more sound than the Knox model) the only unknown parameters appearing in the model are A' and D'. These parameters need to be fitted on the plate-height data, as their values are generally only known as an order of magnitude. The resulting fitting curves in Figure 6.3 readily show that the traditional modeling approach fails to accurately predict and describe the plate-height curves obtained for the TSM structures. This poor fitting quality is

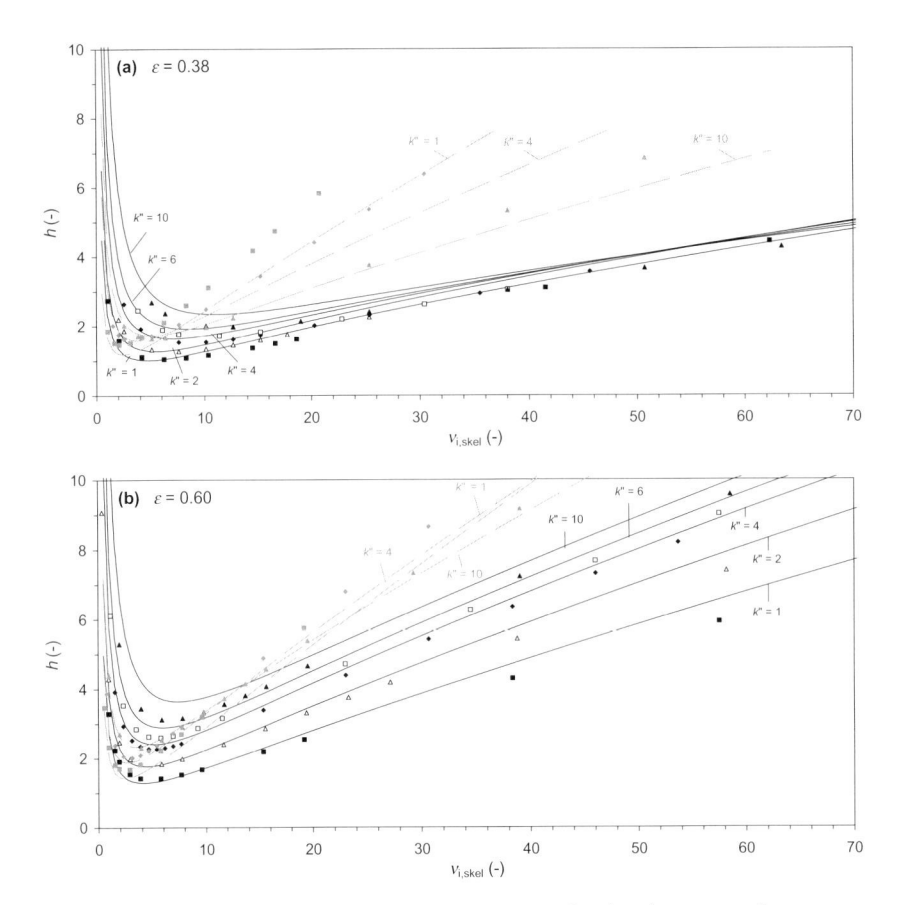

Figure 6.3 d_{skel}-based reduced van Deemter plot obtained for different combinations of the ε, k'' and D_{skel}: (a) $\varepsilon = 0.38$ and (b) $\varepsilon = 0.60$; $D_{skel} = 5 \times 10^{-10}\,m^2\,s^{-1}$ (black curves) and $D_{skel} = 1 \times 10^{-10}\,m^2\,s^{-1}$ (gray dashed curves) with $k'' = 1$ (full square), $k'' = 2$ (open triangle), $k'' = 4$ (full diamond), $k'' = 6$ (open square) and $k'' = 10$ (full triangle). The general plate-height model (6.36) was fitted to the numerically simulated data points using the traditional modeling approach (parameter values in Tables 6.1 and 6.2). The Giddings equation (6.40) was used to describe h_{ax}, wherein the D_{eff} values are obtained from Equation 6.18, with $\gamma = 0.60$ in case of $\varepsilon = 0.38$ and $\gamma = 0.72$ in case of $\varepsilon = 0.60$.

Table 6.1 Best-fit constants and ERR-values for the different fitting cases considered for the case of the TSM with $\varepsilon = 0.38$. The crosses denote the parameters that were actually fitted (least square method).

	Nonporous skeleton		Porous skeleton	
	Knox	Giddings	Traditional approach	Fitted to plate-height data
A	*0.057	–	–	–
A'	–	*0.20	*3.2	*0.66
D'	–	*0.060	*0.069	*0.033
γ_d	–	–	0	*6.1
γ_c	–	–	2.08	*0.75
n	–	–	1/3	1/3
Sh_{skel}	–	–	8	*7.36
ERR (%)	7.1	9.3	11.4	1.2

Table 6.2 Best-fit constants and ERR values for the different fitting cases considered for the case of the TSM with $\varepsilon = 0.60$. The crosses denote the parameters that were actually fitted (least square method).

	Nonporous skeleton		Porous skeleton	
	Knox	Giddings	Traditional approach	Fitted to plate-height data
A	*0.10	–	–	–
A'	–	*0.40	*15	*2.5
D'	–	*0.050	*0.062	*0.059
γ_d	–	–	0	*2.28
γ_c	–	–	1.52	*1.07
n	–	–	1/3	1/3
Sh_{skel}	–	–	8	*7.41
ERR (%)	4.4	7.2	7.4	1.6

also reflected in the high value of the average linear error of all considered data points (Tables 6.1 and 6.2), defined as:

$$\text{ERR} = \frac{1}{n*m} \sum_{i=1}^{m} \sum_{j=1}^{n} \frac{|h_{i,j} - h_{i,j,\text{fit}}|}{h_{i,j}} \tag{6.43}$$

There are several reasons why the traditional literature approach is unable to describe the TSM band-broadening curves.

First, the $\alpha = 4$ value originating from the infinitely long cylinder assumption adopted in nearly all monolithic silica modeling studies [12, 49, 51] is incorrect because it neglects the fact that large parts of the monolithic material, that is, that in the node points, is shielded from the outside. As a consequence, a monolithic skeleton has a significantly smaller specific surface than an infinitely long cylinder. This was recognized for the first time been by Skudas et al. [4]. Studying the specific outer surface of a series of real silica monolith samples using a variety of characterization methods, they found that α is rather equal to 3 instead of to 4. Using accurate geometrical measurement methods, a value of $\alpha = 2.74$ can be calculated for the more densely packed TSM ($\varepsilon = 0.38$), and a value of $\alpha = 3.21$ for the less densely packed TSM ($\varepsilon = 0.60$). The latter has a smaller fraction of its material occupying the node points, hence explaining that its α value lies closer to that of an infinitely long cylinder [46].

Provided that the α value of the TSM strongly deviates from that of an infinitely long cylinder and considering that Sh_{part} is also a shape factor appearing in the expression for h_C, one could suspect that the latter will also deviate from the value that is obtained for an infinitely long cylinder. Therefore, Sh_{part} should be treated as a freely changeable variable and should be fitted to the band-broadening data.

Another reason for the poor fit observed in Figure 6.3 is the use of the semiempirical Wilson–Geankoplis expression to describe the mass transfer in the mobile zone. Using this equation (or the other Sh_{tp} correlations like the Kataoka equation or the penetration model) to model the h_{Cm}-band broadening in a packed bed of spheres as well as in monolithic columns (as is done in the literature [13, 51]) would mean that the Wilson–Geankoplis expression is independent of the shape of the interstitial space. Physical intuition, however, suggests a strong dependency of Sh_{tp} on the geometry. Using the more general expression for Sh_{tp} (6.24), this dependency on the geometry is incorporated into the values of two shape factors γ_d and γ_c. Given the complex structure of monoliths however, their values are not known and need to be fitted to the plate-height data.

Also, the use of the questionable time-weighted average expression for the calculation of D_{eff} results in fitting errors. It has been shown that the best way to accurately calculate D_{eff} is by performing peak-parking measurements [36], either experimentally or through numerical simulations on reconstructed computer models of the silica monolith structure.

When the above-mentioned corrections are applied to the modeling strategy (use of the calculated α value and the exact D_{eff} value as obtained through peak-parking measurements, use of A', D', γ_d, γ_c and Sh_{part} as freely variable fitting constants and $n = 1/3$), the fitting quality is strongly improved. As shown in Figure 6.4, the numerically derived plate-height data are well fitted using only one set of parameters for all the different degrees of retention and intra-skeleton diffusion coefficients (ERR values are given in Tables 6.1 and 6.2). Hence, provided that the values of D_{mol} and D_{skel} are well known, the general plate-height model can be used to closely predict the influence of retention and intraskeleton diffusivity on the band broadening. Rough approximations on the values of the geometrical constants should therefore be avoided, and ideally, their values should be determined by fitting them to the plate-height data or through numerical simulation on reconstructed computer models.

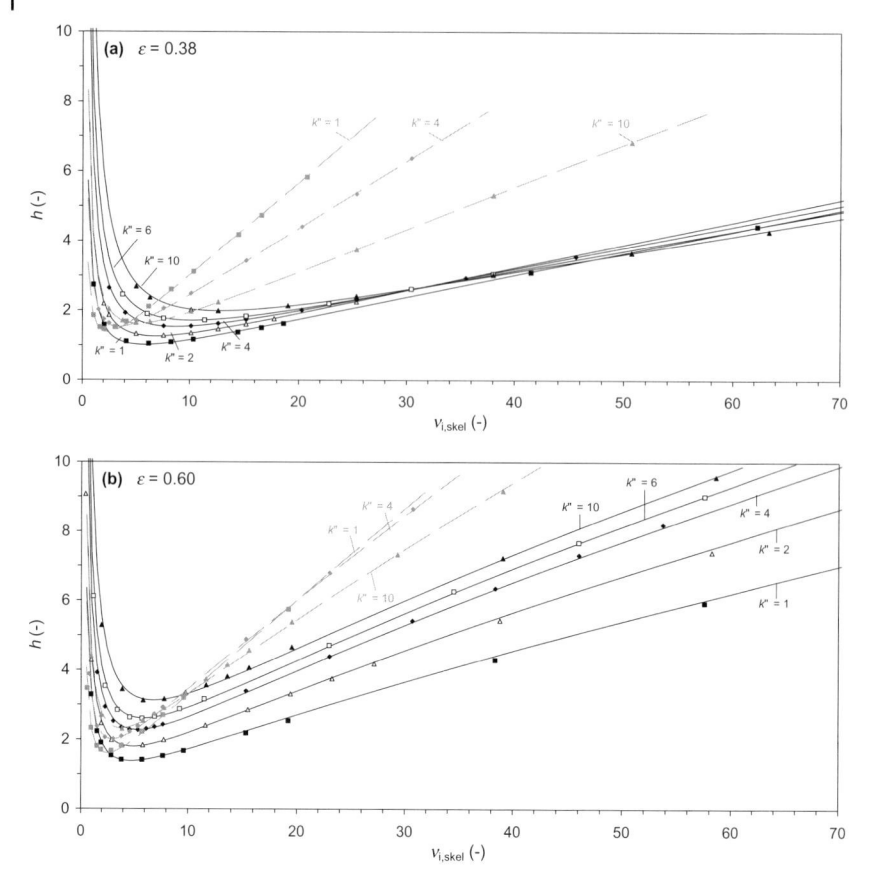

Figure 6.4 d_{skel}-based reduced van Deemter plot with the same legend as in Figure 6.3, but the fitted curves are obtained by freely fitting the geometrical constants (parameter values in Tables 6.1 and 6.2). The Giddings equation (6.40) was used to describe h_{ax}, wherein the D_{eff} values are obtained from peak-parking measurements.

6.4 Conclusion

In this chapter, it has been shown that the general plate-height model of liquid chromatography can be tuned to predict how the band broadening in monolithic packing structures will depend on the species retention factors and the species diffusion coefficients in the mobile and stationary zone. This general plate-height model of liquid chromatography has been established by pioneering authors such as Giddings, van Deemter and Lapidus and Amuindson, who all proposed variants of the same basic expression. However, before the model can be used, the values for the geometrical constants (α, Sh_{tp}, Sh_{skel} and the eddy-dispersion parameters) appearing in it need to be determined separately, either through numerical simulation on reconstructed computer models of the structure or through experimental plate-height measurements using test components with well-known diffusion coefficients. Also, the B term appearing in the general plate-height model could be predicted in advance and its value should be determined experimentally, either by peak parking or measurements at very low flow rates.

Acknowledgments

F.D. gratefully acknowledges a research grant from the Research Foundation – Flanders (FWO Vlaanderen).

Symbols

A, A', B, D, D'	dimensionless constants appearing in the H_{ax} expressions (see Equation 6.16)
C	concentration (mol/m^3)
d	diameter (m)
D_{diff}	diffusion coefficient (m^2/s)
D_{eff}	effective diffusion coefficient (m^2/s)
D_{mol}	molecular diffusion coefficient (m^2/s)
D_{skel}	intraskeleton diffusion coefficient (m^2/s)
ΔP	transcolumn pressure drop (Pa)
F	flow rate (m^3/s)
h	reduced plate height (-)
H	plate height (m)
H_{ax}	axial dispersion contribution to plate height (m)
H_{Cs}	intraskeleton mass-transfer resistance contribution to plate height (m)
H_{Cm}	through-pore mass-transfer resistance contribution to plate height (m)
k_f	mass-transfer coefficient (m/s)
k'	phase-retention factor (-)
k''	zone-retention factor (-)
k_0''	zone-retention factor of nonretained species (-)
K_{vi}	permeability based on interstitial velocity (m^2)
l_{skel}	length of the cylindrical building blocks of the TSM structure (m)
L	column length (m)
m	amount of species (kg or mol)
n	empirical exponent in Sh expression
S	cross-sectional surface area of column (m^2)
Sh	Sherwood number (-)
t	time (s)
u_i	interstitial velocity (m/s)

u_0 velocity of a nonretained component (m/s)
V volume (m³)

Greek Symbols

α geometrical constant related to the specific surface, as defined in Equation 6.29 (-)
ε external porosity (-)
ε_{int} intra-skeleton porosity (-)
η dynamic viscosity (Pa s)
v_i reduced interstitial velocity (-)
γ_c, γ_d geometrical factor, as defined in Equation 6.24 (-)
τ tortuosity (-)
Ω skeleton surface area (m²)

Subscripts

ax axial
eq equilibrium
dom domain (sum skeleton and flow-through pore)
i interstitial
m mobile phase
ref reference
skel skeleton
tp through-pore
tp,H hydraulic pore diameter
tp,ϕ through-pore, based on permeability K_{vi}

References

1. Svec, F., and Frechet, J.M. (1992) Continuous rods of macroporous polymer as high-performance liquid chromatography separation media. *Anal. Chem.*, **64**, 820–822.
2. Minakuchi, H., Nakanishi, K., Soga, N., Ishizuka, N., and Tanaka, N. (1996) Octadecylsilylated porous silica rods as separation media for reversed-phase liquid chromatography. *Anal. Chem.*, **68**, 3498–3501.
3. Hjerten, S., Liao, L.J., and Zhang, R. (1989) High-performance liquid-chromatography on continuous polymer beds. *J. Chromatogr.*, **473**, 273–275.
4. Skudas, R., Grimes, B.A., Thommes, M., and Unger, K.K. (2009) Flow-through pore characteristics of monolithic silicas and their impact on column performance in high-performance liquid chromatography. *J. Chromatogr. A*, **1216**, 2625–2636.
5. Siouffi, A.M. (2006) About the C term in the van Deemter's equation of plate height in monoliths. *J. Chromatogr. A*, **1126**, 86–94.
6. Knox, J.H. (1999) Band dispersion in chromatography–a new view of A-term dispersion. *J. Chromatogr. A*, **831**, 3–15.

7. Meyers, J.J., and Liapis, A.I. (1999) Network modeling of the convective flow and diffusion of molecules adsorbing in monoliths and in porous particles packed in a chromatographic column. *J. Chromatogr. A*, **852**, 3–23.
8. Minakuchi, H., Nakanishi, K., Soga, N., Ishizuka, N., and Tanaka, N. (1997) Effect of skeleton size on the performance of octadecylsilylated continuous porous silica columns in reversed-phase liquid chromatography. *J. Chromatogr. A*, **762**, 135–146.
9. Minakuchi, H., Nakanishi, K., Soga, N., Ishizuka, N., and Tanaka, N. (1998) Effect of domain size on the performance of octadecylsilylated continuous porous silica columns in reversed-phase liquid chromatography. *J. Chromatogr. A*, **797**, 121–132.
10. Tanaka, N., Kobayashi, H., Nakanishi, K., Minakuchi, H., and Ishizuka, N. (2001) Monolithic LC columns. *Anal. Chem.*, **73**, 420A–429A.
11. Tanaka, N., Kobayashi, H., Ishizuka, N., Minakuchi, H., Nakanishi, K., Hosoya, K., and Ikegami, T. (2002) Monolithic silica columns for high-efficiency chromatographic separations. *J. Chromatogr. A*, **965**, 35–49.
12. Miyabe, K., and Guiochon, G. (2002) The moment equations of chromatography for monolithic stationary phases. *J. Phys. Chem. B*, **106**, 8898–8909.
13. Miyabe, K., Cavazzini, A., Gritti, F., Kele, M., and Guiochon, G. (2003) Moment analysis of mass-transfer kinetics in C-18-silica monolithic columns. *Anal. Chem.*, **75**, 6975–6986.
14. Tallarek, U., Leinweber, F.C., and Seidel-Morgenstern, A. (2002) Fluid dynamics in monolithic adsorbents: Phenomenological approach to equivalent particle dimensions. *Chem. Eng. Technol.*, **25**, 1177–1182.
15. Leinweber, F.C., Lubda, D., Cabrera, K., and Tallarek, U. (2002) Characterization of silica-based monoliths with bimodal pore size distribution. *Anal. Chem.*, **74**, 2470–2477.
16. Leinweber, F.C., and Tallarek, U. (2003) Chromatographic performance of monolithic and particulate stationary phases–hydrodynamics and adsorption capacity. *J. Chromatogr. A*, **1006**, 207–228.
17. Lapidus, L., and Amundson, N.R. (1952) Mathematics of adsorption in beds. VI. The effect of longitudinal diffusion in ion exchange and chromatographic columns. *J. Phys. Chem.*, **56**, 984–988.
18. van Deemter, J.J., Zuiderweg, F.J., and Klinkenberg, A. (1956) Longitudinal diffusion and resistance to mass transfer as causes of non ideality in chromatography. *Chem. Eng. Sci.*, **5**, 271–289.
19. Giddings, J.C. (1961) Role of lateral diffusion as a rate-controlling mechanism in chromatography. *J. Chromatogr.*, **5**, 46–60.
20. Giddings, J.C. (1965) *Dynamics of Chromatography Part 1*, Marcel Dekker, New York.
21. Miyabe, K., and Guiochon, G. (2000) A kinetic study of mass transfer in reversed-phase liquid chromatography on a C18-silica gel. *Anal. Chem.*, **72**, 5162–5171.
22. Desmet, G., and Broeckhoven, K. (2008) Equivalence of the Different C_m- and C_s-term expressions used in liquid chromatography and a geometrical model uniting them. *Anal. Chem.*, **80**, 8076–8088.
23. Gritti, F., and Guiochon, G. (2006) General HETP equation for the study of mass-transfer mechanisms in RPLC. *Anal. Chem.*, **78**, 5329–5347.
24. Haynes, H.W., and Sarma, P.N. (1973) Model for application of gas-chromatography to measurements of diffusion in bidisperse structured catalysts. *AIChE J.*, **19**, 1043–1046.
25. Huber, J. (1969) High efficiency high speed liquid chromatography in columns. *J. Chromatogr. Sci.*, **7**, 85–90.
26. Horvath, C., and Lin, H.-J. (1976) Movement and band spreading of unsorbed solutes in liquid-chromatography. *J. Chromatogr.*, **126**, 401–420.
27. Horvath, C., and Lin, H.-J. (1978) Band spreading in liquid-chromatography–general plate height equation and a method for evaluation of individual plate height contributions. *J. Chromatogr.*, **149**, 43–70.

28. Kucera, E. (1965) Contribution to theory of chromatography linear non-equilibrium elution chromatography. *J. Chromatogr.*, **19**, 237–248.
29. Guiochon, G., Golshan-Shirazi, S., and Katti, A.M. (1994) *Fundamentals of Preparative and Nonlinear Chromatography*, Academic Press, Boston.
30. Hong, L., Felinger, A., Kaczmarski, K., and Guiochon, G. (2004) Measurement of intraparticle diffusion in reversed phase liquid chromatography. *Chem. Eng. Sci.*, **59**, 3399–3412.
31. Kaczmarski, K., and Guiochon, G. (2007) Modeling of the mass-transfer kinetics in chromatographic columns packed with shell and pellicular particles. *Anal. Chem.*, **79**, 4648–4656.
32. Miyabe, K. (2008) Evaluation of chromatographic performance of various packing materials having different structural characteristics as stationary phase for fast high performance liquid chromatography by new moment equations. *J. Chromatogr. A*, **1183**, 49–64.
33. Grimes, B.A., Lüdtke, S., Unger, K.K., and Liapis, A.I. (2002) Novel general expressions that describe the behavior of the height equivalent of a theoretical plate in chromatographic systems involving electrically-driven and pressure-driven flows. *J. Chromatogr. A*, **979**, 447–466.
34. Wilson, E.J., and Geankoplis, C.J. (1966) Liquid mass transfer at very low Reynolds numbers in packed beds. *Ind. Eng. Chem. Fundam.*, **5**, 9–14.
35. Gzil, P., Vervoort, N., Baron, G.V., and Desmet, G. (2004) A computational study of the porosity effects in silica monolithic columns. *J. Sep. Sci.*, **27**, 887–896.
36. Desmet, G., Broeckhoven, K., De Smet, J., Deridder, S., Baron, G.V., and Gzil, P. (2008) Errors involved in the existing B-term expressions for the longitudinal diffusion in fully porous chromatographic media – part I: computational data in ordered pillar arrays and effective medium theory. *J. Chromatogr. A*, **1188**, 171–188.
37. Broeckhoven, K., Cabooter, D., Lynen, F., Sandra, P., and Desmet, G. (2008) Errors involved in the existing B-term expressions for the longitudinal diffusion in fully porous chromatographic media – Part II: Experimental data in packed columns and surface diffusion measurements. *J. Chromatogr. A*, **1188**, 189–198.
38. Wilke, C.R., and Chang, P. (1955) Correlation of diffusion coefficients in dilute solutions. *AICHE J.*, **1**, 264–270.
39. Scheibel, E.G. (1954) Liquid diffusivities. *Ind. Eng. Chem.*, **46**, 2007–2008.
40. Li, J., and Carr, P.W. (1997) Accuracy of empirical correlations for estimating diffusion coefficients in aqueous organic mixtures. *Anal. Chem.*, **69**, 2530–2536.
41. De Malsche, W., Gardeniers, H., and Desmet, G. (2008) Experimental study of porous silicon shell pillars under retentive conditions. *Anal. Chem.*, **80**, 5391–5400.
42. Incropera, F.P., and De Witt, D.P. (1985) *Fundamentals of Heat and Mass Transfer*, 2nd edn, John Wiley & Sons, Inc., New York.
43. Kataoka, T., Yoshida, H., and Ueyama, K. (1972) Mass transfer in laminar region between liquid and packing material surface in the packed bed. *J. Chem. Eng. Jpn.*, **5**, 132–136.
44. Bird, R.B., Stewart, W.E., and Lightfoot, E.N. (1962) *Transport Phenomena*, John Wiley & Sons, Inc., New York.
45. Richardson, J.F., Harker, J.H., and Backhurst, J.R. (2002) *Coulson & Richardson's Chemical Engineering*, vol. 2, 5th edn, Elsevier Butterworth-Heinemann, Oxford.
46. Detobel, F., Gzil, P., and Desmet, G. (2009) *J. Sep. Sci.*. doi: 10.1002/jssc.200900222
47. Vervoort, N., Gzil, P., Baron, G.V., and Desmet, G. (2003) A correlation for the pressure drop in monolithic silica columns. *Anal. Chem.*, **75**, 843–850.
48. Vervoort, N., Gzil, P., Baron, G.V., and Desmet, G. (2004) Model column structure for the analysis of the flow and band-broadening characteristics of

silica monoliths. *J. Chromatogr. A*, **1030**, 177–186.

49. Gritti, F., and Guiochon, G. (2009) Mass transfer kinetic mechanism in monolithic columns and application to the characterization of new research monolithic samples with different average pore sizes. *J. Chromatogr. A*, **1216**, 4752–4767.

50. Gritti, F., Piatkowski, W., and Guiochon, G. (2003) Study of the mass transfer kinetics in a monolithic column. *J. Chromatogr. A*, **983**, 51–71.

51. Guiochon, G. (2007) Monolithic columns in high-performance liquid chromatography. *J. Chromatogr. A*, **1168**, 101–168.

7
Comparison of the Performance of Particle-Packed and Monolithic Columns in High-Performance Liquid Chromatography

Georges Guiochon

7.1
Introduction

Although the concept of monolithic columns was suggested by Knox in the early 1970s, the first practical implementation appeared in the 1990s with monolithic columns becoming commercially available in 1999. During the last ten years, no significant changes were made to these initial offerings. So, the practical impact of the revolutionary concept of monolithic columns, the first and only significant change in the structure of chromatographic columns since their design by Tswett [1–3], remains insignificant. Nearly five years ago, column technology began a new cycle of rapid evolution, following a long period of dormancy. This renaissance is largely due to the commercial reaction involving a well-established technology challenged by the advent of a fundamentally new approach, the advantages of which could threaten it and, possibly, lead to its complete replacement by the newcomer. Paradoxically, the most important practical consequence of the advent of monolithic columns so far has been the development of many new, high-performance, fine particles. A new generation of packing media offers chromatographers a variety of attractive choices, with solid, shell, and fully porous particles having diameters between 1.5 and 3 µm, hence offering higher column efficiency and/or faster analyses. The outcome of this progress has led to a significant increase in analytical throughput observed in many industrial laboratories.

The advent of monolithic silica standard- and narrow-bore columns and of several families of polymer-based monolithic columns could considerably change the potential perspectives of the HPLC field, but so far it has exhibited little effect. The new type of column bed is attracting considerable intellectual interest but little investment in practice due to a lack of competition and a dirth of innovative attractive commercial offerings. To the initial 100×4.6 mm Chromolith column of 1999 (Merck, Darmstadt, Germany), only a few models, were added, having 25 and 50 mm in length or 10 and 25 mm id, all with the same mesopore and through-pore-size distributions. A narrow-bore 150×0.1 mm column is also available. No monolithic column giving a higher efficiency, if at the cost of a lower

Monolithic Silicas in Separation Science. Edited by K.K. Unger, N. Tanaka, and E. Machtejevas
© 2011 WILEY-VCH Verlag GmbH & Co. KGaA, Weinheim
ISBN: 978-3-527-32575-7

permeability, has been made available. Competition is barred by the intellectual property protection that also precludes the overcoming of technical difficulties. As a consequence and because most scientists have been unable to master the production of efficient monolithic columns, there is a lack of systematic comparisons between the two types of column beds available, their different characteristics, their performance, and their potentialities. In contrast, the literature contains many comparisons between the specific performance of one of two particular monolithic columns (Chromolith or Onyx columns) and the performance of a wide variety of columns packed with many given materials. Although instructive, these reports do not provide conclusions that could be generalized.

7.2
Basic Columns Properties

Chromatographic columns use two phases, a solid adsorbent, the stationary phase, and a fluid solvent, the mobile phase. The mobile phase flows along the stationary phase (open tubular columns) or percolates through a bed of the adsorbent that makes the stationary phase (packed and monolithic columns). The separation is due to a difference in the migration velocities of the mixture components that equilibrate between the two phases, their molecules moving along the column at the mobile-phase velocity when they are in solution and at a velocity that is naught if they are adsorbed. A critical property of any implementation of chromatography is the kinetics of exchange between the two phases involved in a separation scheme. This kinetics must be fast to achieve high separation efficiencies or fast separations [4].

Accordingly, there should be two distinct pore networks in a chromatographic column, a network of through-pores through which the stream of mobile phase percolates and a network of mesopores inside the adsorbent used as the stationary phase. Admittedly, there are a few exceptions: in open tubular (OTC) or wall-coated open tubular (WCOT) columns, the through-pore network consists in the single, open channel at the column center while the stationary phase consists of a layer of adsorbent etched in the internal wall of the column tube or coated on this wall. Some columns are packed with solid particles, covered by a rugose layer of adsorbent that has practically no mesopores. The specific surface area of these adsorbents is very low, orders of magnitude lower than that of usual adsorbents. Generally, however, the stationary phase is made of a porous adsorbent containing a complex network of mesopores (average size between ca. 50 or 60 Å and 120 to 500 Å and often a rather wide distribution, depending on the material). The specific surface area of these mesopores is typically a few hundreds of square meters per gram. Such a large area is useful to achieve both a significant retention without requiring a high adsorption energy and a sufficient saturation capacity. These two properties are required to obtain the high-efficiency columns needed in modern analyses and to accommodate samples sufficiently large to permit trace analyses.

The kinetics of equilibration between the two phases of a chromatographic system, also called the mass-transfer kinetics in the column, depends on the distance over which the molecules of sample components must diffuse to reach the absorbent surface from the center of the mobile phase stream and conversely, from the surface back to the stream. This time depends obviously on the distance over which equilibrium must be reached, hence on the average dimensions of the through-pores and of the agglomerates of mesopores that constitute the particles of packed columns or the porons of monolithic columns. Since the latter are usually far from being spherical in shape, we admit that their smallest dimension controls the equilibration kinetics.

We discuss the different characteristics of the through-pores in chromatographic columns, their porosity and permeability, and the methods of determination of these characteristics. Finally, we describe the methods used to estimate or assess column efficiency and the approaches to compare efficiency data for widely different column-bed morphologies.

7.2.1
Total, External and Internal Column Porosity

The determinations of the column total, external, and internal porosities raise a number of difficulties, in part related to the ambiguity of the definitions of these different parameters.

7.2.1.1 Definition of the Total Column Porosity
This porosity is the fraction of the geometrical volume of the column tube that is occupied by the mobile phase contained inside the column. It is difficult to measure accurately because the mobile phase itself is adsorbed on the stationary phase with which it is in equilibrium and the partial molar volume of a compound adsorbed on a surface is usually smaller than its partial molar volume in the bulk liquid. Different methods have been developed to measure the total column porosity. They must take this possible source of error into account.

7.2.1.2 Measurement of the Total Column Porosity
Due to adsorption of the monolayer of mobile phase in contact with the adsorbent surface, the total porosity is not exactly the difference between the weight of the column filled with the mobile phase minus the weight of the empty column, divided by the product of the mobile phase density and the column geometrical volume ($\pi D_c^2 L/4$). This method is delicate because it is needed to make sure that the column is really empty of mobile phase without affecting its integrity when attempting to dry it. The two pycnometric methods (see below) are improvements of this crude approach.

The simplest and most popular method consists in measuring the retention volume of an inert compound, not adsorbed on the stationary phase. The only difficulty with this method, but a serious one, is to find a compound that is not adsorbed on the stationary phase. The most popular probes used in RPLC are thiourea and uracil. Both are retained to some extent, although the use of a strong mobile phase like pure methanol or acetonitrile or the operation of the column at high temperature could help. It was noted, however, that the total porosity thus measured depends on the composition of the mobile phase used (due to adsorption of the organic modifier).

Other, more sophisticated, methods of measurements of the column porosity are solvent pycnometry, helium pycnometry, mercury intrusion porosimetry [5]. These methods are also used to determine separately the external and internal porosities and to measure the pore-size distributions of porous materials.

In solvent pycnometry, the column is filled with a low-density solvent (methanol or acetonitrile) and weighed. Then, this solvent is replaced with a higher density solvent (dichloromethane, chloroform) then weighed again. The total porosity is derived from the differences between these two weights and between the densities of the two solvents. Care should be taken to measure the temperature of the columns and correct the solvent densities. It is generally assumed that the corrections for adsorption of the solvents compensate, which is probably correct for adsorption on RPLC stationary phases. In helium pycnometry, the adsorbent is placed in a vacuum chamber. After vacuum is achieved, the chamber is connected to a second chamber of known volume, containing helium under a known pressure. The total porosity is determined from the known volumes of the two chambers and the initial and final helium pressure. The method assumes that helium is not adsorbed, which is true of almost all the adsorbents used in chromatography, at room temperature. Due to the difficulty in achieving vacuum along a column filled with a bed that has a very low permeability (see Section 7.2.2), this method is not practical to measure the porosity of a column. It can be used to determine the porosity of a sample of porous material but that of a column is usually lower due to the compression of the bed during its packing.

In mercury intrusion porosimetry, mercury is pressurized into an appropriate container containing a known weight of the adsorbent [6]. To allow mercury to penetrate all the pores without other impediments than the one resulting from its contact angle, the sample is evacuated prior to any measurement. Then, mercury is pushed into the chamber under a known pressure and the volume of mercury entering the chamber is noted. After equilibrium has been reached, the pressure is increased by successive increments, following an exponential scale. The relationship between the radius, r_p, of a pore opening and the pressure, P, needed to fill it is given by the Washburn equation [7]:

$$P = -\frac{2\gamma}{r_p}\cos\theta \tag{7.1}$$

where γ is the surface tension of mercury (474 mN/m or dyne/cm at 25 °C) and θ its contact angle on the surface of the adsorbent studied. In the absence of a better estimate, this angle is assumed to be 130°. Thus, under a pressure of 1 MP (i.e. 10 kg/cm² or 142 psi), pores of 0.61 µm are filled. The contact angle is a function of the temperature and the pressure. During the measurement, the pressure is raised to approximately 410 MP (ca. 60 000 psi or 4100 kg/cm²). Because mercury is compressible, the temperature of the sample increases, by approximately 30 °C. As a consequence, the contact angle varies systematically during the measurements, which introduces a systematic error that cannot be corrected easily but is well reproducible for equipments of a given model. The variation of temperature is nearly independent of the sample because the heat capacity of mercury far exceeds that of the sample. The temperature dependence of the contact angle depends on the sample but is probably close for different packing materials (particularly for the stationary phases used in RPLC).

7.2.1.3 Definitions of the Column External and Internal Porosities

The external porosity of the column is the volume fraction of the column occupied by the mobile phase contained in the through-pores. It is the volume fraction of the channels offered to the stream of mobile phase. The internal porosity is the volume fraction of the column that is occupied by the micropores. It would appear that the sum of these two fractions should be equal to the column total porosity. This is not so for two reasons, one physical, one due to the actual definitions used. First, there is a degree of overlap that depends on the material studied between through-pores and the mesopores. Streamlets of the mobile phase may flow through the largest mesopores. Although small, the surface area of the particles may adsorb some of the molecules of the sample. Also, the methods used to measure these two porosities may show, for certain beds at least, an overlap between the pore-size distributions corrresponding to the through-pores and the mesopores and the common part must be apportioned, which is somewhat arbitrary. Furthermore, the definition of the internal porosity is not the same in chromatography and in chemical engineering [8]. In the former case, the internal porosity is the fraction of the whole column volume occupied by the mesopores. In the latter case, it is the fraction of the particle volume occupied by the mesopores.

7.2.1.4 Measurement of the Column External and Internal Porosities

These measurements can be made by mercury intrusion porosimetry, nitrogen sorptometry, and inverse size-exclusion chromatography [5, 9]. However, the new pore-blocking method [10] is far easier to use than mercury intrusion porosimetry or nitrogen sorptometry to measure the internal porosity of a column and more precise and accurate than inverse size-exclusion chromatography [11]. It consists in measuring the total porosity of a column from the retention time of a non-retained compound, then blocking all the mesopores by filling them with nonane, and washing off the nonane from the external porosity with a stream of water passing 5 to 10 columns volumes. This method was recently developed [10] and has not been used often enough yet to be considered as fully validated but it seems highly promising. Some results regarding a few research monolithic rods are available [12].

Recently, Unger *et al.* reviewed different conventional methods of measurement of the column porosities and compared the results that they provide on a number of monolithic rods of different total porosities (between 0.708 and 0.919) and external porosity (between 0.478 and 0.671) [13]. The methods used include mercury porosimetry, permeability measurements and the analysis of SEM photographs. Differences between the values of the skeleton diameter provided by these three methods were in substantial agreement, with differences of the order of 3%. The main result of this investigation was to show the close correlation between the surface to volume ratio of the skeleton of monolithic rods and their efficiency (see Section 7.5).

7.2.2
Column Permeability

The column permeability characterizes the degree of obstruction that the column bed opposes to the stream of mobile phase that the analyst wants to percolate through this column. To force this percolation, a pressure must be applied to the column head. This pressure is proportional to the mobile-phase viscosity, the column length, and the mobile-phase velocity required. It is also inversely proportional to the column permeability [4]. The mobile phase flow velocity is related to the column characteristics through the Darcy equation [14, 15]

$$\Delta P = \frac{\eta u_F L}{k_{p,F}} \qquad (7.2)$$

where ΔP is the pressure difference between column inlet and outlet, η is the viscosity of the mobile phase, u_F is the superficial velocity or ratio of the column flow rate and the cross-section area of the column tube, and $k_{p,F}$ is the superficial velocity-based column permeability [16]. This velocity is the ratio of the flow rate F_v and the cross-sectional area of the column tube, $\pi D_c^2/4$ with D_c column inner diameter. Note that the correct definition of the permeability is based on the use of the superficial velocity [17], not of the chromatographic velocity, u_0 with $u_0 = L/t_0$ where t_0 is the hold-up time. These two velocities are related through $u_F = \varepsilon_T u_0$. From Equation 7.2, the permeability has the dimension of a surface area.

7.2.2.1 Permeability of Packed Columns
For packed beds, the column permeability is related to the average particle size by the Kozeny–Carman equation [17]

$$k_{p,F} = k_F d_p^2 = \frac{d_p^2}{180} \frac{\varepsilon_e^3}{(1-\varepsilon_e)^2} \qquad (7.3)$$

There is some uncertainty as to the actual value of the numerical coefficient (180). This correlation is valid only for densely packed beds. The permeability would be larger for expanded beds but those beds are rarely used in chromatography because they are unstable.

7.2.2.2 Permeability of Monolithic Columns
Due to the much larger external porosity of these columns, Vervoort *et al.* thought that it was improbable that the Kozeny–Carman correlation that had been

developed for packed beds would apply to monolithic columns [18]. These authors used a computational fluid dynamics program (FLUENT) to calculate the permeability of a model of monolithic columns and to derive a similar correlation. They choose as a model of the monolithic column structure a regular tetrahedral skeleton column in which the silica skeleton is supposed to be made of bars of identical length and diameter oriented as the four diagonals of a regular tetrahedron [18]. Their calculations of the dependence of the flow velocity on the applied inlet pressure suggest the following equation:

$$k_{p,F} = \frac{d_s^2}{55} \left(\frac{\varepsilon_e}{1-\varepsilon_e} \right)^{1.55} \quad (7.4)$$

where d_s is the average size of the skeleton of the monolithic column. From the data supplied by the authors, it seems that the numerical calculations were made assuming that the rods have a diameter of 2 μm and a length of 5 μm, resulting in an external porosity of the monolithic rod of 0.863. It seems somewhat strange to correlate the permeability of a monolithic column and the average size of the skeleton rather than the average size of the through-pores. One of the important characteristics of monolithic columns is the rather wide range of external porosity that can be achieved by changing the experimental conditions of their preparation [19]. Admittedly, the Kozeny–Carman equation is also based on the use of the particle size diameter, that is, the size of the solid elements that are obstructing the flow stream. However, there is a strong correlation for packed beds between the average particle size and the average diameter of the through-pores in the packed bed [17]. There is a similarly strong correlation between the size of the rods in the tetrahedral skeleton model and its external porosity. Whether such a correlation extends to actual monolithic columns is questionable. We feel that the permeability should be correlated with the average size of the through-pores that control the size of the channels available to the mobile-phase stream. The size of the skeleton is important because it controls the amount of porous silica available for retention, hence the specific surface area of silica and the mechanical strength of the monolith. Skudas et al. [13] have prepared a series of monolithic columns for which they have measured the external porosities and the average through-pore sizes. Figure 7.1 shows a plot of the permeability calculated from Equation 7.4 versus the value of the average through-pore size. Although the range of external porosity explored (0.385 to 0.551) is relatively narrow, the correlation appears to be reasonable. Unfortunately, no values of the actual permeabilities of these columns are available.

The results of this equation were later compared to experimental results, using a series of silica rods of approximately 6 mm diameter, with skeleton sizes between 2.20 and 8.06 μm, pore sizes between 2.90 and 9.35 μm (both sizes determined by laser scanning confocal microscopy), and external porosities between 0.467 and 0.664 [20]. The possible influence of a wide size distribution of the silica filaments or porons, the intricacies of which form the monolith was investigated. For this purpose, the average and variance of the skeleton size were determined from SEM photographs of numerous slices of the rod, using a line method. The

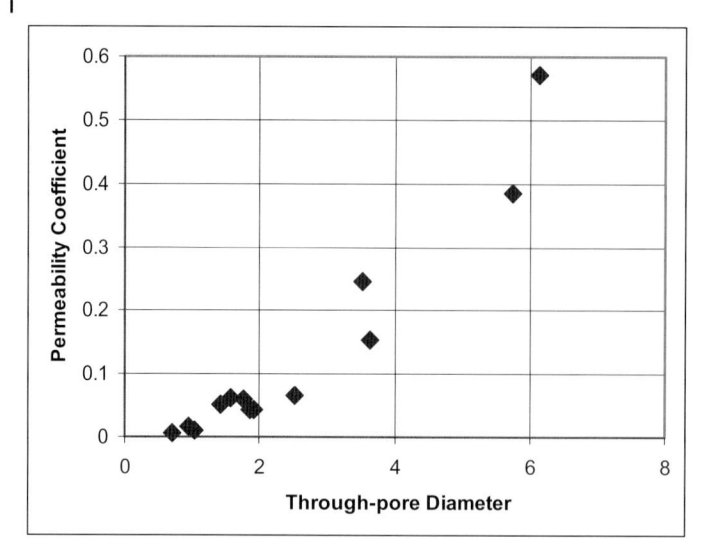

Figure 7.1 Relationship between the column permeability ($k_{p,F}$), its total porosity, and the average through-pore size (see Equation 7.4).

computational program was modified by enlarging the size of the basic cell and introducing random variations of the rod sizes. The average and variance of the skeleton size in the model were determined by applying to a 3D image of the model the same line method as used on the SEM photographs. Finally, a correction was applied to the experimental data to account for the influence of the internal porosity (arbitrarily estimated to 0.50 of the porous silica) on the chromatographic velocity and convert to the superficial velocity.

In all cases, the flow rates in both the model and the actual monolithic rods followed Darcy law. An approximate relationship between the average through-pore size, d_p, the average skeleton size, d_s, and the external porosity was derived

$$\frac{d_p}{d_s} = \sqrt{\frac{32}{55}\left(\frac{1}{1-\varepsilon_e}\right)^{1.55}} \tag{7.5}$$

It seems, however, that the numerical parameters in Equation 7.5 underestimate the amplitude of the variation of d_p/d_s with ε_e. The experimental data do not support Equation 7.4, the column permeability being nearly twice the one predicted. In conclusion, it is suggested to report the permeability of monolithic column to the average skeleton diameter, using the following correlation

$$K = \frac{d_p^2}{C}\left(\frac{\varepsilon}{1-\varepsilon}\right)^{0.21} \tag{7.6}$$

with C being an adjustable parameter. Unfortunately, the data points reported for the rods prepared are quite scattered, covering a range of permeability in excess of a factor 10 at constant external porosity [20], leaving a certain uncertainty regarding the validity of Equation 7.6.

Saito et al. proposed the use of Happel's equation to calculate the permeability of a monolithic column [21]. This equation is based on a model of a bed in which two concentric spheres of radii α and β serve as the Happel model for a random assemblage of spheres moving relatively to a fluid [22]. Although at first glance such a model would seem to be more appropriate for packed beds, it gives reasonable results for monolithic rods. The permeability is written

$$K_{\text{Happel}} = \frac{9}{2}(1-\varepsilon_e)\frac{3+2(1-\varepsilon_e)^{5/3}}{3-9/2(1-\varepsilon_e)^{1/3}+9/2(1-\varepsilon_e)^{5/3}-3(1-\varepsilon_2)^2}\varepsilon_e\left[1+\frac{1-\varepsilon_e}{\varepsilon_e}\varepsilon_1\right] \quad (7.7)$$

The last term in this equation is needed to correct the chromatographic velocity (at which a peak of unretained compound migrates along the column) and report the permeability to the interstitial velocity [21]. Finally, it is important to note that Rumpf and Gupte have measured the velocity of liquid flow through beds packed with spherical particles specially prepared and having porosities between 0.35 and 0.68, in a wide range of velocities (Reynolds numbers between 1×10^{-2} and 1×10^2) [23]. Their experimental results confirm the validity of Equation 7.3 under these experimental conditions. Therefore, the use of Equation 7.3 by the many scientists who investigated the properties of monoliths seems to be justified.

7.2.3
Column Efficiency

The column efficiency is traditionally defined as the width of an elution band relative to its retention time. It is characterized by the number of theoretical plates of the column or by its height equivalent to a theoretical plates. Let N be the efficiency of a column of length L and H the height equivalent to a theoretical plate of this column, we have by definition

$$H = \frac{L}{N} = \frac{L}{16}\left(\frac{w}{t_R}\right)^2 \quad (7.8)$$

where t_R is the retention time of the band considered and w its baseline width. Alternate equations assuming a Gaussian profile for the band and using bandwidth at fractional heights are often used. This is an empirical definition that does not characterize the column studied but the particular combination of that column, the instrument on which it is set, and the test compound used. The true column efficiency characterizes the contribution to the bandwidth due only to the sources of band broadening operating in the column and the actual bandwidth should be corrected for the contribution of the chromatograph [24]. Unfortunately, this later contribution is often far from negligible, especially with modern high-efficiency columns packed with fine particles. A correction is necessary to determine the true performance of the column. However, there is no simple way to correct for the instrument contribution but to determine the first two moments of the band eluted from the column and those of the same sample recorded by the instrument when the column is replaced with a zero volume connection [11, 19]. The first two moments, μ_1 and μ_2^*, of a peak are given by

$$\mu_1 = \frac{\int_0^\infty C(t, L) t \, dt}{\int_0^\infty C(t, L) \, dt} \tag{7.9}$$

$$\mu_2^* = \frac{\int_0^\infty C(t, L)(t - \mu_1)^2 \, dt}{\int_0^\infty C(t, L) \, dt} \tag{7.10}$$

where $C(t, L)$ is the recorded profile of the peak at the column exit ($z = L$). The first moment, μ_1 is the closest estimate of the peak retention time when its profile is not symmetrical. It has been shown that the actual contribution of the column to the two moments of the band, μ_1 and μ_2^*, is the difference between the moment derived from the peak eluted from the column and the corresponding moment derived from the peak profile obtained when no column is fitted to the instrument [24, 25]. The column HETP is given by

$$H = \frac{\mu_2^* L}{\mu_1^2} \tag{7.11}$$

The main advantages of this definition are that it is valid whatever the shape of the elution peak and that the effects of the instrument contributions to band broadening are properly corrected for. Its main disadvantage is a certain lack of precision due to the difficulty in choosing the appropriate times when the integration of Equations 7.9 and 7.10 must be started and ended. This leaves experimentalists facing the unpleasant choice of using precise but quite inaccurate or relatively imprecise but accurate numbers for the column efficiency. Regrettably, many elect to choose the former and few elect the latter option [11, 12].

7.2.3.1 The HETP Equation
Traditionally most HETP data are fitted to the Knox equation [26] that correlates experimental HETP measurements and the corresponding mobile-phase velocity.

$$H = \frac{B}{u_0} + A u_0^{1/3} + C u_0 \tag{7.12}$$

where u_0 is the chromatographic velocity ($u_0 = L/t_0$, with t_0 the hold-up time of the column) and A, B, and C are numerical parameters. This equation is not a fundamental equation, it has no actual ground in physical chemistry. It is an empirical correlation and the parameters A, B, and C have only a qualitative meaning. They characterize the extent of band broadening due to axial diffusion (B), to eddy diffusion (A), and to the resistances to mass-transfer kinetics (C). They do not measure any of the multiple parameters that are related to these phenomena and combine to account for the extent of band broadening. It is not possible to derive any quantitative information regarding which properties of a chromatographic bed are responsible for the band broadening. Serious investigations of the mechanism of band broadening, use a more general and more correct definition of the column plate number that account for the elution profiles not being Gaussian [24]. However, for our purpose here and because most experimental data available on monolithic columns use the traditional approach and Equation 7.12, we will base our discussions on the results of this approach.

7.2.3.2 Reduced HETP and Reduced Velocity

The efficiency of classical, packed columns scales as the average size, d_p, of the particles of the packing material used. From this observation, the reduced column HETP and the reduced mobile-phase velocity were defined

$$h = \frac{H}{d_p} \tag{7.13}$$

$$v = \frac{ud_p}{D_m} \tag{7.14}$$

where D_m is the molecular diffusivity of the solute studied in the bulk mobile phase. These two parameters have been widely used in the study of the efficiency of columns packed with fine particles, between 1.5 and 5 μm that have been produced during the last five years. Using these parameters, Equation 7.12 becomes

$$h = \frac{b}{v} + av^{1/3} + cv \tag{7.15}$$

where the coefficients a, b, and c are independent of the average size of the packing material and, to some extent, of other experimental conditions like the nature of the mobile phase. The values of the parameter b are close to 0.8 and those of a are generally close to 1 for well-packed columns.

In principle, this definition could be extended to monolithic columns, replacing d_p in Equations 7.13 and 7.14 with d_s. Since Vervoort et al. showed that the permeability of a monolithic column is probably proportional to d^2 [18], while Unger showed that the efficiency of a monolithic column is proportional to d_s [13, 27], the analogy is complete. The major problem in extending to monolithic columns the definition of reduced HETP and reduced velocity is in the need for an accurate determination of the average size of the skeleton and of its distribution. The only approach to determine the skeleton size is the complex line method described by Vervoort et al. [20], which is tedious, time consuming, and requires SEM photographs of numerous slices of the monolith. The problem of comparing the chromatographic performance of monolithic and packed columns was solved otherwise, through the use of the column impedance.

7.2.4
Column Impedance

The column impedance is a dimensionless performance index that relates the hold-up time of a column and its efficiency [28] to the mobile phase viscosity and the inlet pressure available.

$$E = \frac{t_0}{N^2}\frac{\Delta P}{\eta} = \frac{H^2 \varepsilon_T}{k_{p,F}} = \frac{h^2 \varepsilon_T}{k_F} \qquad (7.16)$$

Equation 7.16 applies to packed and monolithic beds. It is often used as a basis for the comparison between the performance of these two types of columns. The impedance increases with increasing minimum HETP of the column and with decreasing column permeability. For packed columns, it is nearly independent of the average diameter of the packed particles (since both H^2 and $k_{p,F}$ are proportional to d_p^2).

Obviously, like H, the impedance depends on the mobile phase flow velocity. It can vary by nearly an order of magnitude in the range of flow rate normally used. So, we will consider the impedance at the optimum flow velocity, for which the column efficiency is maximum, hence H and E are minimum. Forgetting this condition and providing values of the column impedance at some unspecified flow velocity leads to meaningless numbers and misleading comparisons.

7.3
Comparison of the Through-Pore Structures and Related Properties

7.3.1
Porosity and Through-Pore Structure

The structure of the through-pore network in packed and monolithic columns appears quite different at the first glance of scanning electron micrographs. A careful examination of the SEM of monolithic columns suggests that the tortuosity and the constriction of the channels available to the mobile phase are not fundamentally different in the two types of bed. From the examination of photographs of two-dimensional sections of a bed, it is difficult to identify the possible channels available for the flow of mobile phase. At any rate, these channels must be tortuous and constricted in monolithic beds and a careful comparison between the characteristics of the streams in both types of beds cannot be made from these photographs.

7.3.1.1 External Porosity of Packed and Monolithic Columns

A few experimental results are collected in Table 7.1. The data regarding the Merck Chromolith column and a column packed with 5 µm Luna particles are from Al-Bokari et al. [29]. The data from the research monolithic samples are from Gritti et al. [12] and from Unger et al. [13]. The total porosity of commercial monolithic columns is nearly 25% larger and their external porosity nearly 80% larger than that of conventional packed columns. The external porosity of the latter is generally between 38 and 42% and may depend to some extent on the rugosity of the surface of the packed particles and on the pressure under which they have been compressed during the packing process. The external porosity of monolithic columns can be between 0.45 and 0.70. In practice, however, the larger

Table 7.1 Porosities of some monolithic and packed columns.

	Chromolith Merck	Research Sample	Monolithic column[b]	Packed column[a]
Total porosity	0.838	0.85	0.79–0.92	0.617
RSD (%)	1.2	–		2.5
External porosity	0.693	0.69		0.386
RSD (%)	1.5	–		6
Internal porosity	0.154			0.238
RSD (%)	0.7			5

a) Unger's research samples [13]
b) Luna particles, Phenomenex, Torrance, CA

value is that of the only columns available to most investigators who do not have access to columns other than those manufactured by Merck.

7.3.1.2 Importance of the Size of the Through-Pores

The two essential parameters that determine the performance of chromatographic columns are the average size and the size distribution of their through-pores; and their external porosity. These parameters, which are closely related, determine the hydraulic resistance of the columns and have a major influence on their efficiencies.

7.3.1.3 Average Size of the Through-Pores in Packed Columns

The size of the through-pores in packed columns is scaled by the average particle size. Assuming a uniform packing of spheres, Bird et al. showed that the friction factors of a stream percolating through the bed and that of a stream flowing at the same velocity along a cylindrical tube are equal if the tube has a hydraulic radius R_h [17]. However, the hydraulic radius of a tube having an actual radius R_0 is $R_h = R_0/2$ [19], so the average size of the through-pores of a packed bed is

$$R_0 = \frac{d_p}{3} \frac{\varepsilon_e}{1-\varepsilon_e} \quad (7.17)$$

This equation was suggested by McHugh [30] and used by Stegeman et al. [31] and by Venema et al. [32]. Stegeman et al. used a bed of nonporous silica spheres ($d_p = 1.4\,\mu m$, $L = 15\,cm$) to carry out measurements of hydrodynamic HPLC. Their packed bed had an external porosity of $\varepsilon_e = 0.380$, a permeability $k_{p,F} = 1.66 \times 10^{-15}\,m^2$, a specific permeability coefficient of 8.5×10^{-4} (consistent with the value of 1×10^{-3}, which is usually accepted), and a permeability coefficient (the experimental value of the numerical coefficient in Kozeny–Carman equation) of 168. Equation 7.17 gives a hydraulic radius of $0.572\,\mu m$, approximately equal to a third of the average particle diameter, which is not quite consistent with the results of the hydrodynamic size exclusion separations performed by these authors [31]. This discrepancy was later explained by the elongation of the large molecules of the linear polymers studied by shear deformation during their migration through the packed bed [32].

Given the external porosity of conventional packed columns ($\varepsilon_e \simeq 0.40$), the average through-pore diameter could be estimated from Equation 7.17 at approximately $d_p/3$.

7.3.1.4 Average Size of the Through-Pores of Monolithic Columns

The macroporous structures of silica monoliths has been studied in detail by the research groups of Nakanishi [33–35] and Tanaka [36] and, later, by Cabrera [37, 38]. The average size of the through-pores was estimated by mercury intrusion porosimetry and found to be close to 1.7 µm, while the external porosity was estimated at 0.65 [35]. These values are consistent with those found by other authors [39–43], which is not surprising since they all used the same silica monoliths.

Saito et al. have compared the performance of mercury intrusion porosimetry and of laser scanning confocal microscopy (LSCM) to measure the geometrical properties of porous silica materials [21]. LSCM provides a 3D image of macropore structures. Various structural parameters can be extracted from these images. While the values provided by both methods agree on the porosity data, the pore-size distributions differ at low porosities. The values measured by LSCM are the same as those obtained by mercury intrusion for $\varepsilon_e = 0.65$ but are 20% larger for $\varepsilon_e = 0.50$. This suggests the presence of a significant fraction of bottleneck-shaped pores. The macropore structure provided by LSCM is qualitatively consistent with the tetrahedral skeleton models of Miyabe et al. [44] and Vervoort et al. [18]. However, significant differences arise. The tetrahedral skeleton model is made of uniform cylinders but actual monoliths have branches of uneven lengths and diameters and are irregular in shape.

Unger et al. [13] showed that the surface to volume ratio of silica monoliths has more influence on their efficiency than the average size of the through-pores, which has no influence on the column efficiency. However, from their data (Table 6), it is possible to plot the surface to volume ratio of the silica monolith rods studied versus their through-pore diameter (Figure 7.2). As expected, the surface to volume ratio of the monoliths is inversely proportional to the through-pore diameter. It seems that many geometrical characteristics of monoliths are strongly correlated.

7.3.2
Column Permeability

7.3.2.1 Permeability of Packed Columns
The works of two generations of chromatographers confirm the validity of Equation 7.3, as long as the external porosity is properly measured, which is not easy, as explained earlier.

7.3.2.2 Permeability of Monolithic Columns
Kele et al. measured the characteristics of six Chromolith columns (average pore size, 2 µm), their chromatographic properties and the relative standard deviations

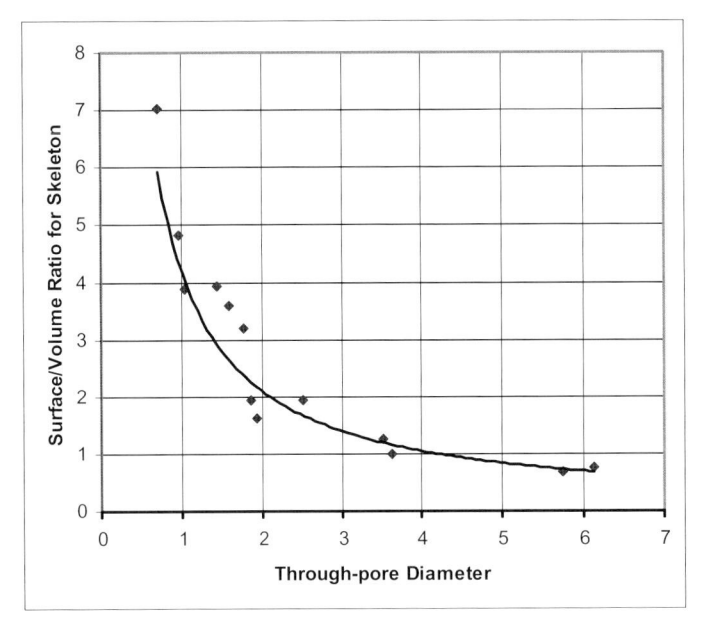

Figure 7.2 Plot of the surface to volume ratio of the skeletons of monolithic rods prepared in ref. [13] versus the through-pore size. Symbols: experimental data; line: inverse proportionality trend.

(RSD) of these data for one column in series of multiplicate measurements and for a set of six columns delivered on the same day [45].

The permeability was measured using three aqueous solutions of methanol having 80, 55, and 30% methanol. Although the RSD of the permeability measured for each column with a given mobile phase composition was between 0.25 and 0.68%, the apparent permeability varied significantly with the mobile phase composition (up to 8%). The average permeability of the six columns was $7.75 \times 10^{-10}\,cm^2$, with a RSD of 6%. All these columns had the same permeability as conventional columns packed with between 8.9 and 9.2 μm particles. Similar measurements by Leinweber and Tallarek using a Chromolith column (average pore size 1.9 μm) lead to an equivalent particle size of 11 μm [39, 41]. Data reported by Tanaka and his group are collected in Table 7.2 [46–48].

The values of the equivalent particle sizes given by Kele et al. [45], Leinweber and Tallarek [41], and Tanaka et al. [46–48] were calculated using Equation 7.3 to relate permeability, external porosity and equivalent particle size. It is interesting to compare the results obtained with those given by Equations 7.4 and 7.6. Using the values of the external porosity of a Chromolith monolithic rod measured by Al-Bokari et al. (0.693 [49]) and of the permeability of these same columns measured by Kele ($7.83 \times 10^{-10}\,cm^2$ [45]), Guiochon derived from Equation 7.3 a value of the equivalent particle size of 2.03 μm [4]. Equation 7.4 gives a value of ds of 1.21 μm, which is significantly lower. This means that many of the values

Table 7.2 Correlation between average pore size, skeleton size, HETP, and permeability of silica monolithic columns.

Macropore size (µm)	Skeleton size (µm)	HETP (µm)	Permeability (10^{-10} cm^2)
8.0	2.0	12.5	130
4.5	2.0	11.9	56
2.8	1.4	10.5	25
2.0	1.5	9.3	19
2.2	1.1	8.7	15
2.0	1.0	7.9	8

of the domain sizes found in the literature are probably overestimated since most were calculated using Equation 7.3.

Both Equations 7.3 and 7.6 illustrate how the expansion of the bed in monolithic columns enhances their permeability. The different forms of the two equations suggest that bed expansion increases considerably the permeability of columns having a low or moderate external porosity (packed columns or dense monolithic columns) but that the effect decreases with increasing porosity (expanded beds, low-density monolithic columns). These differences have been explained in part by Saito et al. [21] who showed that their data are in better agreement with the prediction of Equation 7.7, provided that the standard deviation of the through-pore-size distribution is properly taken into account. The LSCM method is the best approach to determine this standard deviation.

7.3.3
Column Radial Homogeneity

Whether packed or monolithic, chromatographic columns are not homogeneous. The central portion of long packed columns is more homogeneous and has a lower HETP than the two end sections [50]. However, the overall effect on the column efficiency remains moderate if not practically negligible. In contrast, the radial heterogeneity has most important consequences. There are radial distributions of the mobile-phase velocity and of the column efficiency. The consequences are a loss of the overall performance of the columns. The reasons for the observed radial heterogeneity are different in packed and monolithic columns. The magnitude of the consequences are also different.

7.3.3.1 Radial Heterogeneity of Packed Columns

The phenomenon was first identified by Knox [51]. Later it was confirmed by the results of numerous investigations [52–56]. The local HETP close to the column wall is several times larger than it is in its center [51, 52, 55]. The distribution of mobile-phase velocities across the column has nearly a cylindrical symmetry. The mobile-phase velocity of slurry packed columns is nearly homogeneous in the central section of the column, which has a cross-sectional area approximately half that of the column [53–56]. Closer to the wall, the velocity increases slightly along a narrow crown, then drops to nearly 80% of the velocity in the center. The consequence of this important radial variation of the velocity is a significant contribution to band broadening. The chromatograms eluted by the different streamlines do not overlay; the peak elution times and their widths increase with increasing distance from the column center [55].

A theoretical discussion of this contribution was recently published [57]. Unfortunately, the practical application of its results is more than difficult. The column model studied neglects the radial dependence of the radial dispersion coefficient. This dependence is significant since it is related to the radial variation of the local HETP. The solution of this model predicts that the overall column HETP increases linearly with increasing column length and tends toward a limit value that increases as the square of the difference in velocities in the center and along the wall of the column and as the square of the column radius. This limits HETP decreases with increasing radial dispersion coefficient, if this coefficient is a constant, which it is not in practice. Although these conclusions seem to be qualitatively consistent with the experimental results they are not helpful for their quantitative interpretation, due to the approximations made.

7.3.3.2 Radial Heterogeneity of Monolithic Columns

Based on a visual interpretation of the SEM photographs of the cross-section of various monolithic columns, Nakanishi and Tanaka concluded that their monolithic rods were radially homogeneous [34, 35]. It is most difficult, however, to visually assess the degree of homogeneity of such photographs, particularly when it has been shown that a systematic variation of 5% of the flow velocity across the column is deleterious for its efficiency. Furthermore, there are serious reasons to suspect a significant degree of heterogeneity of the pore-size distribution. The reactions involved during the formation of monolithic rods are exothermal, causing the column to experience radial thermal gradients for some time during its formation. Temperature is known to affect the average through-pore size [34]. Eventually, the rod shrinks and the interface between it and the tube in which it forms snaps, causing nonlinear, inelastic strain. Finally, the rod is wrapped in shrinkable Teflon and clad in a PEEK tube. All these steps cause a significant degree of radial heterogeneity [19].

Recently, Cabrera demonstrated the heterogeneity of monolithic rods by recording the elution-band profiles of injections made with a syringe in different points of the column inlet [58]. Using microelectrodes, Abia et al. recorded the elution profiles of a conventional injection of p-benzoquinone in a number of different locations across the column exit section of a semipreparative (10 mm id) and a conventional (4.6 mm id) monolithic rods [59–61]. In both cases, significant, systematic differences were found between the mobile-phase velocity and the column efficiency in the center of the column and along its wall. The local HETP of the semipreparative column was about 12 μm in the column center and varied from ca. 12 to 25–30 μm along the column wall. Its distribution was not cylindrical symmetric. The mobile-phase velocity was approximately 4% larger along the column wall than in its center (in contrast with what is observed for packed columns) [59]. Surprisingly, an analysis of the radius dependence of the four constants of the Knox equation (7.12) showed that the B and C terms are nearly constant all across the column, while the A term increases linearly from the center to the wall, nearly four times [60]. This large A term is consistent with the observations generally made regarding the efficiency of monolithic columns (see later) [62].

For the analytical monolithic rod, the mobile-phase velocity was higher close to the column wall and the local efficiency lower than in the column center. Although important, the relative variations in the radial direction of both the local mobile-phase velocity and the local efficiency were smaller than for the semi-preparative column [61]. Although nearly constant in the core region, the column HETP is nearly 25% higher close to the wall than in the column center.

7.3.3.3 Consequences of Column Radial Heterogeneity

If the radial distribution of the flow velocity is not flat, the migration of the band is not described by piston flow. The band that was injected as a flat distribution in a plane perpendicular to the column axis is warped. The part of the band in the core region migrates at a different average velocity than the part close to the wall region. In a packed column, the concentrations in the wall region migrate more slowly than those in the wall region [54, 55]. In monolithic columns, the converse is true [59–61]. The consequence is that the side of the bands do not reach the detector at the same time as its center, causing an increase in the apparent bandwidth, hence a loss in the efficiency measured for the column [56].

7.4
Thermodynamic Properties

These properties are essentially related to the chemical properties of the adsorbent surface and to its specific surface area. The former is most similar for columns packed with porous silica particles and for monolithic columns containing porous silica, especially when both materials were chemically bonded using the same reagents. The differences are related to minor changes in the experimental conditions under which the adsorbents are prepared, so we should expect differences in relative retention factors similar to those observed between materials prepared by different manufacturers [45]. The differences between the surface areas of the adsorbents per unit volume of the column are expected to be more important between packed and monolithic columns.

7.4.1
Retention

Kele *et al.* measured the retention factors of a large number of compounds and their efficiency on large batches of several commercial RPLC columns and investigated the short- and long-term reproducibility of the performance of these columns [45, 63–67]. Later, Gritti et al. determined the reproducibility of the single-component equilibrium isotherms of several compounds on RPLC columns, including the six monolithic columns previously used by Kele [45].

7.4.1.1 Retention Factors

Kele et al. measured the reproducibility of the retention times and retention factors of thirty compounds on a set of six monolithic columns [45] and on sets of numerous columns packed including some of the most important commercial brands of particles (Symmetry, Waters, Milford, MA [64]; Kromasil, Eka Chemicals, Bohus, Sweden [65],; Luna, Phenomenex, Torrance, CA [66]; and Vydac 218TP C18, Vydac, The Separation Group, Hesperia, CA [67],). The reproducibility of the columns was characterized by the relative standard deviations (RSD) of the retention factors of these thirty compounds on one column, measured first after a few hours, then after several months, and by the reproducibility of the data measured on the column set [63]. Values of these last RSDs were less than 2% for most compounds, with the exception of caffeine (4%), pyridine (5%), 2,2-dipyridyl (4.5%), propranolol (4%), amitriptyline (5%), and procainamide (13%). The long-term reproducibilities of these retention factors on a given column were much better, with a RSD lower than 0.2%, except for propranolol (1.5%), amitriptyline (1.5%), procainamide (6.5%) and benzylamine (1.5%).

The protocol developed by Kele et al. [63] selected 30 compounds divided into five groups. The first group include eight neutral compounds and phenol. The second group include five basic compounds, two aromatics (also included in the first group) and ethyl benzoate. The compounds of the third group are supposed to characterize the hydrogen-bonding capacity of the packing material; they include pyridine and two chelate-forming compounds. Besides aromatic compounds, the fourth group includes two strong aromatic bases, propranolol and amitriptyline. Finally, the compounds of the last group are benzyl amine, benzyl alcohol, benzoic acid and phenol. The values of the retention factors of all these compounds were reported. The retention factors of the compounds of the first and of the second groups are close on Symmetry, Kromasil and Luna; they about three times smaller on Chromolith and relatively close on Vydac and Chromolith.

Gritti et al. compared the adsorption equilibrium of 3-phenyl 1-propanol, 4-tert-butylphenol, butylbenzene, and butyl benzoate on a 100×4.6 mm Chromolith column and on a packed 150×3.9 mm Symmetry column, both C_{18} bonded and endcapped [68]. These two columns have similar characteristics (close specific surface areas and bonding densities). For each compound, the isotherm model best accounting for the data was the same on both columns, the solute polarity determining the class of this model. For 3-phenyl 1-propanol and 4-tert-butylphenol, it was a Langmuirian isotherm; for butylbenzene, an anti-Langmuir convex-downward isotherm model, a result confirming earlier findings by Cavazzini et al. [69]; the isotherm of butyl benzoate exhibited a nearly linear behavior, depending on the methanol concentration of the mobile phase. A slightly convex downward isotherm was obtained at high methanol concentrations, while the best fitting was obtained with a liquid–solid extended multilayer B.E.T. isotherm model at low concentrations. These models were carefully validated. For all four compounds, similar values of the adsorption–desorption constants were found on each column, underlining the closeness of the adsorption energies on the two columns. In contrast, the adsorption capacity of the monolithic column was found to be approximately 1.4 times greater than that of the packed column.

Leinweber and Tallarek reached the same conclusion when comparing the isotherms of caffeine on a Chromolith column and a particle packed column [41]. The monolithic column has a somewhat lower adsorption capacity than the packed column but this decrease is proportionally less than that of the mass of silica in the column. The surface concentration of the monolith is approximately the same as that of the particles when they are in equilibrium with solutions having the same concentration. The mass of silica per unit volume of column is lower with the monolith column but this is in part compensated by the skeleton of the rod having a significantly higher internal porosity than the particles.

7.4.1.2 Reproducibility of Retention Factors and Isotherms

Gritti *et al.* acquired single-component adsorption isotherm data by frontal analysis for phenol, aniline, caffeine, o-toluidine, p-toluidine and propylbenzoate on one Chromolith column, using methanol:water solutions as the mobile phase [70]. The adsorption–energy distributions were also derived, when possible, and used for the selection of the best isotherm model [71]. These data were modeled for best agreement between the experimental data points and the most suitable adsorption isotherm model. The six compounds have widely different isotherms, four being convex upward (i.e. Langmuirian), the other two having at least one inflection point. Overloaded band profiles corresponding to different sample sizes were recorded on six monolithic columns belonging to the same manufactured lot [72]. These columns were the same as those used by Kele *et al.* [45].

7.4.2
Column Loadability

The saturation capacities measured on packed RPLC columns having a similar surface chemistry are of the same order of magnitude as those measured on the monolithic columns [70, 72]. The column loadability is then proportional to the surface area of adsorbent in the column. Because the external porosity of the monolith columns is larger than that of packed columns, the volume of silica that they contain is lower, of the order of twice lower. This means that the surface area of the adsorbent would also be halved and so would be the column loadability. This is not inconsistent with the retention factors (also proportional to the total surface area of the porous silica) being approximately three times lower than those observed on the classical Kromasil, Luna, and Symmetry columns [45, 64–66]. The process of preparation of porous particles of silica is sufficiently different from that of preparation of the silica monoliths to explain this difference in total surface area.

7.5
Kinetic Properties and Column Efficiency

The efficiency of monolithic columns has been studied by many authors, including Tanaka *et al.* [36], Cabrera *et al.* [37, 38], Kele *et al.* [45], Tallarek *et al.* [39, 41], Unger *et al.* [13]. Most of these investigators used the same Chromolith columns and their results are consistent. A few were able to use rods having different external porosities and/or average pore size (or skeleton diameter). Many authors observed that the minimum plate height obtained with monolithic columns is compared to that of columns packed with 10 µm particles [41, 45, 73, 74]. This agreement is not surprising between authors who used Chromolith columns, which are highly reproducible [45]. It is less expected from those who made narrow-bore columns themselves.

In spite of numerous alternatives [75], it has been recognized that the most effective way to compare the performance of packed and monolithic columns consists in comparing their HETP curves (see Equation 7.12), their Poppe plots [76] and, possibly, their impedance (see Equation 7.16).

Vervoort *et al.* [77] applied the tetrahedral skeleton model previously developed by Miyabe [44] to calculate the influence of the average size of the domain of a monolithic rod on the column efficiency. Using a computational fluid-dynamics software, they calculated a value of 120 for the minimum possible impedance and a minimum value of 0.8 for the reduced plate height (based on the domain size, d_{dom} [78]). This value is the same as the minimum value of the reduced HETP of open tubular columns and close to what is considered as the limit efficiency for packed columns, that is, the efficiency measured by NMR, when all the effects of the radial heterogeneity of the packed bed are eliminated [79]. The fit of the efficiency values so calculated in a wide enough range of velocities to the Knox equation (see Equation 7.12) gave predicted values of the coefficients B and C that are in the same range as those usually measured and reported for actual monolithic columns. However, the numerical values predicted by this model for the coefficient A were ten-fold smaller than those typically derived from experimental data.

Cledera-Castro *et al.* optimized the experimental conditions for the RPLC separation of a mixture of 11 phenolic compounds that are considered as important pollutants. The optimum experimental conditions were somewhat different for the several columns studied. The analysis time for baseline separation of all the phenols were between 17 and 22.5 min for the packed columns. In contrast, it was only 3 min for the Chromolith column. This short analysis time is due to the slow decrease of the column efficiency when the flow rate is increased and to the high column permeability that permits operation at a flow rate that is nearly three times higher than the packed columns [80].

7.5.1
Axial Dispersion

When Nakanishi *et al.* compared the performance of a monolithic and a packed column, they obtained some intriguing results [81]. The packed column had a lower minimum HETP than the monolithic column, with values of ca. 12.5 and 14 μm, respectively. However, the optimum velocity of the former, at ca. 0.8 mm s^{-1} was less than half that of the latter. Furthermore, the HETP of the monolithic column increased more slowly with increasing flow velocity. It is 15 μm at 3 mm s^{-1} and 17 μm at 5 mm s^{-1}, values that are lower than the HETP of the packed column (19 and 30 μm, respectively). McCalley observed similar results when he compared the efficiencies of benzene, aniline, 4-ethyl aniline and pyridine on a Chromolith and several commercial packed columns [82]. This suggests that a fit of the HETP data to the Knox equation would yield a larger value of the A coefficient and a smaller value of C for the monolithic than for the packed column.

In a study of the efficiency of monolithic columns, Kele et al. showed that the values of the A term for thiourea ($k^1 = 0.0$), butylbenzene ($k^1 = 2.0$), and triphenylene ($k^1 = 4.3$) on a Chromolith column were 5.9, 7.4, and 9.6, respectively, after correction for the extracolumn band-broadening contribution [45].

These results are consistent with the analyses of both Vervoort et al. [77] and Billen et al. [83] who showed that current monolithic columns probably have a large variance of the size distribution of their domains and that preparing more homogeneous monoliths would markedly improve their efficiency. They agree also with the consequences of our earlier observations regarding the large degree of heterogeneity found for the radial distribution of the domain size [59–61]. Mass-transfer kinetics seems to be exceptionally fast in monolithic columns. We hope that those who prepare monolithic columns adjust the process of preparation of silica monoliths to improve this aspect of the morphology of the monoliths and to reduce the extent of the eddy diffusion contribution to the HETP. Preparing more homogeneous monoliths would probably result in improved performance without reducing the average through-pore size [12].

Gritti et al. studied the relationship between the column efficiency and mobile-phase velocity for phenol, toluene, acenaphthene, amylbenzene, and insulin on three research grade monolithic columns having different average pore sizes ($\approx$150, 225, and 350 Å, respectively) and the same dimensions (100 × 4.6 mm). The contribution of the B term was found to be negligible compared to that of the A term, even at very low reduced velocities. At moderate velocities ($1 < v < 3$), the contribution of the A term decreases with increasing average mesopore size and molecular diffusivity of the compound studied, due to faster mass transfer across the column. The heterogeneity of the distribution of the eluent velocities from the column center to its wall is the main source of efficiency loss [12]. The relative velocity difference between the rod center and its wall is of the order of 2% in a 100 × 4.6 mm commercial monolithic column. Larger values were obtained for the new research-grade rods, between 3.3 and 6.7%.

7.5.2
Mass-Transfer Kinetics

In a recent review, Siouffi reviewed the values of the C term derived from HETP data that had been obtained by a dozen different authors working with nineteen monolithic rods and eight narrow-bore monolithic columns [62]. The data pertain mostly to aromatic compounds, including paraben [84], but also to angiotensin and insulin [40]. The values of C obtained for nonretained compounds are small, less than 1 ms. Those for aromatic compounds are between 1 and 3 ms and increase with increasing retention factor. Similar values were obtained for other low molecular weight compounds. The values reported for angiotensin (MW = 1180) and insulin (MW = 5807) are larger, ca. 3 and 10 ms, respectively, on a rod column with 12.5 nm average mesopore size, and 6 ms for insulin on a 20 nm mesopore-size rod [40]. These values are systematically almost twice as small for monolithic columns as for packed columns with similar phase systems [62]. This explains why numerous authors have reported faster separations on monolithic columns than on packed columns.

In their study of the efficiency of a series of monolithic columns of different average pore sizes, Gritti et al. found that, at high reduced velocities ($v > 5$), the C term of the monolithic columns is controlled by film mass-transfer resistance between the eluent circulating in the large through-pores and the eluent stagnant inside the thin porous skeleton [12]. The experimental Sherwood number measured on the monolith columns increases from 0.05 to 0.22, while the adsorption energy increases by nearly 6 kJ/mol. Stronger adsorption leads to an increase in the value of the estimated film mass transfer coefficient when a first-order film mass-transfer rate is assumed ($j \propto k_f \Delta C$). The average pore size and the trans-skeleton mass transfer have no (<0.5%, small molecules) or little (<10%, insulin) effect on the overall C term.

At moderate mobile-phase velocities the efficiencies of silica monolithic columns are comparable, some times even lower than those of packed columns but they increase far more slowly with increasing mobile-phase velocities, allowing very fast analyses [80]. Nováková et al. showed that the RPLC analysis of ketoprofen in topical ketoprofen gels was three times faster on a Chromolith columns than on several conventional columns packed with 5 μm particles [85]. Guillarme et al. compared the analytical performance of a monolithic column and of conventional columns packed with 1.7, 1.9, and 5 μm particles. They found that the precision and accuracy of the quantitative analyses of lidocaïne were nearly the same. The analysis times were the same on the monolithic column and the column packed with 1.7 μm particles if the latter was operated at an inlet pressure of 400 bar; however, if the inlet pressure of the packed column is raised to 1000 bar, it is twice as fast as the monolithic column [86]. Similarly, Kelly et al. showed that replacing a conventional RPLC column packed with 5-μm particles with an Onyx C18 monolithic column permits the achievement of a 10 times faster separation of DNA bases and nucleosides [87].

In principle, because their permeability is so much larger than that of packed columns of comparable efficiencies, monolithic columns should permit the achievement of extremely high efficiencies, hence of very difficult separations in a relatively short period of time. Unfortunately, silica monolithic columns are available only as short rods. Stringing column rods into long series is theoretically possible but significant losses of efficiency are observed due to back-mixing in

Table 7.3 Characteristics and impedance of a few packed columns.

	D_p (μm)	K-C constant	ε_e	ε_T	$k_F \times 10^4$	h	E
Kinetex	2.5	180	0.372	0.542	7.1	1.1	924
Halo	2.7	270	0.391	0.532	8.42	1.4	1178
Zorbax	3.5	200	0.426	0.515	11.73	2.0	1756
Luna	3	190	0.383	0.645	7.38	2.0	3495
Luna	5	210	0.372	0.630	6.87	1.9	3310
Atlantis	3	170	0.380	0.695	8.40	2.0	3309

the connectors. While a 10-cm long rod column gave 11 200 theoretical plates for hexylbenzene, strings of two, four and six gave only 18 000, 32 000, and 41 000 theoretical plates, respectively. It took 14 such columns to generate peaks exhibiting 108 000 theoretical plates (with a 117-atm inlet pressure), a 30% loss [38]. Long monolithic columns can be produced only if their inner diameter is narrow, below 0.5 mm; a wider rod deforms and bends to a degree during the drying process, in the same time as it shrinks [48, 88, 89]. This practically limits the length of commercial columns to 10 cm. Narrow-bore columns exhibiting extremely high efficiencies, up to a million plates or more, can be produced and used for difficult separations. At present, the use of such columns appears to be limited since they are not commercially available.

7.5.3
Column Impedance

As explained earlier, the impedance is a dimensionless parameter, easy to calculate, that permits a comparison of the performance of monolithic and packed columns. Values of the impedance of several packed columns are listed in Table 7.3. They were derived from experimental data reported by Gritti et al. [90].

For packed columns, the impedance is usually close to 3000 (see Table 7.3), but it depends somewhat on the external and the total column porosities. There are some remarkable exceptions, however. Conventional columns packed more loosely than average (e.g., Zorbax in Table 7.3) or having exceptionally low reduced HETP (e.g., Halo and Kinetex in Table 7.3) may have an impedance as low as 1000 to 1700. It seems that the performance of packed columns has dramatically increased since the advent of monolithic columns. It is the best service that the development of monolithic columns has so far rendered to HPLC: they induced among the manufacturers of particles and the packers of columns the fear of competition that always has been the most effective engine in the drive to progress.

For monolithic columns, the impedance is often much lower, with values usually between 250 and 800 [45, 47, 89]. For Chromolith, using the values of the external porosity measured by Al-Bokari et al. (0.693 [49]), of the permeability ($k_{p,F} = 7.83 \times 10^{-10}$ cm^2 [45]) and of the average minimum HETP for several compounds measured by Kele ($H \approx 9$ μm [45]), the impedance is 715. This value is consistent with other results [47, 89]. This value is barely 50% lower than that corresponding to the Halo column, which might provide an explanation for the lack of commercial success of the former column.

Recent work suggests that monolithic columns having still lower impedance could be achieved by properly adjusting the parameters of the preparation of silica monolithic rods [91]. Using the tetrahedron skeleton model [18, 44] and a computational fluid-dynamics software, Vervoort *et al.* [77] calculated a theoretical minimum value of 120 for the impedance of perfectly regular monolithic columns. This value would constitute a minimum for monolithic columns. No lower values have yet been reported.

Finally, for open tubular columns, an impedance of ca. 32 is expected, with a significant dependence on the retention factor. This value seems to be the lowest limit achievable.

7.5.4
Kinetic Properties

The mass-transfer kinetics of butylbenzoate was measured by Gritti *et al.* on a Chromolith C_{18}-bonded column, elution with an aqueous solution of methanol, using a perturbation method, which measures the HETP of peaks obtained as responses to the injection of small pulses of solute on a concentration plateau [92]. The equilibrium isotherm of butylbenzoate, previously determined by frontal analysis, is well accounted for by a liquid–solid extended multilayer B.E.T. isotherm model. The equilibrium data derived from the pulse method are in excellent agreement with those of frontal analysis in the accessible concentration range (0 to $8 g/dm^3$). Plots of the HETP of small pulses, injected on eight different plateau concentrations, were acquired in a wide range of mobile-phase flow velocities. The axial dispersion and the mass-transfer kinetic coefficients were derived from these data. The mass-transfer kinetics of butylbenzoate depends strongly on the plateau concentration. At low and moderate velocities, axial dispersion controls the HETP. Eddy diffusion and axial diffusion in the macropores control band dispersion. At high velocities, the results show that the mass-transfer kinetics is mainly controlled by pore diffusivity inside the monolith skeleton and is fast. The film mass-transfer contribution is small. Processes involving adsorptive interactions between the solute and the stationary phase, for example, surface diffusion and adsorption/desorption kinetics, combine in series to the external mass-transfer kinetics and to effective pore diffusivity. The column efficiency decreases only slowly with increasing mobile-phase velocity [62, 93].

7.6
Conclusions

Monolithic columns hold considerable promise, which will remain on hold until more useful versions are made available. Lack of commercially available improved monolithic columns during the last decade has caused this approach to be

Table 7.4 Comparison of monolithic and particle-packed columns.

Characteristics	Current monolithic columns[a]	Potential monolithic columns[b]	Current particle-packed columns	Potential particle-packed columns[c]
External porosity	0.70	0.50 to 0.90	0.37 to 0.43	0.37 to 0.43
Domain size (µm)	2 to 3	0.5	1.5 to 5	1
Permeability cm^2	7.83×10^{-10}	5×10^{-12}	8×10^{-12}	7×10^{-13}
Efficiency (HETP, µm)	9–10	2	3.5 to 14	1.1[d]
Impedance	750 to 8000	120	1100 to 3000	1000 to 1500

a) The only silica monolithic column available is the Chromolith.
b) Based on the assumption that radially homogeneous monoliths with low domain sizes could be manufactured.
c) Based on the assumption that 1-µm particles similar to Halo (ATM, Wilmington, DE) in packing characteristics could be made.
d) Performance of the new Kinetex-C$_{18}$ columns from Phenomenex (September 2009).

overwhelmed by advances in particle chromatography. Intellectual property rights play a key role in the immediate future of this separation media. Merck KGaA (Darmstadt, Germany) owns two nearly contemporaneous patents, one on the principle of the use of a porous rod of ceramic as a column for chromatography [94], the other on the production of porous silica monoliths [95, 96]. Although these patents date from 1993 and will end soon, a more recent patent may be used to extend the protection [97]. Finally, the production and sale of narrow-bore monolithic columns are protected by two patents that are 10 years younger [98, 99]. Although it appears that Merck has yet to license production and research on monoliths, there is an arrangement under which the Onyx column offered by Phenomenex (Torrance, CA) is practically identical to the 10-year-old Merck Chromolith.

Current circumstances under which monoliths are tightly restrained is reminiscent of the treatment of the Golay patent covering open tubular columns. Instead of licensing the idea to others who could assist in the research, production and commercialization of a variety of columns, Perkin Elmer wasted the opportunity to participate in scientific advances and to receive important financial compensation by sitting on the patent and waiting for the world to beat a path to its door. The real money was made later, by several small companies.

Once the field of monolithic columns becomes available to the scientific community either through controlled licensing or the expiration of patent protection, there will be considerable advances in this fertile area. Table 7.4 compares the performance of monolithic and packed columns that are currently available commercially and those of columns that could possibly be produced if current research projects eventually succeed.

Acknowledgments

This work was supported in part by grant CHE-06-08659 of the National Science Foundation. The author acknowledges fruitful, informative discussions with Karin Cabrera (Merck KGaA, Darmstadt, Germany), Tivadar Farkas (Phenomenex, Torrance, CA), Attila Felinger (University of P'ecs, Hungary), Fabrice Gritti (UTK, Knoxville, TN, USA), the late Marianna Kele (Waters Corp., Milford, MA), Kanji Miyabe (Toyama University, Japan), Uwe Neue (Waters Corp., Milford, MA), Antoine Siouffi (Universit'e Paul C'ezanne, Marseille, France), Nobuo Tanaka (Kyoto Institute of Technology, Kyoto, Japan).

Symbols

A	numerical parameter in the plate height equation, characterizes eddy dispersion
B	numerical parameter in the plate height equation, characterizes axial diffusion
C	numerical parameter in the plate height equation, characterizes the rate of mass transfers
C	adjustable parameter in Equation 7.6
D_c	column diameter
D_m	molecular diffusivity of the solute studied in the bulk mobile phase
d_c	column inner diameter
d_{dom}	average size of a domain
d_p	diameter of a particle
d_s	average size of the skeleton of the monolithic column
F_v	flow rate
H	column height equivalent to a theoretical plate
h	reduced height equivalent to a theoretical plate
k_F	chromatographic velocity-based column permeability
K_{Happel}	permeability in the model of Happel [22]
k^1	retention factor
$k_{p,F}$	superficial velocity-based column permeability
L	column length
P	pressure
r_p	radius of a particle
t_0	hold-up time
t_R	retention time of the band considered
R_h	hydraulic radius
u_F	superficial velocity or ratio of the column flow rate and the cross-sectional area of the column tube
u_0	chromatographic linear velocity of the mobile phase ($u_0 = L/t_0$)
w	baseline width of a peak

Greek Symbols

γ surface tension of mercury (474 mN/m or dyne/cm at 25 °C)
ΔP pressure difference between column inlet and outlet
ε_e external porosity of the column
ε_i internal porosity of the column
ε_T total porosity of the column
η viscosity of the mobile phase
θ contact angle of mercury on the surface of the adsorbent studied
μ_1 first-order moment of a peak
μ_2^* second-order moment of a peak
ν reduced velocity of the mobile phase

References

1. Tswett, M.S. (1906) *Ber. Deut. Botan. Ges.*, **24**, 316.
2. Tswett, M.S. (1906) *Ber. Deut. Botan. Ges.*, **24**, 384.
3. Tswett, M.S. (1910) *Khromofilly V Rastitel'nom Zhivotnom Mire [Chromophylls in the Plant and Animal World]*, Izd. Karbasnikov, Warzaw, Poland, partly reprinted in 1946 by the publishing house of the Soviet Academy of Science, A. A. Rikhter and T. A. Krasnosel'skaya, Eds.
4. Guiochon, G. (2006) *J. Chromatogr. A*, 1126, 6.
5. Guan, H., Guiochon, G., Coffey, D., Davis, E., Gulakowski, K., and Smith, D.W. (1996) *J. Chromatogr. A*, **736**, 21.
6. Orr, C. (1969) *Powder Technol.*, **3**, 117.
7. Washburn, E.W. (1921) *Phys. Rev.*, **17**, 273.
8. Guiochon, G., Felinger, A., Katti, A.M., and Shirazi, D. (2006) *Fundamentals of Preparative and Nonlinear Chromatography*, Elsevier, Amsterdam, The Netherlands.
9. Grimes, B.A., Skudas, R., Unger, K.K., and Lubda, D. (2007) *J. Chromatogr. A*, **1144**, 14.
10. Cabooter, D., Lynen, F., Sandra, P., and Desmet, G. (2007) *J. Chromatogr. A*, **1157**, 131.
11. Gritti, F., and Guiochon, G. *AIChE. J.*, in press.
12. Gritti, F., and Guiochon, G. *J. Chromatogr. A*, 2009, **1216**, 4752–4767.
13. Skudas, R., Grimes, B.A., Thommes, M., and Unger, K.K. (2009) *J. Chromatogr. A*, **1216**, 2625.
14. Darcy, H. (1856) *Les Fontaines Publiques de la Ville de Dijon*, Dalmont, Paris.
15. Guiochon, G., Felinger, A., Katti, A.M., and Shirazi, D. (2006) *Fundamentals of Preparative and Nonlinear Chromatography*, Elsevier, Amsterdam, The Netherlands, Ch 5.
16. Neue, U.D. (1997) *HPLC Columns. Theory, Technology, and Practice*, John Wiley & Sons, Inc., New York, NY.
17. Bird, R.B., Stewart, W.E., and Lightfoot, E.N. (1962) *Transport Phenomena*, John Wiley & Sons, Inc., New York, NY.
18. Vervoort, N., Gzill, P., Baron, G.V., and Desmet, G. (2003) *Anal. Chem.*, **75**, 843.
19. Guiochon, G. (2007) *J. Chromatogr. A*, **1168**, 101.
20. Vervoort, N., Saito, H., Nakanishi, K., and Desmet, G. (2005) *Anal. Chem.*, **77**, 3986.
21. Saito, H., Nakanishi, K., Hirao, K., and Jinnai, H. (2006) *J. Chromatogr. A*, **1119**, 95.
22. Happel, J. (1958) *AIChE J.*, **4**, 197.
23. Rumpf, H., and Gupte, A.R. (1971) *Chem. Ing. Tech.*, **43**, 367.
24. Guiochon, G., Felinger, A., Katti, A.M., and Shirazi, D. (2006) *Fundamentals of Preparative and Nonlinear Chromatography*, Elsevier, Amsterdam, The Netherlands.

25. Dose, E.V., and Guiochon, G. (1990) *Anal. Chem.*, **62**, 1723.
26. Knox, J.H., and Saleem, M. (1969) *J. Chromatogr. Sci.*, **7**, 614.
27. Unger, K.K., Skudas, R., and Schulte, M.M. (2008) *J. Chromatogr. A*, **1184**, 393.
28. Bristow, P.A., and Knox, J.H. (1977) *Chromatographia*, **10**, 279.
29. Al-Bokari, M., Cherrak, D., and Guiochon, G. (2002) *J. Chromatogr. A*, **975**, 275.
30. McHugh, A.J. (1984) *CRC Crit. Rev. Anal. Chem.*, **15**, 1963.
31. Stegeman, G., Oostervink, R., Kraak, J.C., Poppe, H., and Unger, K.K. (1990) *J. Chromatogr.*, **506**, 547.
32. Venema, E., Kraak, J.C., Poppe, H., and Tijssen, R. (1996) *J. Chromatogr. A*, **740**, 159.
33. Nakanishi, K., and Soga, N. (1992) *J. Non-Cryst. Solids*, **139**, 1.
34. Nakanishi, K., and Soga, N. (1992) *J. Non-Cryst. Solids*, **139**, 14.
35. Minakuchi, H., Nakanishi, K., Soga, N., Ishizuka, N., and Tanaka, N. (1996) *Anal. Chem.*, **68**, 3498.
36. Tanaka, N., Ishizuka, N., Hosoya, K., Kimata, K., Minakuchi, H., Nakanishi, K., and Soga, N. (1993) *Kuromatogurafi*, **14**, 50.
37. Cabrera, K., Wieland, G., Lubda, D., Nakanishi, K., Soga, N., Minakuchi, H., and Unger, K. (1998) *TrAC*, **17**, 50.
38. Cabrera, K. (2004) *J. Sep. Sci.*, **27**, 843.
39. Tallarek, U., Leinweber, F.C., and Seidel-Morgenstern, A. (2002) *Chem. Eng. Technol.*, **25**, 1177.
40. Leinweber, F.C., Lubda, D., Cabrera, K., and Tallarek, U. (2002) *Anal. Chem.*, **74**, 2470.
41. Leinweber, F.C., and Tallarek, U. (2003) *J. Chromatogr. A*, **1006**, 207.
42. Cavazzini, A., Gritti, F., Kaczmarski, K., Marchetti, N., and Guiochon, G. (2007) *Anal. Chem.*, **79**, 5972.
43. Gritti, F., Cavazzini, A., Marchetti, N., and Guiochon, G. (2007) *J. Chromatogr. A*, **1157**, 289.
44. Miyabe, K., and Guiochon, G. (2002) *J. Phys. Chem. B*, **106**, 8898.
45. Kele, M., and Guiochon, G. (2002) *J. Chromatogr. A*, **960**, 19.
46. Ishizuka, N., Kobayashi, H., Minakushi, H., Nakanishi, K., Hirao, K., Hosoya, K., Itegami, T., and Tanaka, N. (2002) *J. Chromatogr. A*, **960**, 85.
47. Motokawa, M., Kobayashi, H., Ishizuka, N., Minakushi, H., Nakanishi, K., Jinnai, H., Hosoya, K., Itegami, T., and Tanaka, N. (2002) *J. Chromatogr. A*, **961**, 53.
48. Tanaka, N., Kobayashi, H., Ishizuka, N., Minakushi, H., Nakanishi, K., Hosoya, K., and Itegami, T. (2002) *J. Chromatogr. A*, **965**, 35.
49. Al-Bokari, M., Cherrak, D., and Guiochon, G. (2002) *J. Chromatogr. A*, **975**, 275.
50. Wong, V., Shalliker, R.A., and Guiochon, G. (2004) *Anal. Chem.*, **76**, 2601.
51. Knox, J.H., Laird, G.R., and Raven, P.A. (1976) *J. Chromatogr.*, **122**, 129.
52. Eon, C.H. (1978) *J. Chromatogr.*, **149**, 29.
53. Baur, J.E., Kristensen, E.W., and Wightman, R.M. (1988) *Anal. Chem.*, **60**, 2334.
54. Farkas, T., Sepaniak, M.J., and Guiochon, G. (1997) *AIChE J.*, **43**, 1464.
55. Farkas, T., and Guiochon, G. (1997) *Anal. Chem.*, **69**, 4592.
56. Guiochon, G., Farkas, T., Sajonz, H.G., Koh, J.-H., Sarker, M., Stanley, B.J., and Yun, T. (1997) *J. Chromatogr. A*, **762**, 83.
57. Broeckhoven, K., and Desmet, G. (2009) *J. Chromatogr. A*, **1216**, 1325.
58. Cabrera, K. (2007) HPLC-2007, pp. Communication L–17:03.
59. Mriziq, K.S., Abia, J.A., Lee, Y., and Guiochon, G. (2008) *J. Chromatogr. A*, **1195**, 97.
60. Abia, J.A., Mriziq, K.S., and Guiochon, G. (2009) *J. Chromatogr. A*, **1216**, 3185.
61. Abia, J.A., Mriziq, K.S., and Guiochon, G. (2009) *J. Sep. Sci.*, **32**, 923.
62. Siouffi, A.-M. (2006) *J. Chromatogr. A*, **1126**, 86.
63. Kele, M., and Guiochon, G. (1999) *J. Chromatogr. A*, **830**, 41.
64. Kele, M., and Guiochon, G. (1999) *J. Chromatogr. A*, **830**, 55.
65. Kele, M., and Guiochon, G. (1999) *J. Chromatogr. A*, **855**, 423.
66. Kele, M., and Guiochon, G. (2000) *J. Chromatogr. A*, **869**, 181.

67. Kele, M., and Guiochon, G. (2001) *J. Chromatogr. A*, **913**, 89.
68. Gritti, F., Piatkowski, W., and Guiochon, G. (2002) *J. Chromatogr. A*, **978**, 81.
69. Cavazzini, A., Bardin, G., Kaczmarski, K., Szabelski, P., Al-Bokari, M., and Guiochon, G. (2002) *J. Chromatogr. A*, **957**, 111.
70. Gritti, F., and Guiochon, G. (2003) *J. Chromatogr. A*, **1003**, 43.
71. Gritti, F., and Guiochon, G. (2004) *J. Chromatogr. A*, **1028**, 105.
72. Gritti, F., and Guiochon, G. (2003) *J. Chromatogr. A*, **1021**, 25.
73. Eeltink, S., Rozing, G.P., Schoenmakers, P.J., and Kok, W.T. (2004) *J. Chromatogr. A*, **1044**, 311.
74. Eeltink, S., Gzil, P., Kok, W.T., Schoenmakers, P.J., and Desmet, G. (2006) *J. Chromatogr. A*, **1130**, 108.
75. Desmet, G., Clicq, D., and Gzil, P. (2005) *Anal. Chem.*, **77**, 4058.
76. Poppe, H. (1997) *J. Chromatogr. A*, **778**, 3.
77. Vervoort, N., Gzill, P., Baron, G.V., and Desmet, G. (2004) *J. Chromatogr. A*, **1030**, 177.
78. Minakuchi, H., Nakanishi, K., Soga, N., Ishizuka, N., and Tanaka, N. (1998) *J. Chromatogr. A*, **797**, 121.
79. Tallarek, U., Bayer, E., and Guiochon, G. (1998) *J. Am. Chem. Soc.*, **120**, 1494.
80. Cledera-Castro, M., Santos-Montes, A., and Izquierdo-Hornillos, R. (2005) *J. Chromatogr. A*, **1087**, 57.
81. Nakanishi, K., Minakuchi, H., Soga, N., and Tanaka, N. (1998) *J. Sol-Gel Sci. Technol.*, **13**, 163.
82. McCalley, D.V. (2002) *J. Chromatogr. A*, **965**, 51.
83. Billen, J., Gzil, P., Baron, G.V., and Desmet, G. (2005) *J. Chromatogr. A*, **1077**, 28.
84. Desmet, G., Clicq, D., Nguyen, D.T.-T., Guillarme, D., Rudaz, S., Veuthey, J.-L., Cervoort, N., Torok, G., Cabooter, D., and Gzil, P. (2006) *Anal. Chem.*, **78**, 2150.
85. Novakova, L., Matysova, L., Solichova, D., Koupparis, M., and Solich, P. (2004) *J. Chromatogr. B*, **813**, 191.
86. Guillarme, D., Nguyen, D.T.-T., Rudaz, S., and Veuthey, J.-L. (2007) *J. Chromatogr. A*, **1149**, 20.
87. Kelly, M.C., White, B., and Smyth, M.R. (2008) *J. Chromatogr. B*, **863**, 181.
88. Kobayashi, H., Kajiwara, W., Inui, Y., Hara, T., Hosoya, K., Ikegami, T., and Tanaka, N. (2004) *Chromatographia*, **60**, S19.
89. Kobayashi, H., Tokuda, D., Ichimaru, J., Ikegami, T., Miyabe, K., and Tanaka, N. (2006) *J. Chromatogr. A*, **1109**, 2.
90. Gritti, F., and Guiochon, G. (2007) *J. Chromatogr. A*, **1166**, 30.
91. Nakanishi, K., Kanamori, K., Saito, H., Morisato, K., and Minakuchi, H. (2006) HPLC-2006, pp. Communication L–0503.
92. Gritti, F., Piatkowski, W., and Guiochon, G. (2003) *J. Chromatogr. A*, **983**, 51.
93. Gritti, F., and Guiochon, G. (2003) *J. Chromatogr. A*, **1021**, 25.
94. Cabrera, K., Sattler, G., and Wieland, G. (1994) Trennmittel (separator), european patent EP 0 686 258 b1.
95. Nakanishi, K., and Soga, N. (1993) Inorganic Porous Column, Japan patent 5-200392.
96. Nakanishi, K., and Soga, N. (1993) Production of Inorganic Porous Body, Japan patent 5-208642.
97. Nakanishi, K., and Soga, N. (1997) Inorganic Porous Material and Process for Making Same, US patent 5,624,875.
98. Lubda, D., and Muller, E. (2003) Method for Producing Monolithic Chromatography Columns, US patent application 2003/0155676 a1.
99. Nakanishi, K., Soga, N., and Minakuchi, H. (2003) Capillary Column Including Porous Silica Gel Having Continuous Through Pores and Mesopores, US patent 6,531,060.

Part Three
Applications

8
Quantitative Structure–Retention Relationships in Studies of Monolithic Materials

Roman Kaliszan and Michał J. Markuszewski

8.1
Fundamentals of Quantitative Structure–Retention Relationships (QSRRs)

Linear free-energy relationships (LFERs) may be regarded as linear relationships between the logarithms of the rate or equilibria constants for one reaction series and those for a second reaction series subjected to the same variation in reactant structure or reaction conditions. Retention parameters can be assumed to reflect the free-energy changes associated with the chromatographic distribution process. Accordingly, a chromatographic column can be treated as a "free-energy transducer", translating differences in chemical potentials of analytes, arising from differences in their structure, into quantitative differences in retention parameters.

The basic methodology of employing LFER to predict differences in pharmacological activity within a series of related agents was proposed by Hansch and Fujita (QSAR–quantitative structure–activity relationships) [1]. Multiple regression analysis was applied in 1977 to chromatographic data (QSRR–quantitative structure–retention relationships) [2]. Later, other chemometric methods of data analysis have been introduced to QSRR. QSRR are now one of the most extensively studied manifestations of LFER.

To undertake QSRR studies two kinds of input data are needed (Figure 8.1). One is a set of quantitatively comparable retention data (dependent variable) for a sufficiently large (for statistical reason) set of analytes. The other is a set of quantities (independent variables) assumed to account for structural differences among the analytes being studied. Through the use of chemometric computational techniques, retention parameters are characterized in terms of various descriptors of analytes (and their combinations) or in terms of systematic knowledge extracted (learnt) from these descriptors [3, 4].

Once statistically significant and physically meaningful QSRR have been obtained, one can exploit them for [5]:

1. prediction of retention of a new analyte;
2. identification of the most informative structural descriptors possessing the highest retention prediction potency;

Monolithic Silicas in Separation Science. Edited by K.K. Unger, N. Tanaka, and E. Machtejevas
© 2011 WILEY-VCH Verlag GmbH & Co. KGaA, Weinheim
ISBN: 978-3-527-32575-7

8 Quantitative Structure–Retention Relationships in Studies of Monolithic Materials

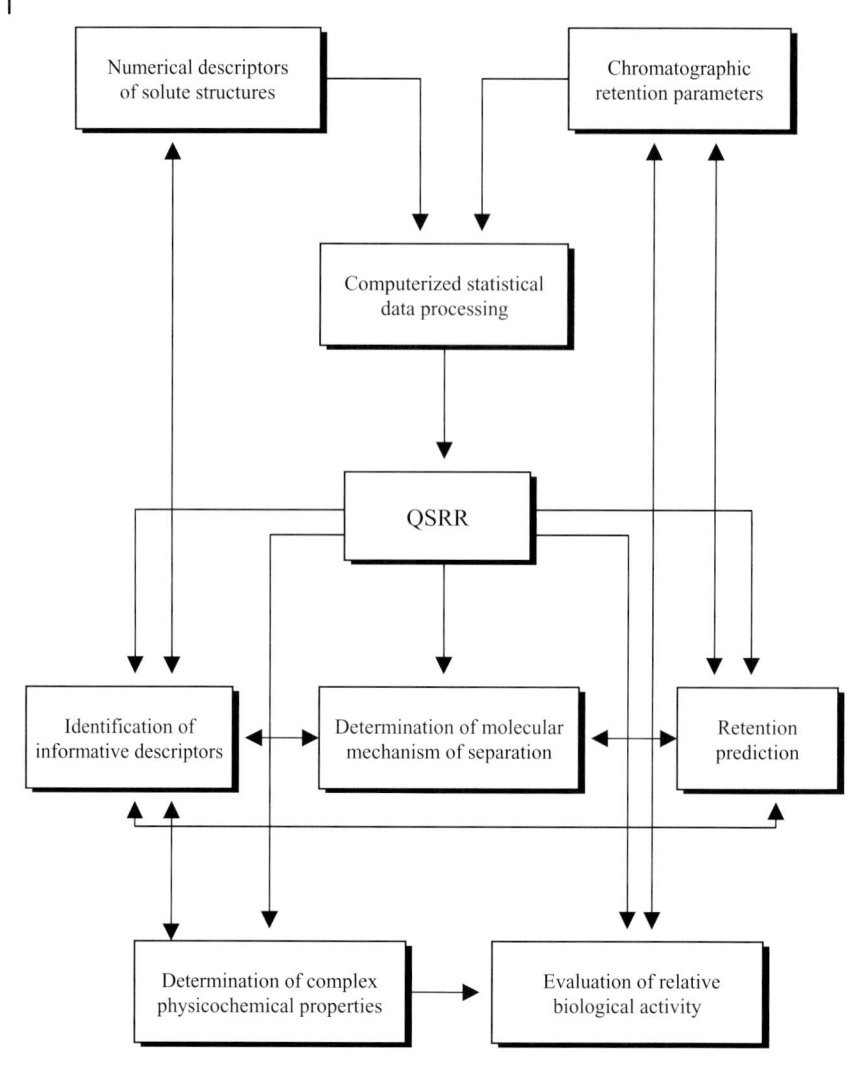

Figure 8.1 Methodology and goals of the quantitative structure–retention relationships (QSRR).

3. getting insight into the molecular mechanism of separation operating in individual chromatographic systems;
4. evaluation of complex physicochemical properties of analytes, for example, their hydrophobicity;
5. prediction of relative biological (pharmacological) activities within a set of drugs and other xenobiotics.

Assuming LFER, a given chromatographic retention parameter ($\log k$, R_M, Kováts index) may be described (statistically) by a set of analyte structural descriptors:

$$\text{Retention Parameter} = f(a_1 x_1, \ldots, a_n x_n) \tag{8.1}$$

The coefficients a_1–a_n for individual n descriptors are calculated by multiple regression. There are computer programs available commercially that serve to derive regression coefficients and to evaluate statistical value of the regression model assumed.

Multivariate methods of data analysis, like discriminant analysis, factor analysis and principal component analysis are often employed if multiple regression methods fail [6]. The most popular chemometric method in QSRR studies is principal component analysis (PCA).

By PCA one reduces the number of independent variables in a data set by finding linear combinations of those variables that explain most of the dependent-variable variability. Commercially available software packages tabulate the component weights and the values of individual principal components. Plots of component weights for each variable (structural descriptor of analytes) are useful in QSRR analysis. Analogously, the scatterplots for the first two principal components illustrate the distribution of objects (analytes) according to their inputs to the principal components.

The translation of molecular structures into numerical descriptors is important not only in QSRR but also to many subdisciplines of chemistry and pharmacology. There is a magnitude of diverse empirical, semiempirical, quantum chemical, topological and other calculation chemistry acquired molecular descriptors [7].

Structural descriptors related to molecular size may be treated as reflecting the ability of an analyte to take part in nonspecific intermolecular interactions (dispersive interactions or London interactions) with the components of a chromatographic system. They are the factors most often found significant in QSRR analysis. The bulkiness parameters are decisive in the description of separations of closely congeneric analytes. For example, carbon number normally suffices to differentiate retention of individual members of a homologous series.

What is more or less intuitively understood as molecular polarity of an analyte is difficult to quantify unequivocally. The descriptors of polarity are expected to account for differences among analytes regarding their dipole–dipole, dipole–induced dipole, hydrogen bonding and electron-pair donor–electron-pair acceptor (EPD-EPA) interactions.

Obviously, geometry-related or molecular shape parameters are difficult to quantify one-dimensionally. Single numbers reflecting molecular shape differences are adequate only in the case of rigid and planar solutes. They become significant in QSRR equations if the range of analytes considered comprises compounds of similar size and polarity.

The multiparameter QSRR based on linear solvation energy relationships (LSER) possess a high predictive power regarding reversed-phase HPLC retention. The model developed by Abraham and coworkers to predict the n-octanol/water coefficient, log P, appears useful also in chromatography [8]:

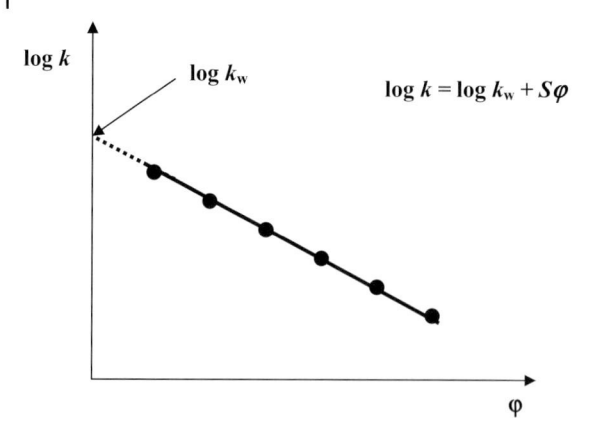

Figure 8.2 Representation of the linear relationships between log k and the volume fraction of the organic modifier, φ, in an aqueous eluent.

$$\log k = c_0 + c_1 V_X + c_2 \pi_2^H + c_3 \sum \alpha_2^H + c_4 \sum \beta_2^H + c_5 R_2 \tag{8.2}$$

where log k denotes logarithm of retention factor calculated from the retention times of the analyte, t_r, and of an unretained marker compound, t_0:

$$k = (t_r - t_0)/t_0 \tag{8.3}$$

The logarithmic form, log k, is used as a hydrophobicity index. The logarithm of retention factor obtained at a given defined mobile phase concentration is called the isocratic log k. Instead of the isocratic log k values, more often log k_w values are used. The log k_w is obtained by extrapolation of the plots of "polycratic" log k values against the concentration of organic modifier in the mobile phase (Figure 8.2). The extrapolation is based on the assumption of the linear relationships between log k and the volume fraction of the organic modifier in a binary aqueous eluent.

In Equation 8.2 V_X is the so-called McGowan's characteristic volume that can be calculated simply from the analyte's molecular structure; π_2^H is dipolarity/polarizability of the analyte that can be determined through gas-chromatographic and other measurements or calculated theoretically; $\sum \alpha_2^H$ is the effective or summation hydrogen-bond acidity; $\sum \beta_2^H$ is the effective or summation hydrogen-bond basicity and R_2 is an excess molar refraction that can be obtained from refractive-index measurements and is an additive quantity. The LSER-based structural descriptors are now available for a large number of compounds.

A typical multiparameter approach to predicting retention of an unknown compound, based on structural features and chromatographic properties of it and other more or less similar compounds, consists of generating a multitude of analyte descriptors that are next regressed against retention data [3, 4]. The structural descriptors are usually derived by computational chemistry methods for the energy-optimized conformations. Software systems have been developed that produce and process hundreds of quantum chemical, molecular modeling, topological and semiempirical additive-constitutive descriptors after sketching the molecule in the computer. Observing all the rules and recommendations for meaningful statistics the minimum number of descriptors (uncorrelated) is selected that are needed to produce a QSRR equation with a good predictive ability. The descriptors that serve eventually to predict retention of new analytes are sometimes of obscure physical meaning. Nevertheless, for several groups of compounds, prediction of retention by means of QSRR is reliable enough for identification purposes, especially when there is no better alternative [5, 9].

The QSRR equations that comprise physically interpretable structural descriptors can be discussed in terms of the molecular mechanisms involved in the chromatographic process.

The proper QSRR strategy aimed at objective characterization of differences in retentive potency of individual chromatographic columns should employ a well-designed set of test analytes [10].

8.2
Quantitative Relationships between Analyte Hydrophobicity and Retention on Monolithic Columns

In QSRR studies often the retention parameters of a series of test analytes are first linearly regressed against the reference log P values from the n-octanol-water partition system, as a standard measure of compounds' hydrophobicity. Good correlations obtained are interpreted as evidence of a partition mechanism operating in the chromatographic system under study.

Hydrophobicity or lipophilicity is understood to be a measure of the relative tendency of an analyte "to prefer" a nonaqueous over an aqueous environment or the tendency of two (or more) analyte molecules to aggregate in aqueous solutions. The partition coefficients of the substances may differ if determined in different organic–water solvent systems but their logarithms are often linearly related. Hydrophobicity is commonly quantified by means of the measured or calculated logarithm of partition coefficient of an analyte between n-octanol and water forming a two-phase system (log P). It is a laborious procedure in which one measures equilibrium concentrations of the analyte in a lipophilic phase (n-octanol) and in an aqueous phase (water or an appropriate buffer solution). The so-called "shake-flask" method is suitable only for nonionized compounds with a strong UV absorption, sufficient solubility in the aqueous phase and log P values in the range from −3 up to +3. Impurities and instability of the analyte can result in unreliable experimental data. In order to overcome the limitations and disadvantages of the "shake-flask" method chromatographic methods have alternatively been used to assess hydrophobicity of organic compounds. These are reversed-phase thin-layer chromatography (TLC) and reversed-phase high-performance liquid chromatography (RP-HPLC) methods [11]. Occasionally centrifugal partition chromatography (CPC) has been employed [5, 12].

The most commonly used stationary phase in reversed phase HPLC is octadecylsilica (ODS) that has an ability to separate a wide variety of analytes. The octadecyl-bonded silica phases have also been commonly employed in hydrophobicity studies. There is similarity between a chromatographic system with hydrocarbonaceous silica stationary phase and the n-octanol/water system. Hence, the outputs from both the systems, that is, logarithms of partition coefficient ($\log P$) and retention parameters (R_M, $\log k$), should correlate.

The relationship between $\log P$ values and RP HPLC retention parameters offers a possibility of not only estimating analyte hydrophobicity by chromatography but also of predicting its reversed-phase chromatographic retention. The general form of that relationship presents the equation:

$$\log k_w = k_1 + k_2 \log P \tag{8.4}$$

where k_1 and k_2 are the coefficients of regression.

The high $\log k_w$ vs. $\log P$ correlations confirm the general similarity of the slow-equilibrium "shake-flask" octanol–water partition system and the fast-equilibrium reversed-phase chromatographic systems. One can presume that the k_2 values (slope) in Equation 8.4 reflect the degree of similarity between the chromatographic and the "shake-flask" systems. The more alike to octanol the solvated stationary phase is the closer to 1 should k_2 be.

Evaluation of lipophilicity parameters for basic compounds using different chromatographic stationary phases was a subject of investigation by Welerowicz and Buszewski [13]. Differences in correlation between the lipophilicity of compounds and experimental chromatographic results obtained in pseudomembrane systems showed a strong influence of stationary phase structure on physicochemical properties. β-Adrenolytic drugs with varying lipophilicity and bioactivity were chosen as test compounds. The stationary phases used for the study were among others monolithic rod-structure C18 phases. The authors reported a linear correlation of retention parameters, $\log k_g$, with $\log P$ at the level of $R^2 = 0.921$ and 0.909 for methanol and acetonitrile, respectively, as organic modifiers in the mobile phase.

$$\log k_{g\,\text{Monolith-C18 (MeOH buffer)}} = 0.374\ (\pm 0.045) + 0.392\ (\pm 0.038)\log P \tag{8.5}$$

$$\log k_{g\,\text{Monolith-C18 (ACN buffer)}} = -0.265\ (\pm 0.069) + 0.554\ (\pm 0.058)\log P \tag{8.6}$$

In RP HPLC, lipophilicity of organic compounds can be measured employing standard HPLC equipment.

In 1997 RP HPLC parameters of hydrophobicity were proposed based on the fast gradient elution HPLC. Krass et al. [14] determined the parameter k_g:

$$k_g = (V_g - V_d - V_m)/V_m \tag{8.7}$$

where V_g is gradient volume; V_d is the equipment dwell volume and V_M is the column dead volume. The authors reported a good correlation between k_g from gradient elution and the standard $\log k_w$ extrapolated from a series of isocratic measurements.

8.2 Quantitative Relationships between Analyte Hydrophobicity and Retention

At the same time Du et al. [15] proposed a chromatographic hydrophobicity index, CHI, derived from retention times, t_r, in a fast gradient reversed-phase HPLC.

The newly introduced pH gradient reversed-phase HPLC method [16] consists in a programmed increase during the chromatographic run of the eluting power of the mobile phase with regards to ionizable analytes. The eluting strength of the mobile phase increases due to its increasing (in case of acidic analytes) or decreasing (basic analytes) pH, whereas the content of organic modifier remains constant. It has been theoretically and experimentally demonstrated [17, 18] that the pK_a and log k_w values can be evaluated based on retention data from a pH gradient run, combined with appropriate data from two organic solvent gradient runs. The gradient HPLC derived log k_w parameters correlate well with analogous parameters determined isocratically as well as with reference lipophilicity parameter log P.

Gradient retention time, t_R, of a set of structurally diversified fifteen test analytes was used as a dependent variable in the QSRR equation describing retention on monolithic column in terms of theoretically calculated partition coefficient, clog P [19]:

$$t_R = 3.489\ (\pm 1.345) + 4.882\ (\pm 0.406)\ \text{clog}\,P$$
$$R = 0.958,\ s = 2.40,\ F = 145,\ p < 2\text{E}{-}08 \tag{8.8}$$

where R is correlation coefficient, F is the value of the Fisher test of significance and p is the significance level of the regression equation.

The high t_R versus clog P correlation confirms the similarity of the slow-equilibrium octanol/water partition system and the fast-equilibrium partition chromatographic process. Bączek et al. [19] suggested a hydrophobicity order of the stationary phases studied based on the slope coefficient in Equation 8.8. The order was as follows: Discovery HS F5 < Inertsil ODS-3 < Supelcosil LC-18 < XTerra MS C18 ≤ LiChrospher 60RP-select B < Chromolith ≤ Nova-Pak C18 < Symmetry C18 ≤ Aqua C18. According to the coefficients of the hydrophobic-subtraction model, Chromolith is similar to the Nova-Pak C18, Aqua C18 and Symmetry C18 columns.

Detroyer et al. [20] studied performance of the recently introduced monolithic silica columns in micellar liquid chromatography (MLC). Micellar methods are utilized in high-throughput QSRR to estimate an indicator of the membrane permeability of drugs, namely, the octanol–water partition coefficient, log P. The monolithic column's ability to function at higher flow rates might be useful to speed up these chromatographic methods estimating the log P. Therefore, the elution behavior of diverse basic pharmaceutical substances was determined on a classical particle-based and a monolithic column, both with and without a micellar medium in the mobile phase. Utilizing principal component analysis (PCA), the extent to which these methods differ in retention characteristics was examined in the context of high throughput determination of log P. Results indicated that combining monolithic columns with micellar media leads to a faster determination of log P and possibly even better permeability predictions. The QSRR involving log *P* and the proposed "fast" monolithic micellar systems appear to be comparable to, sometimes even better than, the classic HPLC methods employing organic modifier/buffer eluents. As molecules from a broader log *P* range can be analyzed with a micellar system, "fast" MLC have been proposed as a faster and potentially better membrane-permeability predicting system.

8.3
QSRR Based on Structural Descriptors from Calculation Chemistry

There is a QSRR approach, which allows for interpretation of the molecular mechanism of retention in terms of structural parameters of analytes obtained from molecular modeling. Test analytes are subjected to geometry optimization that employs the molecular mechanics, semiempirical methods and sometime *ab initio* calculations. There are a number of structural descriptors readily produced by commonly available molecular modeling software [7]. However, it is often difficult to assign defined physical meaning to individual descriptors obtained [5, 21].

The calculation chemistry derived parameter most often used in QSRR calculations is the solvent-accessible molecular surface area, SAS (Figure 8.3). That

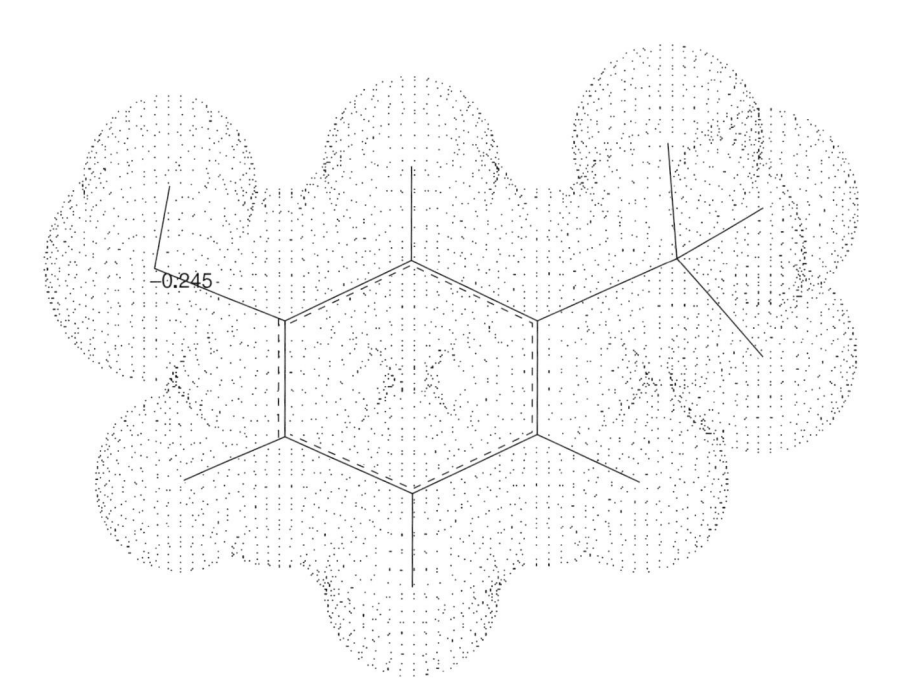

Figure 8.3 Solvent-accessible molecular surface area, SAS, and maximum excess of electrons on an indicated atom in the molecule, δ_{min}, on the example of 3-trifluoromethylophenol.

molecular descriptor is assumed to account for the strength of dispersive interactions between the analyte and the molecules forming the chromatographic system. Dipole moment, μ, or square of the dipole moment, μ^2, account for the differences among the analytes regarding their dipole–dipole or dipole–induced-dipole interactions. Maximum excess of electrons on an atom in the molecule, δ_{min}, may reflect the ability of analytes to form local dipoles. Energies of the lowest unoccupied molecular orbital, E_{LUMO}, and the highest occupied molecular orbital, E_{HOMO}, explain the differences in tendency of analytes to form the electron-pair donor/electron-pair acceptor (EPD/EPA) complexes with other molecules or their ability to take part in charge-transfer interactions.

Exemplary QSRR equation characterizing a C18 silica stationary phase, *Hypersil ODS*, and describing log k_w obtained in methanol–water eluents in terms of δ_{min}, μ^2 and SAS is presented below:

$$\log k_w = -1.758(\pm 0.606) + 5.223(\pm 1.345)\, \delta_{min} - 0.106(\pm 0.025)\, \mu^2 +$$
$$0.018(\pm 0.002)\, SAS \tag{8.9}$$
$$n = 23;\ R = 0.961;\ s = 0.452;\ F = 77$$

Equation 8.9, employing the molecular modeling-based parameters, makes good physical sense although it has a slightly lower predictive potency than similar equations comprising the LSER-based parameters [22].

Analyzing Equation 8.9 one notes that the net positive effect to retention is due to the analyte bulkiness descriptors, SAS. The dispersive attractions of analyte are stronger from the side of the bulky hydrocarbon ligand of the stationary phase than from the side of the small molecules of the eluent. The net effect to retention provided by μ^2 is negative because the dipole–dipole and dipole–induced-dipole attractions are stronger between the analyte and the polar molecules of the eluent than between the same analyte and the nonpolar hydrocarbon ligand of the stationary phase. The positive sign of the coefficient at the δ_{min} term (the value of δ_{min} is negative) means that local dipoles are more easily formed between the analyte and the polar molecules of the eluent than with the chemically inert hydrocarbons of the stationary phase.

To gain insight into the molecular mechanism of the separation of xanthines on the Chromolith RP-18e columns QSRR were derived describing retention parameters determined at different pH [23]. Molecular descriptors of the protonated forms of the xanthines were applied in the QSRR analysis for retention data corresponding to the citrate (pH 2.00) and the acetate (pH 4.66) buffered systems. The "best" QSRR equations obtained were:

$$\log k\ (\text{pH } 2.00) = 0.1575(\pm 0.0187)\, \text{Refr} - 0.1218(\pm 0.0417)\, E_{HOMO} - 8.0874$$
$$R = 0.9367;\ F = 36.972 \tag{8.10}$$

$$\log k\ (\text{pH } 4.66) = 0.1104(\pm 0.0132)\, \text{Refr} - 0.0643(\pm 0.0295)\, E_{HOMO} - 5.2971$$
$$R = 0.9337;\ F = 35.234 \tag{8.11}$$

where Refr denotes molar refractivity and E_{HOMO} is energy of the highest occupied molecular orbital.

Equations 8.10 and 8.11 show that retention of xanthines on the Chromolith RP-18e phase at low pH depends on the bulkiness (as expressed by Refr) and electronic properties (represented by E_{HOMO}) of the analytes. This seems reasonable for the reversed-phase type of chromatographic separations [24]. As expected for partition chromatography, dispersive interactions between the analyte and the bulky ligand of the stationary phase are stronger than those between the same analyte molecule and the small molecules of the mobile phase, whereas polar interactions prove to be stronger between a solute and polar molecules of mobile phase than between the analyte and the nonpolar ligand of the stationary phase. The authors of Ref. [22] concluded that in their studies there were identified the structural factors that affect the retention of xanthines on the Chromolith RP-18e columns at different pH, thus illustrating the applicability of semiempirical calculations to the explaining of retention mechanisms operating in a specific chromatographic system.

A study reported by Bączek et al. [19] was aimed at quantitative comparison of retention properties of modern stationary phases, such as the monolithic column, for RP HPLC. The multiple regression equations obtained are based on the three structural parameters from molecular modeling (i.e. total dipole moment, electron excess charge of the most negatively charged atom and water-accessible molecular surface area). These equations make good physical sense:

$$t_R = -0.261\,(\pm 2.5) - 2.678\,(\pm 0.374)\,\mu + 22.044\,(\pm 6.060)\,\delta_{min} + 0.080\,(\pm 0.006)\,A_{WAS}$$
$$R = 0.986,\ s = 1.52,\ F = 128,\ p < 8E-09$$

(8.12)

The value of the coefficient at A_{WAS} is positive, in accordance with a positive contribution to retention of the nonspecific analyte–stationary phase interactions, characterized by the this parameter. These interactions require a close contact of the interacting molecules fragments. Larger values of A_{WAS} indicate a larger surface area of the stationary phase hydrocarbon moiety that is accessible to the analyte.

The specific, polar intramolecular interactions are characterized by the coefficients at μ and δ_{min} parameters. Negative values of the coefficients at the μ parameter suggest that the net effect to the retention of dipole–dipole (and dipole–induced-dipole) attractions between the analytes and the stationary phase, on one hand, and the components of the eluent, on the other hand, is negative. This can be explained by a stronger attraction between the total dipole of the analyte and the total and/or fragmental dipoles (both permanent and induced) of the polar molecules of the eluent, as compared to the respective interactions between the analytes and the nonpolar alkyl chains of the stationary phases. The values of the coefficient at μ and hence, the polarity of the stationary phase, increase in the order: Symmetry C18≤Aqua C18≤Nova-Pak C18≤Chromolith <XTerra MSC18≤Supelcosil LC-18 < Inertsil ODS-3 < LiChrospher 60RP-select B< Discovery HS F5.

A similar rationalization applies to the coefficient at δ_{min} parameter in Equation 8.12. The coefficient at δ_{min} has a positive sign because the δ_{min} values (electron deficiencies) are negative. Thus, the more charged an atom is, the higher is the absolute value of the coefficient at δ_{min} term, and the less retained the analyte is. Lower values of coefficient at δ_{min} can be interpreted as indicating stronger local (fragmental) dipole–dipole interactions and/or the formation of the electron-pair-donor–electron-pair-acceptor complexes between the analyte and the stationary phase, with regards to analogous interactions with the eluent. According to the value of coefficient at δ_{min}, the stationary phases can be ordered as follows with decreasing polarity: Symmetry C18 >XTerra MS C18≥Inertsil ODS-3≥Nova-Pak C18 > Discovery HS F5 > LiChrospher 60RP-select B≥Aqua C18 > Chromolith > Supelcosil LC-18. The coefficient at δ_{min} can also be interpreted as reflecting the ability of analytes to take part in hydrogen-bonding interactions with free silanols of the stationary phases support material.

8.4
LSER on Monolithic Columns

In predicting the chromatographic retention the QSRR model based on the solvatochromic comparison method and the linear solvation energy relationships has frequently been employed [25].

Full Abraham equations of the type of Equation 8.2 have a good predictive power regarding retention. However, relatively high standard deviations of some regression coefficients cause the lack of statistical significance in QSRR equations of some LSER-based parameters of analytes. For that reason modified equations have been proposed [22], which retain only the terms significant above the 95% significance level. The most significant for description of reversed-phase retention appeared hydrogen-bond basicity (β_2^H) and McGowan volume (V_X) of analytes. The third significant parameter in LSER-based QSRR is either dipolarity/polarizability (π_2^H) or hydrogen-bond acidity (α_2^H). Exemplary QSRR equations derived for retention parameters determined on the *Hypersil ODS* stationary phase are as follows:

$$\log k_w \text{ (MeOH–water)} = 0.303(\pm 0.257) - 0.840(\pm 0.163)\, \pi_2^H - 2.309(\pm 0.203)\, \beta_2^H + 3.408(\pm 0.168)\, V_X \qquad (8.13)$$

$n = 23; R = 0.989; s = 0.241; F = 288$

$$\log k_w \text{ (ACN–water)} = 0.678(\pm 0.150) - 0.810(\pm 0.116)\, \alpha_2^H - 2.960(\pm 0.117)\, \beta_2^H + 1.935(\pm 0.118)\, V_X \qquad (8.14)$$

$n = 23; R = 0.991; s = 0.168; F = 345$

The following interpretation of the physical sense of these equations is proposed. The coefficient at the McGowan's volume term is positive. It suggests that the attractive dispersion interactions (London interactions) between the analyte and the bulky hydrocarbon ligand of the stationary phase are stronger than the same nonspecific attractive interactions between the analyte and small molecules (water, methanol, acetonitrile) of the eluent.

The net effect of attractive interactions of a hydrogen-bond acceptor analyte with hydrocarbonaceous stationary phase, on one hand, and with the polar molecules of the eluent, which are efficient hydrogen-bond donors, on the other hand, is negative. The negative sign at coefficient with β_2^H confirms this.

The term reflecting solute dipolarity/polarizability, π_2^H, is significant if methanol is used to derive log k_w. Its negative sign reflects the higher strength of dipole–dipole and dipole–induced-dipole attractive interactions between the analyte and the polar molecules of the eluent as compared to the same type of interactions between the analyte and the nonpolar stationary phase.

If acetonitrile was used to determine log k_w the respective LSER-based QSRR equations comprised the hydrogen-bond acidity term, α_2^H, instead of π_2^H. Again, the sign is negative, as expected because the hydrogen bonding of analyte with the polar eluent serving as a hydrogen bond acceptor is stronger than with the hydrocarbonaceous stationary phase.

Lesellier and West [26] presented a QSRR analysis based on the retention factors of aromatic compounds to characterize the different chromatographic behavior of octadecylsiloxane-bonded stationary phases of all sorts: classical, protected against silanophilic interactions or not, containing polar groups (endcapping groups or embedded groups) and monolithic stationary phase material. Varied chemometric methods were used to enlighten the differences between the 27 phases tested. The LSER system constants for each chromatographic system were obtained by multiple linear regression analysis for the logarithms of the measured retention factors. In case of monolithic column the equation has a form:

$$\log k = -1.197 + 0.508E - 0.411S - 0.367A - 0.348B + 0.383V$$
$$n = 27, R^2 = 0.965, SE = 0.045, F = 145.5$$
(8.15)

where E is the excess molar refraction, S is the solute dipolarity/polarizability, A and B are the solute overall hydrogen-bond acidity and basicity, V is the McGowan characteristic volume. Obtained in the equation positive values for E and V coefficients, indicate that an increase in volume and in the polarizability of the solute induces an increase in retention on the monolithic phase. These terms are both related to dispersive interactions as an increase in volume and in polarizability both favor high dispersive interactions with the octadecyl chains. Negative values of coefficients S, A and B, indicate that polar, acidic and basic solutes have greater interactions with the mobile phase than with the stationary phase thus are less retained on these stationary phases. Principal component analysis based on the solvation parameter model enclosed monolithic column in the group of the classical, endcapped or protected ODS columns among five different groups of stationary-phase materials.

According to the hydrophobic-subtraction model studied by Bączek et al. [19], reversed-phase columns were characterized by five selectivity parameters derived from the LSER theory. The hydrophobic-subtraction model applied to a series of diverse test analytes chromatographed on nine modern stationary phases showed no advantage over the less complex QSRR models employing calculated logarithm of octanol–water coefficient, clog$P$, or a set of three descriptors from molecular modeling. Actually, the QSRR based on the hydrophobic-subtraction model appeared statistically questionable because of marked intercorrelations between the descriptors.

8.5
Concluding Remarks

Quantitative structure–chromatographic retention relationships (QSRRs) can be readily derived for retention data determined on monolithic columns. This may be due to a high reliability of retention parameters obtained on such columns. By QSRR a partition chromatography mechanism was demonstrated as prevailing in the separation process. All the three standard types of QSRR equations, that is, those relating retention parameters to hydrophobicity of analytes, those employing linear salvation energy relationships (LSER)-based parameters and those utilizing structure descriptors from molecular modeling, showed similarity of the mechanism of separations to that observed on conventional reversed-phase materials. A good retention prediction potency of the QSRR characterizing monolithic columns allows for reliable predictions of retention for any structurally defined analyte and hence may help to rationally optimize chromatographic separations.

References

1. Hansch, C., and Fujita, T. (1964) *J. Am. Chem. Soc.*, **86**, 1616.
2. Kaliszan, R., and Foks, H. (1977) *Chromatographia*, **10**, 346.
3. Kaliszan, R. (1987) *Quantitative Structure-Chromatographic Retention Relationships*, John Wiley & Sons, Inc., New York.
4. Kaliszan, R. (1997) *Structure and Retention in Chromatography. A Chemometric Approach*, Harwood Academic Publishers, Amsterdam.
5. Kaliszan, R. (2007) *Chem. Rev.*, **107**, 3212.
6. Put, R., Xu, Q.S., Massart, D.L., and Vander Heyden, Y. (2004) *J. Chromatogr. A*, **1055**, 11.
7. Todeschini, R., and Consonni, V. (2000) *Handbook of Molecular Descriptors*, Wiley-VCH Verlag GmbH, Weinheim, Germany.
8. Abraham, M.H. (1993) *Chem. Soc. Rev.*, **22**, 73.
9. Bodzioch, K., Dejaegher, B., Baczek, T., Kaliszan, R., and Vander Heyden, Y. (2009) *J. Sep. Sci.*, **32**, 2075.
10. Al-Haj, M.A., Kaliszan, R., and Nasal, A. (1999) *Anal. Chem.*, **71**, 2976.
11. Pliška, V., Testa, B., and Van de Waterbeemd, H. (1996) *Lipophilicity in Drug Action and Toxicology*, VCH, Weinheim.
12. Heberger, K. (2007) *J. Chromatogr. A*, **1158**, 273.
13. Welerowicz, T., and Buszewski, B. (2005) *Biomed. Chromatogr.*, **19**, 725.
14. Krass, J.D., Jastorff, B., and Genieser, H.G. (1997) *Anal. Chem.*, **69**, 2571.
15. Du, C.M., Valko, K., Bevan, C., Reynolds, D., and Abraham, M.H. (1998) *Anal. Chem.*, **70**, 4228.
16. Wiczling, P., Markuszewski, M.J., and Kaliszan, R. (2004) *Anal. Chem.*, **76**, 3069.
17. Wiczling, P., Kawczak, P., Nasal, A., and Kaliszan, R. (2006) *Anal. Chem.*, **78**, 239.
18. Wiczling, P., Waszczuk-Jankowska, M., Markuszewski, M.J., and Kaliszan, R. (2008) *J. Chromatogr. A*, **1214**, 109.
19. Baczek, T., Kaliszan, R., Novotna, K., and Jandera, P. (2005) *J. Chromatogr. A*, **1075**, 109.

20. Detroyer, A., Heyden, Y.V., Reynaert, K., and Massart, D.L. (2004) *Anal. Chem.*, **76**, 1903.
21. Wang, C., Skibic, M.J., Higgs, R.E., Watson, I.A., Bui, H., Wang, J., and Cintron, J.M. (2009) *J. Chromatogr. A.*, **1216**, 5030.
22. Kaliszan, R., van Straten, M.A., Markuszewski, M., Cramers, C.A., and Claessens, H.A. (1999) *J. Chromatogr. A*, **855**, 455.
23. Baranowska, I., and Zydroń, M. (2002) *Anal. Bioanal. Chem.*, **373**, 889.
24. Markuszewski, M., Krass, J.D., Hippe, T., Jastroff, B., and Kaliszan, R. (1998) *Chemosphere*, **37**, 559.
25. Forgács, E., and Cserháti, T. (1997) *Molecular Bases of Chromatographic Separations*, CRC Press, Boca Raton.
26. Lesellier, E., and West, C. (2007) *J. Chromatogr. A*, **1149**, 345.

9
Performance of Silica Monoliths for Basic Compounds. Silanol Activity
David V. McCalley

9.1
Introduction

Reversed-phase HPLC (RP-LC) continues to be the most widely used technique for routine analysis and research in the pharmaceutical industry due to its many advantages, such as compatibility with aqueous samples, ease of use with gradient elution, reproducibility of retention times, versatility of the mechanism, which can be altered by changing the mobile phase or additives, and long experience with the use of the technique, allowing prediction of appropriate analysis conditions for a given sample [1]. Monolithic columns, which have now been commercially available for about 10 years, have enhanced the choice of RP-LC techniques available to analysts for separation of pharmaceutical and other types of samples. The large majority of pharmaceuticals, and in addition many compounds of biomedical significance, contain ionizable basic groups. According to Wells [2], some 75% of pharmaceuticals are bases. Thus, many of these compounds are protonated over the usual pH range of operation of RP-LC (for most columns ~pH 2.5–7.5), and may undergo detrimental ionic interactions with ionized silanol groups that exist on the surface of these columns.

$$BH^+ + SiOM^+ \rightarrow SiOBH^+ + M^+$$

where BH^+ is the protonated form of a base B and M^+ represents the mobile-phase buffer cation [3]. The average pK_a of silanols was reported as 7.1, but it is clear that some silanols are considerably more acidic, caused by the presence of metal impurities in the silica [4], and thus remain ionized even at the lowest pH values that can be used with most columns. The tailing and poor efficiency that is a consequence of such interactions are not simply due to the presence of a mixed retention mechanism, with hydrophobic interactions occurring alongside an ionic retention mechanism exhibiting slow kinetics. It is likely that the stereochemistry of ionized silanols, being buried deep beneath the hydrophobic ligands, and/or a multiplicative or synergistic hydrophobic/ionic retention mechanism is responsible [5]. Such problems can be reduced by endcapping procedures, which reduce the number of underivatized silanols. More importantly, the use of

Monolithic Silicas in Separation Science. Edited by K.K. Unger, N. Tanaka, and E. Machtejevas
© 2011 WILEY-VCH Verlag GmbH & Co. KGaA, Weinheim
ISBN: 978-3-527-32575-7

pure "Type B" silicas that have low metal content and thus less acidic silanols, considerably ameliorates these problems [6]. Nevertheless, a further problem with the analysis of ionized solutes that occurs even on highly pure phases with very low numbers of acidic silanols is overloading that can cause dramatic losses in column efficiency for sample masses 50–100 times less than for neutral solutes [7].

In this chapter, some of the results obtained for basic solutes on monolithic silica columns, particularly with regard to these effects, will be reviewed.

9.2
Reproducibility of Commercial Monoliths for Analysis of Bases

Kele and Guiochon investigated the repeatability and reproducibility of retention data and band profiles on six batches of the commercial C18 chemically bonded silica monolith from Merck [8]. A total of 30 neutral, acidic and basic test compounds were chromatographed on a series of six Chromolith C18 columns, each from a different production batch. An important conclusion of this work was that the columns exhibited a high degree of reproducibility, being better, or closely matching the values obtained previously on five brands of particle-based columns. This result is noteworthy, as it suggests that for monolith and particle-based columns (at least from major manufacturers), the reproducibility is sufficient such that tests performed on one, or a limited number of columns, are likely to be quite representative of their average properties. The conclusion was remarkable in that each monolith is the result of a single discrete process, where it is individually produced in its own mold (even when considering columns produced at the same time or in the same "batch"), unlike the situation for particle-based columns, where a large number can be packed from a homogenized batch of material.

The testing procedures of Kele and Guiochon were comprehensive, including the weak bases aniline, N,N-dimethylaniline, o-, m-,p-toluidine and pyridine tested in simple methanol–water mixtures without the addition of buffer [8]. The stronger bases benzylamine and procainamide were tested in methanol-phosphate buffer at pH 2.7, and amitriptyline and propranolol in methanol-phosphate buffer pH 7.0. The long-term repeatability of the retention times of the tests carried out in unbuffered solutions on the same column were excellent for the basic compounds (RSD <0.1%), which was not always found by the same authors for some particle-packed columns, indicating a very high degree of stability of the surface chemistry of the monolith. The batch to batch reproducibility of retention times in these tests was also good, with the greatest variation for pyridine (RSD~3%). For tests of the stronger bases in buffered mobile phases, retention reproducibility was not so good, but batch to batch retention reproducibility was still < 4.5%. Nevertheless, reproducibility of peak-shape measurements for basic solutes was poorer. While reproducibility of the tailing factor was 1.2–2.6% for neutral compounds, it was 3.4–22.6% for basic solutes.

9.3
Activity of Monoliths towards Basic Solutes

Surprisingly few publications have evaluated in detail the potential of silica monoliths for analysis of basic solutes. Some publications show successful applications of the separation of basic solutes, but without provision of peak-shape data: for example, the separation of 5 β-blocking drugs, which are typically basic compounds has been reported but with no such details [9]. An early publication by Fekete and coworkers [10] presented a detailed comparison of the performance of five particle-packed RP-LC columns with that of the Merck monolith for the analysis of the antibiotics, ciprofloxacin and vancomycin, which both contain basic groups. The mobile phase was acetonitrile-phosphate buffer at pH 3.0. The column efficiency for Ciprofloxacin was higher for the monolith (around 90 000 plates/m) than any of the 5-μm packed columns, with an asymmetry factor of 1.5, a value that was about average for the group of compounds studied. For vancomycin, the efficiency was around 37 000 plates/m, the second highest in the group, with a asymmetry factor around 1.3. However, triethylamine was added to the mobile phase in each case, in order "to diminish silanol activity and metal-impurity effects". Particularly for ciprofloxacin, the authors concluded that the high plate number was evidence for there being "only hydrophobic interaction between the solute and the stationary phase". In such a case, the commercial monolith was expected to give higher efficiency, as according to Leinweber et al., it has an efficiency comparable to 3-μm particulate columns at least with nonpolar solutes [11]. While it was claimed that the concentration of triethylamine used in the mobile phase was not sufficient to block all the silanol interactions of some of the columns tested [10], clearly the use of blocking agents will reduce the differences observed between various stationary phases and is not generally recommended for column testing. Furthermore, this procedure has fallen somewhat out of favor as such additives may give rise to slow equilibration of the column with the mobile phase, they may be difficult to remove from the column and thus permanently alter its properties, can cause interference with detection, and may even chemically react with some sample constituents [3].

Kele and Guiochon [8] reported peak-shape data for all solutes used in their tests. Peak asymmetry was measured according to the United States Pharmacopeia tailing factor (T_f), determined from the peak width at 5% of height divided by twice the width of the leading edge of the peak. All of the test compounds in the study (which included simple neutral compounds) gave tailing to some degree but for 24 out of 34 compounds, the tailing factor was below 1.4. Propranol ($T_f \sim 3.7$) and amitriptyline ($T_f = 3.2$) gave poor results for the test in methanol-phosphate buffer at pH 7, although tailing factors for procainamide and benzylamine in the acidic mobile phase were considerably lower (~1.5). The authors found some tailing even for neutral compounds. They also investigated high and low loads of some of the basic solutes tested, and ruled out overload as being the cause of this tailing. They concluded that two possible causes remain: a column-bed heterogeneity, and/or slow kinetics of mass transfer (see above).

It is interesting that Guiochon and coworkers later compared the radial heterogeneity of a conventional particle-packed column, a fused-core particle column, and the commercial monolith using an electrochemical microdetector to determine accurately the radial distribution of the mobile-phase velocity [12]. They suggested that the strain resulting from the shrinkage of the monolithic rod during the process of its production could cause the silica layer close to the wall to become deformed, leading to slightly greater porosity in the wall region compared with the center. Such a result could explain the peak asymmetry even for neutral compounds they found measured in the earlier study. Nevertheless, they claimed that the decrease in local efficiency from the core to the wall region was about 25% for the monolith, compared with 35–50% for the particle-packed columns, which showed a different radial velocity pattern. Thus, it is not certain whether the radial heterogeneity of monoliths could be the cause of the tailing of neutral compounds, and clearly this cannot explain the greater tailing experienced for bases than for neutral solutes.

McCalley [13, 14] compared the performance of the commercial monolith with a number of Type-B particle-packed columns for the analysis of basic compounds. Weak bases chromatographed in unbuffered mobile phases, and stronger bases in buffered mobile phases at low and intermediate pH were studied. Considerable additional comparative data for particle-packed columns tested with basic compounds under similar mobile-phase conditions is also available from publications by the same author [15–17]. The asymmetry factor (A_s) of peaks, defined as the width of the trailing edge of the peak divided by the width of the front edge were initially measured for test solutes in unbuffered mobile phases. This parameter is different from the T_f as utilized by Kele and Guiochon, however, it is possible to relate these two quantities for the purpose of comparison of results between the two research groups.

$$T_f = (a+b)/2a \qquad (9.1)$$

where a is the leading, and b is the trailing edge of the peak.

$$A_s = b/a \qquad (9.2)$$

Simple algebra shows that if measured at the same fraction of the peak height,

$$A_s = 2T_f - 1 \qquad (9.3)$$

Thus a USP tailing factor of 2 would imply an asymmetry factor of 3, if both were measured at 5% of the height of the peak. As the magnitude of the asymmetry factor is probably in most cases smaller at the more commonly used 10% of the peak height, the apparently greater severity of the A_s measurement is moderated somewhat. However, comparison with experimental results [18] shows this relationship is approximately true even when measurements are made at different % peak height. Papai and Papp [19] claimed that if the asymmetrical peak can be modeled with an exponentially modified Gaussian function, the following relationship exists between T_f and A_s:

9.3 Activity of Monoliths towards Basic Solutes

Table 9.1 Asymmetry factors for weak bases and benzene on Chromolith and particle-packed columns. Mobile phase: acetonitrile–water (40:60,v/v) at 2 ml/min. Data from [13]

Column	Pyridine	Aniline	4-Ethylaniline	benzene
Inertsil ODS3 (5 µm)	1.3	1.2	1.2	1.1
Chromolith RP-18e	2.1	1.6	1.7	1.4
SymmetryShield (3.5 µm)	2.1	1.1	1.1	1.0
Purospher star RP18e (3 µm)	2.6	1.3	1.4	1.2
Inertsil ODS 3 (3 µm)	1.4	1.2	1.2	1.1

$$T_f = 0.6\, A_s + 0.4 \tag{9.4}$$

implying a calculated value of $A_s = 2.7$ for $T_f = 2.0$, which is not too different from the value generated by Equation 9.3. McCalley found A_s of 1.4 for benzene in acetonitrile-water (Table 9.1) suggesting $T_f = 1.2$ using Equation 9.4, in agreement with the value of $T_f = 1.3$ quoted by Guiochon [8]. Thus, both groups noted a greater tailing on the monolith than is normally obtained for simple neutral compounds on particle-packed columns. The asymmetry factors reported by McCalley for the weak bases (see Table 9.1) were also somewhat worse than asymmetries obtained for these solutes on 3-µm and 5-µm particle-packed columns. Maximum column efficiencies for the weak bases on the monolith at optimum flow were in between those given by 3-µm and 5-µm particle columns. Detailed comparisons of peak shape were made between the monolith and particle-packed columns with 9 more strongly basic compounds at pH 2.7 (using potassium phosphate, ammonium phosphate, formic acid and trifluoroacetic acid as alternative buffers) and at pH 7.0 (using potassium phosphate, ammonium phosphate, ammonium acetate and ammonium carbonate) [14]. Mean column efficiencies in ACN-potassium phosphate buffer measured using the half-height and Dorsey–Foley procedures, together with the asymmetry factors are shown in Table 9.2. The mass of injected solute was 0.2 µg in each case in order to avoid overloading effects (see below). The Dorsey–Foley equation is useful in that it takes into account the asymmetry of the peaks in the measurement of column efficiency [20]:

$$N_{df} = 41.7 \frac{(t_r / w_{0.1})^2}{(A_s + 1.25)}$$

At pH 2.7, the monolith gave the highest efficiency (117 600 plates/m) when measured by the half-height method. However, when measured by the Dorsey–Foley procedure, the efficiency is shown to be more modest than two of the 5-µm

Table 9.2 Comparison of average column efficiencies and asymmetry factors for 9 different probe compounds on a monolithic and 3 particle-packed columns (5 μm) using (a) ACN-potassium phosphate buffer pH 2.7; (b) using ACN-potassium phosphate buffer pH 7.0, 1.0 ml/min. Efficiency in plates/m. Data from [14]

Column	N (half-height)	N (Dorsey–Foley)	A_s
(a)			
Symmetry C18	89 600	72 600	1.34
Discovery C18	90 900	65 300	1.33
Symmetry 300	66 700	47 600	1.47
Chromolith	117 600	53 600	2.01
(b)			
Symmetry C18	23 500	9 300	4.26
Discovery C18	73 300	36 600	1.98
Eclipse XDB C8	50 300	21 700	4.19
Chromolith	25 600	6 900	4.33

particle-packed columns due to the considerably higher peak asymmetry (average = 2.01). At pH 7.0, the monolith gave the lowest number of plates using the Dorsey–Foley equation ($N = 6900$/m) due in part to the high average asymmetry of the peaks (4.33). This value is in very good agreement with the tailing factors for propranolol and amitriptyline reported under similar conditions by Kele and Guiochon (see above). Clearly all of the columns tested (all modern Type-B silicas) show considerably worse performance at pH 7.0 compared with pH 2.7, due to interaction with silanols that are increasingly ionized as the pH is raised. The results obtained for the monolith can also be compared with similar data for 7 additional particle-packed columns obtained under the same test conditions [15–17], which would indicate that the Dorsey–Foley efficiency of the monolith is about average compared with a selection of 5-μm particle size Type-B columns when tested with strong bases at acid pH, but compares unfavorably with the same particle-packed columns when evaluated at pH 7.0.

Tanaka and coworkers reported an improved endcapping method for monolithic silica columns [21]. They first prepared the silylating reagent octadecyldimethyl-N,N-diethylaminosilane (ODS-DEA, $C_{18}H_{37}Si(CH_3)_2N(C_2H_5)_2$) by the reaction of octadecyldimethylchlorosilane ($ODSMe_2Cl$) and diethylamine in toluene. The purified reagent was passed as a plug (20% solution) through bare PEEK clad silica Chromolith columns, maintained in a water bath at 60 °C. Endcapping of the columns was accomplished by passing a solution of hexamethyldisilazane (HMDS) in toluene or trimethylsilylimidazole(TMSI) in acetonitrile through the column, again at 60 °C. As noted by Felix and coworkers from Merck [22], care is necessary for these *in situ* reactions of a PEEK clad monolith. Aromatic solvents such as toluene or xylene are compatible with PEEK, while chlorinated solvents can induce swelling of the polymer, particularly during heating, loosening the cladding from the monolith rod. An additional problem is that PEEK-clad columns cannot be heated much above 60 °C, a considerably lower temperature than is normally used in silylation reactions carried out for preparation of particulate phases (often prepared by refluxing in toluene, b.p. 111 °C). The coverage of the silica with all types of silane is poor at these lower temperatures [22]. Details of the preparation of the commercial monolith have not been disclosed by the manufacturers. Alternatively, it is possible that the monolith is derivatized and end-capped first, then subsequently encased in PEEK. This procedure might also have drawbacks, in that the reagent might not penetrate the monolith as efficiently as small particles, resulting in uneven or poor coverage. Svec and Huber suggest that the porous silica rod is indeed prepared first, encased within a PEEK tube, and then functionalized with a bonded C-18 chemistry *in situ* to create the RP column [23]. Tanaka and coworkers [21] monitored the success of the *in situ* end-capping reaction through the retention factor of phenol (k decreased after one reaction and thereafter remained constant) and caffeine (k decreased gradually to reach a steady value after 4 endcapping reactions with HMDS, or 1 reaction with TMSI). TMSI was shown to react with additional silanols on a surface previously endcapped with trimethylchlorosilane (TMCS), and a single endcapping with TMSI was superior to five treatments with TMCS. Indeed, analysis of the strong bases procainamide and N-acetylprocainamide showed strongly tailing peaks using methanol-phosphate buffer pH 7.6 on a column endcapped with HMDS, but clearly superior results when endcapped with TMSI. Furthermore, pyridine eluted after phenol on a column endcapped with HMDS, but before phenol with a TMSI-endcapped column, indicating that secondary retention interactions of pyridine with unreacted silanols had been reduced in the latter procedure. The endcapping with TMSI was also shown to be effective on commercial monoliths that were already endcapped by the manufacturer. Reduced retention of caffeine relative to phenol was obtained, and considerable improved peak symmetry in the procainamide test at pH 7.6. The authors obtained good peak shapes for two other strongly basic probes, amitriptyline and imipramine at pH 7.6, using their ODS-DEA derivatization procedure followed by TMSI endcapping. However, it appeared that for these more searching probe compounds, satisfactory analysis could not be achieved on early batches of commercially derivatized monoliths subsequently endcapped with TMSI, although later batches of commercial monolith treated in the same way showed improved results. It appears therefore that some improvements in the commercial production process might have been implemented since the production of the first columns, of the type tested by Kele and Guiochon [8] and McCalley [13, 14]. The authors also suggested that for compounds like amitriptyline and imipramine, their in-house ODS-TEA/TMSI procedure produced better peak shapes, but results for pyridine were superior on TMSI-treated Chromolith RP18e already endcapped by the manufacturer. Clearly, the interesting study of Tanaka and coworkers seems to allow significant improvements in the performance of the commercial monolith for the analysis of basic compounds, using rather simple procedures.

9.4
Contribution of Overload to Peak Shapes of Basic Solutes

While interaction with ionized silanol groups is clearly a cause of tailing for ionized bases, overload can also be a major contributor to poor peak shape. McCalley demonstrated that overloading for both ionized basic, but also acidic solutes, occurred very readily compared with neutral solutes [7, 24] resulting in sample capacities 50–100 times smaller than for neutral solutes. Overloading effects increase with decreasing ionic strength of the buffer, and are thus worse for mass-spectrometer-friendly eluents such as formic acid buffers than for the phosphate buffers. The column used [7] was a highly inert hybrid inorganic-organic phase on which there may be effectively no ionized silanols at low pH [25]. In addition, similar overloading occurs with purely polymeric columns [26]. Thus, it seems overloading effects can be independent of the silanol interactions described previously. It is possible that overloading can be attributed to mutual repulsion of ionized solute ions on the column surface reducing its capacity [7], or to the overloading of a small number of strong sites, which as yet remain identified [27].

According to Felix [22], the commercial monolith contains 30% less silica than in a classical particle-packed column of the same dimensions, due to its greater porosity. In this case, it is not surprising that the monolith was shown to be more easily overloaded by strongly basic solutes such as diphenhydramine and nortriptyline in comparison to particle-based columns of the same physical dimensions [24]. Thus, care is necessary especially for monoliths to avoid introduction of large quantities of ionized solute.

9.5
Van Deemter Plots for Commercial Monoliths

As can be deduced from the review of Sioufﬁ, papers dealing with the effect of flow on efficiency for monoliths with basic solutes are not numerous [28]. Figure 9.1 shows Van Deemter plots in acetonitrile–water (40:60, v/v) for the Merck monolith (4.6 mm ID) over the flow-rate range 0.5–5 ml/min, for benzene and 3 rather weakly basic compounds pyridine, aniline and 4-ethylaniline [13]. The strongest of these bases is pyridine with a pK_a of ~5.2 in water although this value is lowered somewhat by the presence of the organic solvent [29]. Although the pH of this unbuffered mobile phase is indeterminate, it seems likely that these solutes are unprotonated under these conditions. Similar solutes have been used by Engelhardt and coworkers to test the suitability of columns for the analysis of basic compounds [30], although the connection between column performance of weak bases in unbuffered mobile phases and strong bases in buffered mobile phases is not always straightforward [15]. The plots based on measurement of plate height (H) from the peak width at half-height are impressive. In particular, the plots are flat in the C term region at high flow. Figure 9.2 shows separation

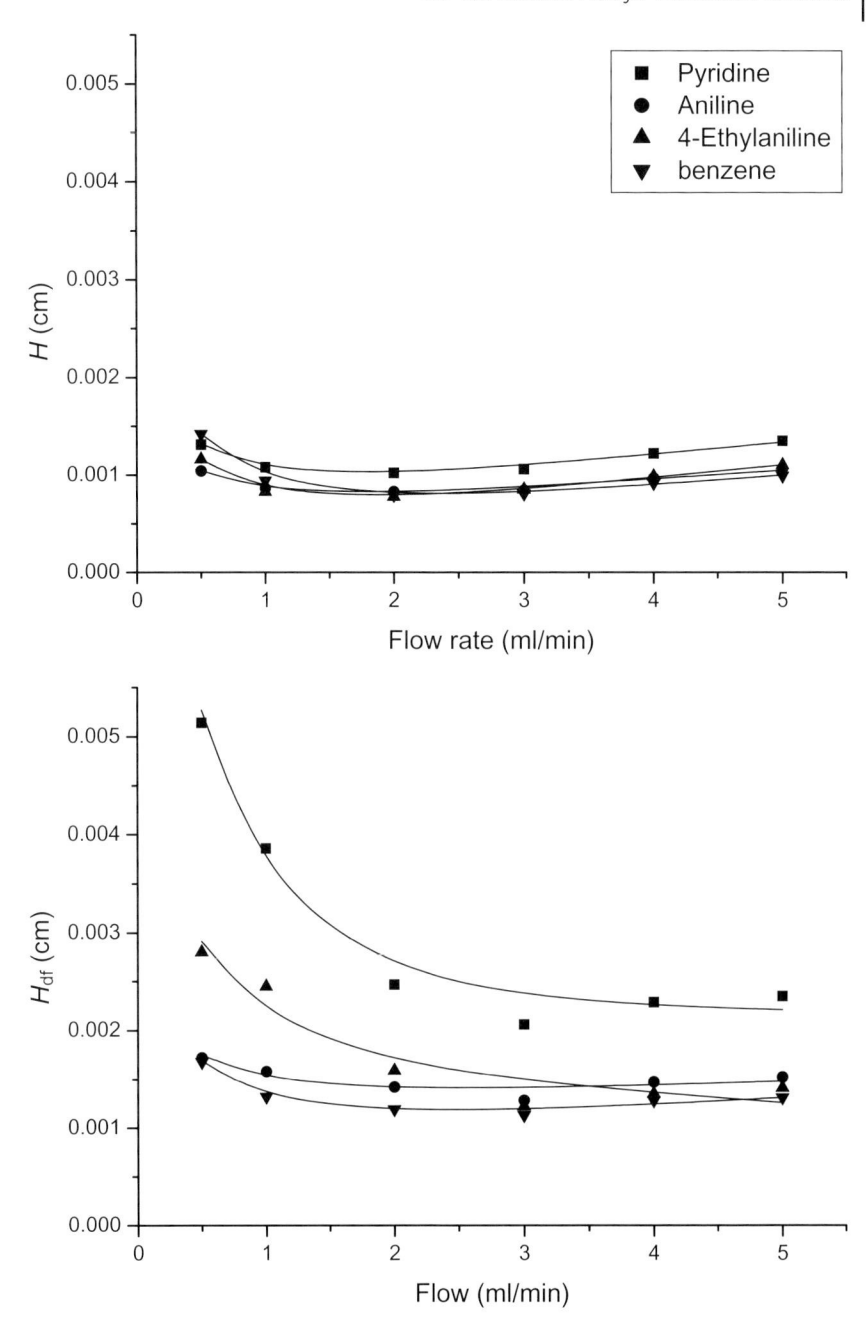

Figure 9.1 Van Deemter plots for weak bases and benzene on Chromolith (10 × 0.46 cm id) using plate height based on N at half-height (H) or the Dorsey–Foley method (H_{df}). Mobile phase acetonitrile-water (40:60, v/v). Injection: 2 µl of 100 mg/l solution. (1 ml/min corresponds to a linear velocity of 0.13 cm/s)

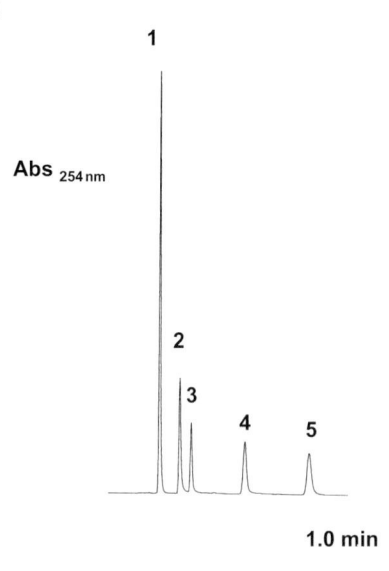

Figure 9.2 Fast separation using Chromolith column. Conditions: as Figure 9.1 with flow rate 5 ml/min. Peak identities: 1 = uracil, 2 = pyridine, 3 = aniline, 4 = 4-ethylaniline, 5 = benzene.

of these test compounds at high flow rate (5 ml/min) in less than 1 min, demonstrating the potential of the monolith to maintain good efficiency in fast analysis. Reviewing the results that have been obtained using silica monoliths (mostly with neutral solutes), Sioufﬁ noted that the commercial monolith exhibited a very low C term due to excellent mass-transfer properties. The values of C obtained by curve fitting of the Van Deemter equation to the plots were in the region of 1.0 ms on average for the basic solutes, and also 1.0 ms for benzene. These are at the more favorable end of the range of values for C published in the review by Sioufﬁ [28]. An additional factor leading to the small C terms of monoliths may well be the low backpressures of the monolith that give rise to only small heating effects in the column. In particulate columns, the center of the column is hotter than the region near the walls due to frictional heating effects, leading to higher linear velocity at the center of the column. This phenomenon can contribute considerably to band spreading especially at high flow rates even in conventional columns [31], and is especially problematic in ultrahigh-pressure LC with sub-2 µm phases. For particulate columns, a reduced plate height of 2 generally indicates a very well-packed column. For the commercial monolith for benzene, the minimum value of H of < 8 µm indicates therefore an excellent column of effective particle size of somewhat less than 4 µm, in line with the measurements of Leinweber et al. [11]. Sioufﬁ emphasized that accurate data for Van Deemter plots should be acquired through use of the second moment of the peak (which reflects the non-Gaussian nature of peaks) rather than peak width at half-height [28]. Van Deemter plots based on measurement of plate height using the Dorsey–Foley efficiency

(H_{df}), which also takes into account the asymmetry of peaks are considerably less optimistic (Figure 9.1). We did not consider the fit of the Van Deemter equation using H_{df} sufficiently good to calculate the C terms for the equation. The poor fit was partially due to poor reproducibility of the measurement of the Dorsey–Foley efficiency from the tailing peaks for the basic solutes, in contrast to the good reproducibility of half-height efficiency measurements. The profiles based on H_{df} still show a flat region at high flow but in addition a steep rise in plate height at lower flows for pyridine and 4-Ethylaniline. This effect was largely due to increased peak asymmetry at low flow for these solutes, as illustrated in Figure 9.3. While the peaks of the basic compounds show improved peak symmetry as the flow rate increases, the asymmetry of the benzene peak remains more or less constant with increasing flow rate. At present, we have no explanation for this phenomenon. This unusual effect throws doubt on the validity of the meaning of the calculated values of the C coefficient of the Van Deemter equation reported above. Tanaka and coworkers [32] found considerable band broadening at low flow rates for monoliths, due to molecular diffusion. They explained this result by the larger values of the obstruction coefficient, leading to higher effective diffusion coefficients in the commercial monolith compared with packed columns, which are explained by the greater porosity of the monoliths. For example, they quoted the porosity of the commercial monolith as 80–85% compared with 60–70% typical of conventional particle-packed ODS columns. They pointed out also that this faster effective diffusion should result in a smaller C term as well as a larger B term.

Figure 9.4 shows similar plots for the commercial monolith with a mixture including the strong bases quinine and nortriptyline using a buffered mobile phase at pH 3.0. The plots of H (from the peak width measured at half-height) are again flat at higher flow rate, with C ~1.3 ms for quinine and ~1.7 ms for nortriptyline. The plot of H_{df} against flow shows similar features to the results in the unbuffered mobile phase. Plate heights for the more strongly basic compounds are somewhat greater, and as shown in Figure 9.3, the asymmetry factor of the strong bases also decreases with increasing flow, as found for the weak bases.

9.6
Performance of Hybrid Capillary Silica Monoliths for Basic Compounds

Some studies have appeared on the use of capillary hybrid silica monoliths for the analysis of basic compounds in the chromatography or electrophoresis mode. Tanaka and coworkers reported the preparation of these hybrids from a mixture of tetramethoxysilane (TMOS) and methyltrimethoxysilane [33]. Hybrid materials were first prepared for particle-packed columns, and for basic compounds have the advantage of lower concentrations of silanol groups on the surface compared with pure-silica columns [34]. Tanaka compared the performance of hybrid monoliths with classical silica capillary monoliths prepared from TMOS alone. Surface modification of the monolithic silica was carried out on the column by feeding a

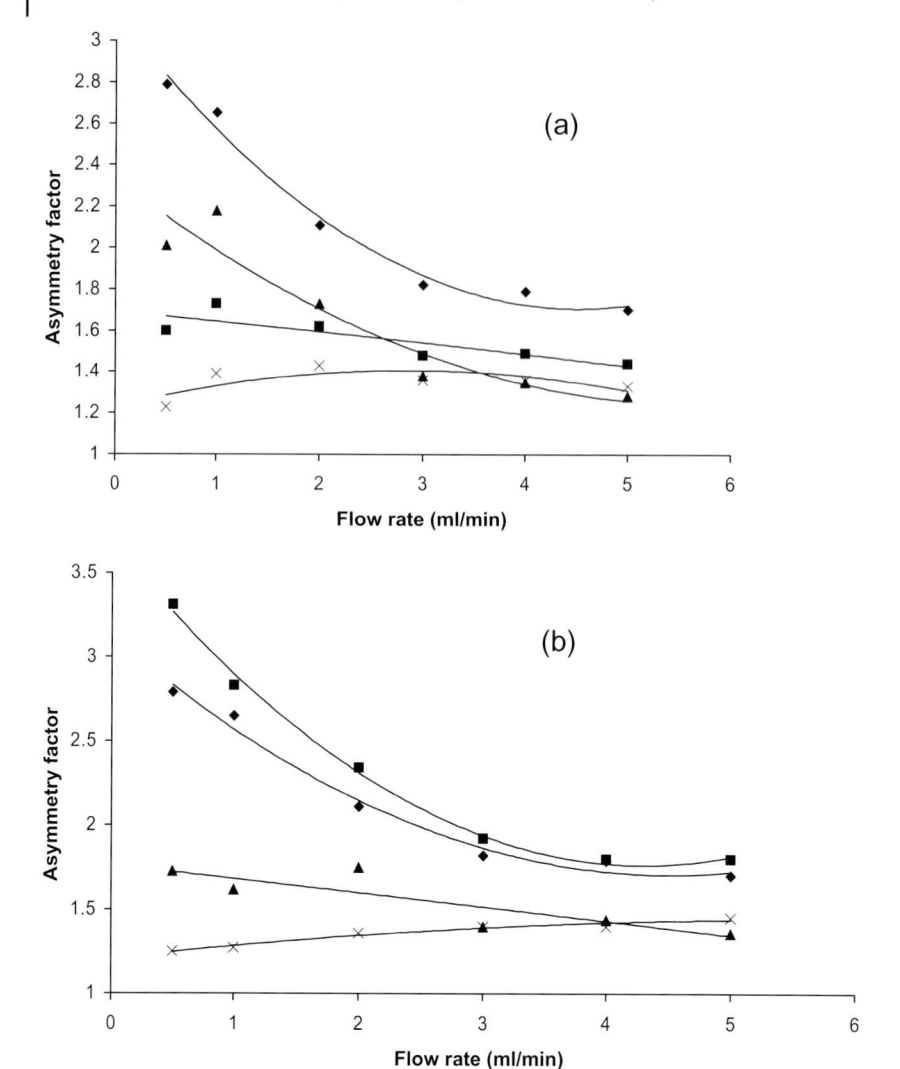

Figure 9.3 Variation of asymmetry factor with flow rate. (a) mobile phase = acetonitrile–water (40:60, v/v). Peak identities: pyridine (diamonds), aniline (squares), 4-Ethylaniline (triangles), benzene (crosses). (b) mobile phase acetonitrile–phosphate buffer pH 3.0 (30:70, v/v). Peak identities: pyridine (diamonds), quinine (squares), nortriptyline (triangles), benzene (crosses). Other conditions as Figure 9.1.

solution of $C_{18}H_{37}Si(CH_3)_2N(C_2H_5)_2$ in toluene (as before) through the column for 3 h at 60°C, although no endcapping was apparently used. The columns were tested with weak bases (anilines) in acetonitrile–water and with stronger bases such as propranolol in acetonitrile–water with 0.1% TFA. Good peak shapes were obtained for anilines on both classical and hybrid columns. Peak shapes for the stronger bases were much improved on the hybrid C18 capillary monolith,

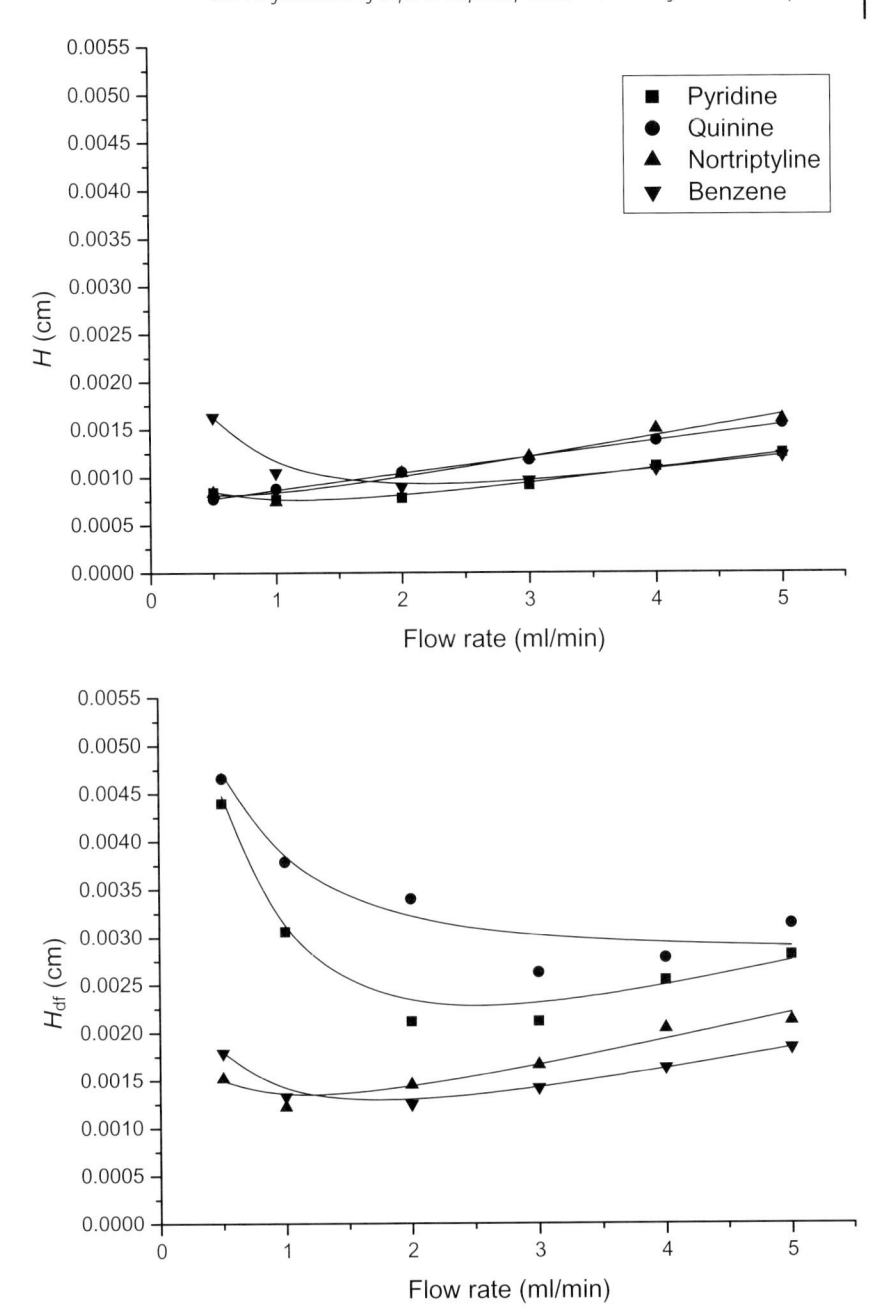

9.6 Performance of Hybrid Capillary Silica Monoliths for Basic Compounds | 185

Figure 9.4 Van Deemter plots for stronger bases, pyridine and benzene. Mobile phase acetonitrile-phosphate buffer pH 3.0 (30:70, v/v). Other conditions as Figure 9.1.

although no quantitative measurement of peak shape was performed. Clearly, the commercial exploitation of a clad hybrid monolith of conventional dimensions (rather than in capillary format) would be of interest. However, it should be noted that the patented technology for the production of hybrid materials (Waters Associates), and that for the production of clad monoliths (Merck), belongs to entirely separate companies.

Tettey and coworkers [35] reported the separation of basic, acidic and neutral analytes by pressure-assisted capillary electrochromatography (CEC). They used a hybrid silica-based monolith, chemically modified with $ODSMe_2Cl$ (by passing a 20% solution through the capillary at 60 °C, followed by endcapping with "HMDS vapor", apparently also at 60 °C. They noted that while the tailing of peaks of basic compounds on silica surfaces is also found in CEC as well as LC, the endcapping with HMDS resulted in near-Gaussian peaks for strongly basic analytes such as nortriptyline, which gave considerable tailing with the nonendcapped material. They reported that this procedure did not result in significant loss in the electro-osmotic flow, whereas endcapping with trimethylchlorosilane (TMCS) gave substantial reduction of the EOF and poor separations. The authors believed that the use of the bulkier reagent HMDS resulted in some uncapped silanols remaining and available to generate the EOF. Column efficiencies of 94 000 plates / meter for benzylamine, and 41 000 plates/meter for nortriptyline were reported in the pressurized CEC mode, but no quantitative measures of peak asymmetry were reported. Xie *et al.* [36] used monolithic (underivatized) silica capillaries in what appears to be the CEC- HILIC mode (using a mobile-phase containing 60% acetonitrile) for the separation of the components of traditional Chinese medicines containing bases and quaternary ammonium compounds. They reported column efficiencies in excess of 250 000 plates/meter with no obvious peak tailing (although no quantitative measures of tailing were used). As shown by other workers using particle-based columns, the HILIC mechanism with bare silica columns is capable of producing excellent peak shapes with strongly basic compounds [37]. Clearly, the mechanism of tailing for bases in RP-HPLC is complex, and does not merely involve silanol interactions, as these are plentiful in the case of HILIC.

9.7
Conclusions

Commercially available clad C18 silica monolith columns show good reproducibility of retention for basic solutes, and acceptable reproducibility of column efficiency. Early versions of these columns show some tailing of weakly basic solutes in unbuffered mobile phases, and for stronger bases in mobile phases buffered at acid pH. The tailing results from silanophilic activity, and can be reduced by the addition of blocking agents such as triethylamine. Due to this tailing, the efficiency of these monoliths, (as shown by use of the Dorsey–Foley calculation) is somewhat reduced for strongly basic solutes in acidic mobile phases, to values comparable with those for 5-μm particle-packed columns, in contrast to performance comparable with 3–4-μm particle columns when tested with neutral solutes. At pH 7, tailing effects with commercial monoliths become more severe. Details of the production process of the monoliths are confidential, and have not been revealed. If silica monoliths are derivatized before encapsulation in PEEK, it is possible that uneven or poor penetration of the reagent into the monolith structure occurs, compared with the derivatization process of the more usual microparticles. Conversely, if the derivatization is carried out *in situ* by passing the reagents through the clad monolith, limitations on the reaction conditions may be imposed by the stability of the cladding. Nevertheless, the use of simple alternative *in situ* derivatization and endcapping reagents can lead to markedly improved performance. Hybrid silica with lower surface silanol concentrations also give considerable improvement in efficiency for bases. These columns have been used successfully in capillary format, both in the HPLC and CEC modes, but are not commercially available in standard column forma.

References

1. Snyder, L.R., Kirkland, J.J., and Glajch, J.L. (1997) *Practical HPLC Method Development*, John Wiley & Sons, Inc., New York.
2. Wells, J.L. (1988) *Pharmaceutical Preformulation: The Physicochemical Properties of Drug Substances*, Ellis Horwood, New York.
3. McCalley, D.V. (2008) *Adv. Chromatogr.*, **46**, 305.
4. Nawrocki, J. (1997) *J. Chromatogr. A*, **779**, 29.
5. Neue, U.D., Tran, K.V., Mendez, A., and Carr, P.W. (2005) *J. Chromatogr. A*, **1063**, 35.
6. Köhler, J., and Kirkland, J.J. (1987) *J. Chromatogr.*, **385**, 125.
7. McCalley, D.V. (2006) *Anal. Chem.*, **78**, 2532.
8. Kele, M., and Guiochon, G. (2002) *J. Chromatogr. A*, **960**, 19.
9. Cabrera, K. (2004) *J. Sep. Sci.*, **27**, 843.
10. Forlay-Frick, P., Nagy Z.B., and Fekete, J. (2002) *J. Liq. Chromatogr. Relat. Technol.*, **25**, 1431.
11. Leinweber, F.C., Lubda, D., Cabrera, K., and Tallarek, U. (2002) *Anal. Chem.*, **74**, 2470.
12. Abia, J.A., Mriziq, K.S., and Guiochon, G.A. (2009) *J. Chromatogr. A*, **1216**, 3185.
13. McCalley, D.V. (2002) *J. Chromatogr. A*, **965**, 51.
14. McCalley, D.V. (2003) *J. Chromatogr. A*, **987**, 17.
15. McCalley, D.V. (1999) *J. Chromatogr. A*, **844**, 23.
16. McCalley, D.V. (1996) *J. Chromatogr. A*, **738**, 169.
17. McCalley, D.V. (1997) *J. Chromatogr. A*, **769**, 169.
18. Dolan, J.W. (2003) *LC.GC Eur.*, **16**, 610.
19. Papai, Z., and Pap, T.L. (2002) *J. Chromatogr. A*, **953**, 31.
20. Foley, J.P., and Dorsey, J.G. (1983) *Anal. Chem.*, **55**, 730.
21. Yang, C., Ikegami, T., Hara, T., and Tanaka, N. (2006) *J. Chromatogr. A*, **1130**, 175.
22. Felix, G., Cabrera, K., Lubda, D., Roussel, J.-M., and Siouffi, A.M. (2004) *Chromatographia*, **60**, 3.
23. Svec, F., and Huber, C.G. (2006) *Anal. Chem.*, **78**, 2100.
24. McCalley, D.V. (2003) *Anal. Chem.*, **75**, 3404.
25. Mendez, A., Bosch, E., Rosés, M., and Neue, U.D. (2003) *J. Chromatogr. A*, **986**, 33.
26. Buckenmaier, S.M.C., McCalley, D.V., and Euerby, M.R. (2002) *Anal. Chem.*, **74**, 4672.

27. Gritti, F., and Guiochon, G. (2009) *J. Chromatogr. A*, **1216**, 3175.
28. Siouffi, A.M. (2006) *J. Chromatogr. A*, **1126**, 86.
29. McCalley, D.V. (1994) *J. Chromatogr. A*, **664**, 139.
30. Engelhardt, H., and Jungheim, M. (1990) *Chromatographia*, **29**, 59.
31. Fallas, M.M., Hadley, M.R., and McCalley, D.V. (2009) *J. Chromatogr. A*, **1216**, 3961.
32. Kobayashi, H., Tokuda, D., Ichimaru, J., Ikegami, T., Miyabe, K., and Tanaka, N. (2006) *J. Chromatogr. A*, **1109**, 2.
33. Kobayashi, H., Kajiwara, W., Inui, Y., Hara, T., Hosoya, K., Ikegami, T., and Tanaka, N. (2004) *Chromatographia*, **60**, S19.
34. Cheng, Y.F., Walter, T.H., Lu, Z.L., Iraneta, P., Alden, B.A., Gendreau, C., Neue, U.D., Grassi, J.M., Carmody, J.L., Gara, J.E.O., and Fisk R.P. (2000) *LC. GC (N. Am.)*, **18**, 1162.
35. Zhang, T., Kadra, I., Euerby, M.R., Skellern, G.G., Watson, D.G., and Tettey, J.N.A. (2008) *Electrophoresis*, **29**, 944.
36. Xie, C.H., Hu, J., Xiao, H., Su, X.Y., Dong, J., Tain, R.J., He, Z.K., and Zou, H.F. (2005) *Electrophoresis*, **26**, 790.
37. McCalley, D.V. (2007) *J. Chromatogr. A*, **1171**, 46.

10
Quality Control of Drugs

Mohammed Taha, Abdelkarem Abed, and Sami El Deeb

10.1
Introduction

During the production and development of drugs, identity, amount, and stability are critical challenges that should be checked through quality control (QC) techniques. Quality control is a procedure intended to ensure that a drug adheres to a defined set of quality criteria. It is performed to insure the assessment of product compliance with stated requirements. Among other instrumental analytical methods applied in QC, HPLC is still the most widely used one. Nowadays, there is a growing need for speeding up the HPLC analysis without much affecting the resolution or facing high system backpressure. To achieve this goal, scientists have focused on fast HPLC separations, where monolithic columns have proven to be one of the most promising developments in this area.

Monolithic columns are attracting attention as a suitable choice for fast HPLC analysis due to their high porosity [1].

To date, hundreds of papers have been published describing the use of monolithic silica columns in various analytical fields that include routine drug analysis and QC [2–5], food and environmental analysis [6, 7], natural products analysis [8] and bioanalysis [9–11]. These wide applications are made possible due to the advantages of monolithic silica columns over conventional particle-packed silica columns [12].

10.2
Analysis of Pharmaceutics

Monolithic silica columns are now largely used for in-process QC. For example, the dissolution test for both selegiline HCl and nimesulide tablets was performed using a Chromolith RP-18e monolithic silica column and acetonitrile-phosphate buffer at pH 7.0 (34:66, v/v) as the mobile phase with a flow rate $4 \, ml \, min^{-1}$, the two compounds were separated in less than 60 s [13]. Monolithic silica columns are also used in pharmaceutical process development, such as reaction

Monolithic Silicas in Separation Science. Edited by K.K. Unger, N. Tanaka, and E. Machtejevas
© 2011 WILEY-VCH Verlag GmbH & Co. KGaA, Weinheim
ISBN: 978-3-527-32575-7

monitoring, column-fraction screening, and analysis of mother liquors and unstable analytes [14].

Regarding the quality control of finished products, monolithic silica columns offer us additional advantages when compared with conventional particle-packed silica columns for routine high-throughput analysis. In one study, a C18, 4.6 mm id monolithic silica column with a length of 10 cm was used to separate and quantitate five beta blockers, atenolol, pindolol, celiprolol, bisoprolol and metoprolol. The separation at a flow rate of 3 ml min^{-1} was 3 times faster than the separation at a flow rate of 1 ml min^{-1}. The efficiency of separation was not sacrificed and column backpressure remained within the acceptable limit, despite increasing the flow rate. By increasing flow rate up to 9 ml min^{-1}, these compounds were separated within less than 1 min [12, 15].

A rapid, sensitive and reproducible HPLC method using C18 monolithic silica column was developed and validated for the analysis of rifampicin and its four related compounds. The total run time was less than 11 min, as opposed to around 60 min when using a C18 particle-packed silica column [16]. In another study, estrogens, 9,11-dehydroestrone and 9,11-dehydroestradiol were separated in less than 1.5 min by using monolithic silica column at a flow rate of 1 ml min^{-1} [17]. A rapid and selective HPLC methods using monolithic silica columns for separation and quantification of amphetamine [18], haloperidol [19] and bacitracin [20] in pharmaceutical preparations were successfully performed. Aboul-Enein *et al.* used monolithic silica columns for fast separation and determination of clopidogrel [21], tadalafil [22], sildenafil citrate [23] and lamivudine [24] at flow rates of 4, 5, 2 and 2 ml min^{-1}, respectively. The analysis time in the four experiments was less than 1 min.

10.3
Natural Products Analysis

Monolithic silica columns were used also in natural products analysis, for example, to determine 16-O-methylcafestol diterpene [25], the characteristic substance for *Robusta coffee* and to determine niaziridin and niazirin [26], which are nitrile glycosides in the leaves, pods and bark of *Moringa oleifera*. They were also used to determine the anticancerous drugs vincristine, vinblastine, catharanthine, and vindoline in the leaves of *Catharanthus roseus* [27].

Due to the need for rapid HPLC separation of ingredients from medicinal plants, conventional particle-packed silica columns were replaced by monolithic silica columns to achieve this purpose. For example, the method for determination of the iridoid glycoside harpagoside in *Harpagophytum procumbens* was successfully transferred from a conventional particle-packed silica column to a monolithic silica column. Therefore, the flow rate could be easily increased from 0.8 ml min^{-1} (conventional particle-packed silica column) to 5 ml min^{-1} (monolithic silica column) and the total run-time reduced from 30 to 5 min without losing resolution or increasing backpressure [28].

10.4
Analysis Speed and Performance

Because fast separation is of great interest today, many studies compared between monolithic silica columns and conventional particle-packed silica columns with regard to speed of analysis. For example, Pan et al. determined chloramphenicol concentration in honey using both a monolithic silica column and a conventional particle-packed silica column. When the same mobile phase and flow rate (1 ml min^{-1}) were used under the same experimental conditions, chloramphenicol eluted from conventional particle-packed silica column in 8 min, while from monolithic silica column it eluded in 4 min, in addition to the improvement in peak shape and signal-to-noise ratio (S/N) [29]. In another study, vardenafil in a pharmaceutical tablets formulation, was identified and quantified by using a conventional C18 and monolithic silica columns with acetonitrile-phosphate buffer mixture (20:80,v/v) as mobile phases. Separation efficiency and resolution were better for the monolithic C18 column at a flow rate of 2 ml min^{-1} than for conventional C18 at a flow rate of 1 ml min^{-1} under otherwise similar separation conditions [30].

Monolithic and conventional particle-packed silica columns were also applied by El Deeb, et al. for determination of propranolol hydrochloride in the presence of its 2 main degradation products, 3-(1-naphthyloxy)propane-1,2-diol and 4-isopropyl-1,7-bis-(1-naphthyloxy)-4-azaheptane-2,6-diol. The separation was investigated on monolithic silica columns at different flow rates ranging from 1 to 9 ml min^{-1} (Figure 10.1). Analysis times were decreased by about 5-fold on monolithic silica columns at a flow rate of 4 ml min^{-1} without significantly scarifying resolution [31].

Another comparison between monolithic silica column and conventional particle-packed silica column was performed by El Deeb, et al. during analysis of pilocarpine hydrochloride in the presence of its degradation products isopilocarpine, pilocarpic acid, and isopilocarpic acid. The separation of pilocarpine from its degradation products was performed on a monolithic silica column at different flow rates from 1 to 9 ml min^{-1} and on a conventional particle-packed silica column at a flow rate of 1 ml min^{-1}. A good resolution was obtained using monolithic C18 column over the conventional C18 column at the same flow rate of 1 ml min^{-1} with a total run time of 9 min compared to 13.5 min for the conventional C18 column. A good resolution and reduction in the total run time to 3.5 min were achieved when the flow rate was increased to 4 ml min^{-1} on monolithic silica column. The precision for both retention time and peak area was equal, or slightly better on monolithic silica column compared to the conventional particle-packed silica column [32].

The performance of monolithic silica columns and conventional C18 columns was compared for two topical preparations, Ketoprofen gel and Estrogel gel. For Ketoprofen analysis, a mixture of acetonitrile-water-phosphate buffer with pH 3.5 (30:68:2, v/v/v) was used as a mobile phase at a flow rate of 5 ml min^{-1}, while a mobile phase of acetonitrile: methanol: water (23:24:53, v/v/v) at a flow rate of 3 ml min^{-1} was used for Estrogel analysis. It was proved that monolith columns,

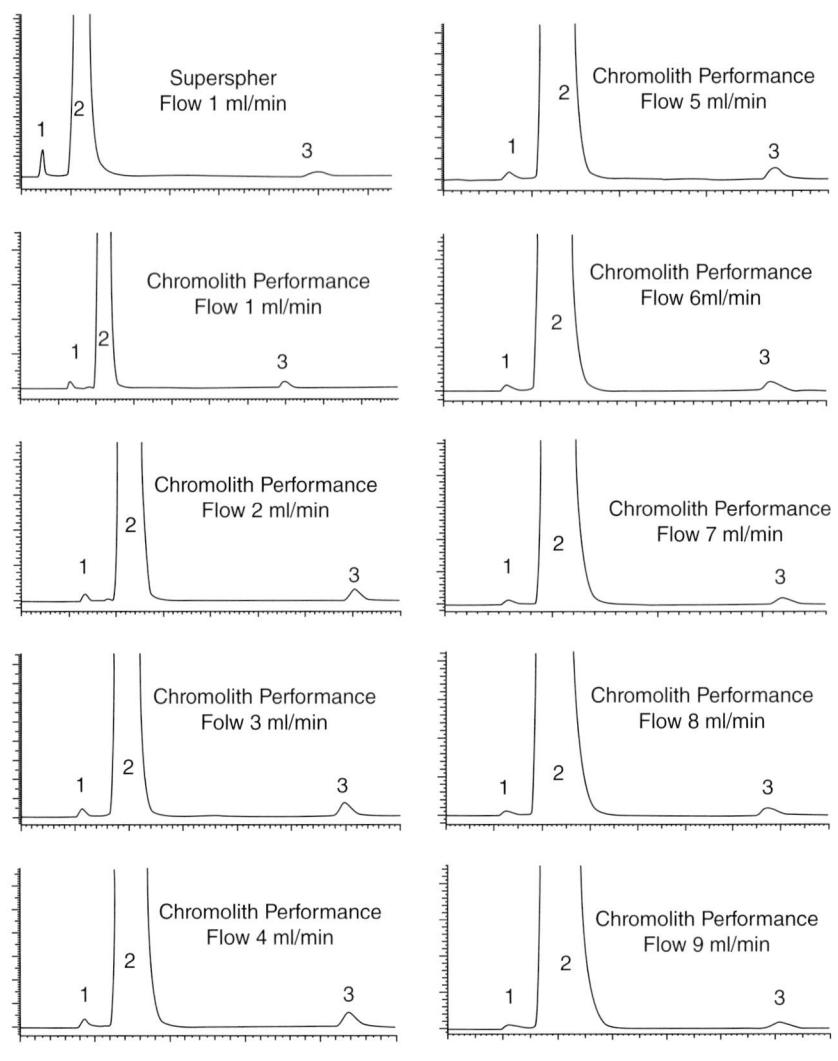

Figure 10.1 Representative chromatograms for propranolol hydrochloride (peak 2) and its 2 degradation products a and b (peaks 1 and 3, respectively) on conventional (Superspher RP-18) and on monolithic (Chromolith Performance RP-18) columns. Mobile phase consists of buffer pH 3.3: acetonitrile (40:60, v/v).

due to their porosity and low backpressure, can save analysis time by about a factor of three with sufficient separation efficiency [33].

Mostly, monolithic silica columns are going to replace the traditional conventional particle-packed silica columns for routine quality control of drugs. They appeared already for official application in USP [34]. In contrast to ultraperformance liquid chromatography (UPLC) as another trend for fast chromatographic analysis, monolithic silica could be used with the conventional HPLC instruments. Accordingly, there is no need to replace a conventional HPLC unit with a new UPLC unit in order to achieve fast chromatographic analysis, just replacing the column will do. So, using monolithic silica columns is a cheap applicable method for fast and efficient analysis that is highly desired for drug quality control. It is important to mention that the consumption of solvents in monolithic silica columns is smaller than conventional particle-packed silica columns even when using high flow rates. This is due to the reduction in total analysis time at high flow rates that counteract the excess use of solvents.

10.5
Method Transfer

Moreover, there were many studies that showed the success of method transfer for small drug molecules from conventional particle-packed silica columns to monolithic silica columns under the same conditions without any modifications. Methods on conventional C18 columns for pilocarpine, propranolol, glibenclamide and glimepiride, and insulin with their degradation or related products were transferred from a conventional (Superspher 100 RP-18e) column to a monolithic (Performance RP-18e) column. In that study, all transfers were successful except for the polypeptide insulin, where the acetonitrile content of the mobile phase was reduced by 0.5% for good resolution [35]. In fact, insulin is a macromolecule that probably interacts with the stationary phase in different ways from small drug molecules. Moreover Mistry *et al.*, suggested the use of polymer monoliths that have smaller surface area for separation of macromolecules [36].

In order to evaluate the ability of a monolithic silica column to separate basic compounds, the separation of four DNA base pairs, adenine, guanine, thiamine and uracil was attempted. The resulting chromatogram shows a successful separation. The four peaks were separated isocratically using a mobile phase consisting of a mixture of methanol 1:1 Mm potassium phosphate buffer with pH 7.2 (2:98, v/v), where the analysis time was 1.3 min [37].

Altria *et al.* used a novel separation method on monolithic silica columns. First, an oil-in-water microemulsion eluent and a conventional C18 column with a flow rate of $1\,ml\,min^{-1}$ were used. Attempts to decrease analysis time were limited due to the high viscosity of the microemulsion that generated relatively high backpressures. When a monolithic silica column was used at a flow rate of $4\,ml\,min^{-1}$, the two formulated products were separated in 90 s and the backpressure reduced 3-fold. This method is used for determination of both nicotine lozenges and naprosyn tablets [38].

Table 10.1 shows a reduction in total analysis time for selected examples when using a monolithic silica column as compared to a conventional particle-packed silica column under the same experimental parameters except for the eluent flow rate.

Table 10.1 Comparison of analysis time for selected examples.

Compound	Conventional C18 Column		Monolithic C18 Column		Reference
	Eluent flow rate	Total run time	Flow rate	Total run time	
Harpagophytum procumbens	0.8 ml min^{-1}	30 min	5 ml min^{-1}	5 min	[28]
Rifampicin and related compounds	1 ml min^{-1}	60 min	2 ml min^{-1}	11 min	[16]
Chloramphenicol in honey	1 ml min^{-1}	8 min	2 ml min^{-1}	4 min	[29]
Pilocarpine and degradation products	1 ml min^{-1}	13.5 min	4 ml min^{-1}	3.5 min	[32]

From the above examples, it was clear that monolithic silica columns could be used successfully to reduce analysis time by isocratically increasing the flow rate without much increasing column backpressure. This fact opened the way to the use of another elution technique that is now known as flow programming. Flow programming is an elution technique that is used to reduce the analysis time for compounds with great differences in elution behavior that produce chromatogram with late eluting peak. It is similar to the use of gradient elution where the elution power of the mobile phase is gradually increased by stepwise modifying its percentile composition. However, flow programming works by gradually increasing the flow rate to increase elution power without changing the mobile-phase composition over the run. Unfortunately, because of the low permeability of the conventional particle-packed silica columns, flow programming could not be applied on it as an intolerable system backpressure will develop at high flow rates that exceeds the maximum allowable value for conventional HPLC systems. Fortunately, this flow programming is applicable on monolithic silica columns where there is high permeability and accordingly a low backpressure will develop at high flow rate that remains beyond the maximum values even at a flow rate of 9.9 ml min^{-1}. In fact, for fast routine quality control, flow programming is advantageous over gradient elution because there is no need for re-equilibrium time after each run. Re-equilibration time is necessary in gradient elution in order to return the column environment to the initial conditions before the next injection. In flow programming the composition of the mobile phase does not change, so the next run of a series for quality control could be started directly without a delay time to re-equilibrate the column to the initial conditions. This could save time and cost for routine quality control. El Deeb, *et al.* have successfully applied flow programming (Table 10.2) using monolithic silica columns and separated a mixture of four

Table 10.2 The flow program used during the separation of compounds of Figure 10.2

Time (min)	Flow rate (ml min^{-1})
0.0	5.0
0.6	6.0
0.7	9.0
1.3	9.9

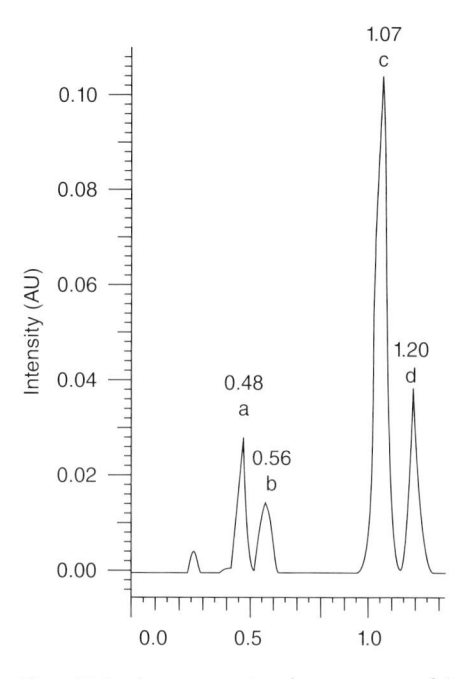

Figure 10.2 A representative chromatogram of the hypoglycemic drugs and related compounds a: related substance a (1), related substance b (2), glibenclamide (3), and glimepiride (4). Retention times are shown above the peaks.

substances, hypoglycemic and related compounds, with a total analysis time less than 80 s (Figure 10.2) [39]. Moreover, the good result repeatability of monolithic silica columns while using flow programming has been approved by Kamniski et al. [40]. Flow programming was successfully applied in the quality control of acyclovir and its major impurity guanine, in acyclovir pharmaceutical creams using 0.2% acetic acid (pH 3.1) as the mobile phase. Guanine was effectively separated in 1.25 min, while the retention time of acyclovir was 2.35 min. Linearity of the assay was validated in the range 0.1–1.0% for guanine and 80–120% for acyclovir. The accuracy and precision of the method were also validated [41].

10.6
Separation of Complex Mixtures

In addition to its efficiency in fast analysis, monolithic silica columns are also efficiently used for complex separations that require long column beds to provide the separation efficiency needed to resolve all the compounds of interest. In conventional particle-packed silica columns, the column length can not be increased by coupling columns together because this can lead to increase column backpressure above the tolerable limit. Contrary to this, in monolithic silica columns, column coupling is one way to increase column bed length and plate numbers for complex separations without being limited by the development of high backpressure due to high porosity and permeability [42].

Two coupled monolithic columns have been used by Rostagno *et al.* for the determination of isoflavones extracted from soybean, the main 12 isoflavones were separated in 10 min at 35 °C by gradient elution using a mobile phase of acidified water (0.1% acetic acid) and methanol at a flow rate of 5 ml min^{-1} [8].

One might think that coupling two columns will improve the separation efficiency of a complex mixture, but the drawback will be a longer time of analysis. However, this could be sometimes overcome by subsequent application of a proper flow program. A good example of using two coupled monolithic silica columns is the separation of a basic mixture of seven alkaloids [43]. The mixture was separated but with poor resolution when using one monolithic silica column and phosphate buffer:methanol (80:20, v/v) as a mobile phase at a flow rate of 3 ml min^{-1}. But when using two coupled monolithic silica columns with the same mobile phase and flow program shown in (Table 10.3), the separation was very efficient with good resolution, as shown in (Figures 10.3a and b). Fortunately, because of the flow programming, the separation on the coupled columns was achieved within the same analysis time obtained with one column [43].

Moreover, column coupling could facilitate the use of two-dimensional HPLC separations by coupling another column of different stationary phase to the monolithic. For example, a mixture of peptide was separated using a fast simple two-dimensional (2D)-HPLC method, in which a polymer-based cation exchange and an octadecylsilylated (C18) monolithic silica columns were used [44].

Table 10.3 The flow program used for the separation of the alkaloid mixture in Figure 10.3b.

Time (min)	Flow rate (ml min^{-1})
0.0	3.0
3.5	3.0
5.0	5.0
5.1	9.0
18	9.0

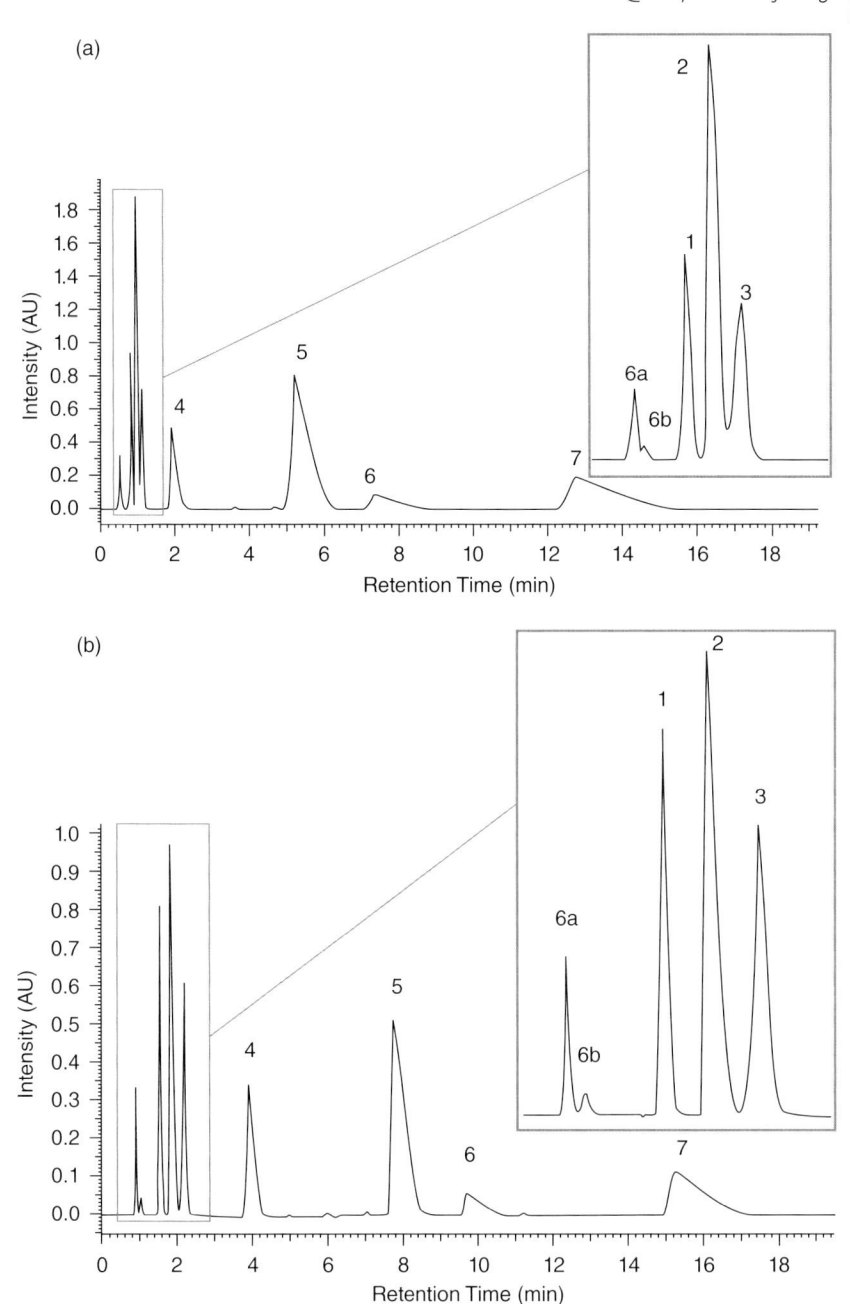

Figure 10.3 (a) Separation of a basic mixture of seven alkaloids. Column: One Chromolith Performance Rp-18e, eluent: phosphate buffer 25 Mm, pH 3:methanol (80:20), flow rate: 3 ml min^{-1}. Compounds are: 1. codeine phosphate, 2. ephedrine HCl, 3. theophiline ethylinediamine, 4. atropine sulfate, 5. yohimbine HCl, 6. butylscopolamine bromide, 7. papaverine HCl, 6a. butylscopolamine impurity a, 6b. butylscopolamine impurity b. (b) Separation of a basic mixture of seven alkaloids. Column: Two connected Chromolith Performance Rp-18e, eluent: phosphate buffer 25 Mm, pH 3:methanol (80:20, v/v). Using the flow program shown in Table 10.3.

10.7
Monolith Derivatives and Versatile Application

Due to the efficiency and diversity of liquid chromatographic applications in separation and identification of compounds, scientists focused on the development of derivatives from monolithic silica to meet this need. Because of the advantages of monolithic silica columns over conventional particle-packed silica columns, many modifications that applied on conventional particle-packed silica columns were also tried on monolithic silica columns to produce a more efficient stationary phase for faster analysis. The modified monolithic silica columns were used for different purposes in many fields such as analysis of compound with certain elements or groups, chiral drug separation, analysis of highly polar compounds and analysis of ionic compounds.

For example, an ether-bonded monolithic silica column for reversed-phase HPLC had been prepared for fast separation. Its chromatographic properties were evaluated by Li *et al.* in separation of nifedipine, nitrendipine and m-nisoldipine. In that study, the column backpressure was found to be weakly influenced by increasing the flow rate [45].

Some studies proved that modified monolithic silica columns can efficiently be used for separation and determination of drugs in biological samples with reduced analysis time and good resolution. An example is the titania-coated monolithic silica stationary phase where TiO_2 nanoparticles are good anticracking and antishrinking agents which do not form Ti–O–Si bonds with SiO_2. This results in the presence of many opening pores in the stationary phase that reduces the probability of cracking, shrinking and blockage of the column [46]. Titania-coated monolithic silica columns also have phospho-selectivity. Accordingly, these columns have been especially used to achieve liquid-chromatographic separation of phosphorus-containing compounds. The titania-coated silica columns exhibited efficient separation with low backpressure and short analysis time [47].

Monolithic silica columns play an important role in separation and determination of chiral drugs. Many articles concerning the development of chiral-based monolithic silica have been published. For example, a chiral monolithic silica columns with covalently modified monolithic silica using 3,5-disubstituted phenylcarbamate derivatives of cellulose and amylase was applied for enantioseparation in capillary liquid chromatography [48].

Bayer *et al.* have synthesized a new chiral monolithic silica column by modification with 2,3-methylated 3-monoacetylated 6-O-tert-butyldimethylsilylated cyclodextrin that bonded onto aminopropyl-derivatized monolithic silica columns. These columns were used to test the enantiomeric separation of 32 chiral compounds from different chemical classes, 14 compounds were resolved into their enantiomers and 7 compounds were successfully stereo-differentiated [49]. Furthermore, chiral monolithic silica column with the macrocyclic antibiotic vancomycin as a chiral selector was prepared for separation and determination of propranolol [50].

Monolithic silica is now modified to produce a sufficient selective stationary phase for highly polar substances. These modified columns are called HILIC (hydrophilic interaction liquid chromatography) mode column or monolithic silica columns with HILIC stationary phase. They are prepared by polymerization of acrylamide on a monolithic silica column with N-(3-trimethoxysilylpropyl) methacrylamide. The modified columns are three times greater in term of permeability and column efficiency for polar substances such as nucleosides, nucleic bases, peptides, proteins, oligosaccharides and carbohydrates compared to a commercially available amide-type HILIC columns [51].

Novel monolithic silica columns with ion exchange and zwitterionic stationary phase were prepared and applied for separation and determination of many acidic and basic drugs. For example, such columns were developed by covalent attachment of phenylalanine to a 3-glycidoxypropyltriethoxysilane silica. Due to the zwitterionic nature of the resulting stationary phase, the net surface charge can be changed by adjusting the pH values of the mobile phase, and the separation mechanisms are related to weak hydrophobic interaction, and weak cation–anion exchange interactions [52].

Moreover, a monolithic-silica-based strong cation-exchange stationary phase was successfully prepared by Chuanhui *et al.* through treatment with 3-mercapto propyltrimethoxysilane. This monolithic silica column with cation-exchange stationary phase has been successfully employed in the separation of beta blockers and alkaloids from traditional Chinese medicines [53].

In other cases, the modifications were done to the chromatographic apparatus itself. For example, a new technique called multisyringe chromatography (MSC) that is based on online coupling of a multisyringe flow system with a monolithic silica column was developed for simultaneous determination of drugs. The system comprised a multisyringe module, three low-pressure solenoid valves, a monolithic C18 column and a diode-array detector. Obando *et al.* used this method for simultaneous determination of hydrochlorothiazide and losartan potassium in tablets. The mobile phase was potassium dihydrogen phosphate (pH 3.1)-acetonitrile:methanol (65:33:2, v/v/v) at a flow rate $0.8\,\text{ml}\,\text{min}^{-1}$. The analysis time was about 400 s and validation results were very good [54].

In addition, Gonzalez-San *et al.* used the same method for determination of three β-lactamic antibiotics, amoxicillin, ampicillin and cephalexin. The mobile phase was methanol:acetic acid-sodium acetate (10:90, v/v) with a flow rate of $2\,\text{ml}\,\text{min}^{-1}$. The conditions selected for MSC separation were compared with those for HPLC system, and similar results were obtained in terms of chromatographic parameters but a difference in retention times was observed [55].

Another modification on the system was performed by Satinsky *et al.* who used a new separation method based on a reversed-phase sequential-injection chromatographic (SIC) technique for simultaneous determination of chloramphenicol and betamethasone in pharmaceutical eye drops. A short monolithic silica column coupled with (SIC) system enabled separation of two compounds in one step. The mobile phase used was acetonitrile-water (30:80, v/v) at a flow rate

0.48 ml min^{-1}. The chromatographic resolution between compound peaks was >2.1 and the analysis time was <8 min under optimal conditions [56].

As is known, biological samples such as plasma, serum, urine, or even whole blood, require time-consuming sample clean up prior to chromatographic analysis. The aim is to eliminate matrix components that block the column and cause rapid decline in HPLC column performance. To solve this problem, pretreatment techniques such as solid-phase extraction (SPE) and liquid–liquid extraction (LLE) were usually applied. Due to high porosity and resistance to blockage, monolithic silica columns played an important role in biological analysis coupled with SPE [57]. An online solid-phase extraction liquid chromatography using a newly developed SPE monolithic silica column was developed and validated for direct analysis of plasma samples containing multiple analytes. This system was developed to increase bioanalysis throughput, where three hundred direct plasma injections were made on one online SPE column without noticeable changes in system performance. Due to the ruggedness and simplicity of this system, generic methods can be easily developed and applied to analyze a wide variety of compounds in a high-throughput manner without laborious offline sample preparation [58, 59].

As an example, ketamine and its two main metabolites, namely, norketamine and dehydronorketamine in human plasma were separated and quantitated using a monolithic silica column. The retention time was reduced by six-fold and the consumption of mobile phase was decreased by two-fold, while the resolution of the analytes remained unaffected [60].

A simple, rapid and sensitive isocratic reversed phase HPLC method using a monolithic silica column has been developed and validated for determination of many drugs and their metabolites in human blood samples. This method was used for determination of gliclazide [61], omeprazole [62] and sotalol [63] by using a Chromolith Performance (RP-18e, 100 mm × 4.6 mm) column. The coefficient of variation for interday and intraday assay was found to be less than 7.0%.

Furthermore, monolithic silica columns and a solid-phase extraction method were used for rapid determination of selected drugs of abuse such as cocaine and its metabolites in human plasma. Cocaine and its metabolites were separated in 5 min at a flow rate of 5 ml min^{-1}. The method was accurate and precise (within-run precision ranged from 1.5 to 12.8% and between-run precision ranged from 0.4 to 12.7%) [64].

Recently, a new extraction method was developed to eliminate the time-consuming pretreatment of the biological samples, this method is called online extraction method (*dilute and shoot method*), which allows for direct injection of biological samples into the column without pretreatment technique to improve automation, and to maintain column performance even at high flow rate due to their high porosity that resists the blockage of the columns and increases sample throughput [57]. The online extraction procedure removes unwanted components that reduce column lifetime and cause detector suppression. Another example is the separation of six drugs, fluoxetine and norfluoxetine, methadone and EDDP, flunitrazepam and norflunitrazepam in human plasma in less than 10 min by direct injection into a Chromolith Flash precolumn (25 × 4.6 mm) [65].

It has been shown that monolithic silica columns can be used with LC-MS systems to obtain a very fast, efficient and highly selective technique deriving benefit from the fast separation of monolithic and selective detection of mass spectroscopy. Practical trials have proved no incompatibility problem from the high flow rates used on monolithic silica columns when coupled to MS. Many successful practical methods were reported using monolithic silica columns in an LC-MS system with and without postcolumn flow split. One can say that a monolithic silica column can be coupled with mass spectrometry (MS) detector despite the fact that it uses a relatively high eluting flow rate. This can usually be achieved by splitting out a part of the eluting liquid, (for example 4:1 splitting) and the remaining part enters the MS to be detected. However, successful methods implying monolithic silica columns coupled to LC-MS system have been also reported without the need for flow split due to high throughput and selectivity, this method has been investigated to increase productivity for quantitative bioanalysis of drugs and their metabolites. The results obtained from this monolithic silica column system were directly compared with the results obtained from a previously validated assay using a conventional C18 column. Both systems have the same sample preparation and mobile phases. The eluting flow rate for the monolithic silica column LC-MS system was 3.2 ml min^{-1} (with 4:1 splitting) and for the conventional LC C18 column system was 1.2 ml min^{-1} (with 3:1 splitting). The monolithic silica column LC-MS system had a run time of 5 min and the conventional C18 column LC system had a run time of 10 min [66]. There are several drugs that were determined in biological fluids by this method, for example, bupropion and its metabolites, hydroxybupropion and threo-hydrobupropion were quantified and separated within 23 s from human, mouse, and rat plasma using a high-throughput LC-MS-MS method. A Chromolith RP-18 (50 mm × 4.6 mm) monolithic silica column and mobile phase consisting of 8 mM ammonium acetate–acetonitrile (55:45, v/v) was used at flow rate 5 ml min^{-1}, and split postcolumn to 2 ml min^{-1}. The monolithic silica column performance as a function of column backpressure, peak asymmetry, and retention time reproducibility was adequately maintained over 864 extracted plasma injections [67].

10.8
Summary and Conclusions

To summarize, monolithic silica columns provide good results during method validation in terms of system suitability, linearity, precision, limits of detection, specificity, stability, and robustness. As a conclusion, one can say that monolithic silica columns either the native silica or the derivatized ones are getting more popular and will in the near future play a big role in the rapid assessment of drug quality control.

References

1. Kobayashi, H., Ikegami, T., Tanaka, N., et al. (2006) Properties of monolithic silica columns for HPLC. *J. Anal. Sci.*, **22**, 491–499.
2. Dear, G., Plumb, R., Mallet, D., et al. (2001) Use of monolithic silica columns to increase analytical throughput for metabolite identification by liquid chromatography/tandem mass spectrometry. *Rapid Commun. Mass Spectrom.*, **15** (2), 152–158.
3. Van Nederkassel, A., Aerts, A., Dierick, A., et al. (2003) Fast separations on monolithic silica columns: method transfer, robustness, and column aging for some case studies. *J. Pharm. Biomed. Anal.*, **32** (2), 233–249.
4. Hsieh, Y., Wang, G., Wang, Y., et al. (2002) Simultaneous determination of a drug candidate and its metabolite in rat plasma samples using ultrafast monolithic column high-performance liquid chromatography/tandem mass spectrometry. *Rapid Commun. Mass Spectrom.*, **16** (10), 944–950.
5. Pihlainen, K., Sippola, E., and Kostiainen, R. (2003) Rapid identification and quantitation of compounds with forensic interest using fast liquid chromatography–ion trap mass spectrometry and library searching. *J. Chromatogr. A*, **994** (1–2), 93–102.
6. Volmer, D.A., Brombacher, S., and Whitehead, B. (2002) Studies on azaspiracid biotoxins. I. Ultrafast high-resolution liquid chromatography/mass spectrometry separations using monolithic columns. *Rapid Commun. Mass Spectrom.*, **16** (24), 2298–2305.
7. Jakab, A., and Forgacs, E. (2002) Characterization of plant oils on a monolithic silica column by high-performance liquid chromatography-atmospheric pressure chemical ionization-mass spectrometry. *Chromatographia*, **56**, S69–S73.
8. Rostagno, M., Palma, M., and Barroso, C. (2007) Fast analysis of soy isoflavones by high-performance liquid chromatography with monolithic columns. *J. Anal. Chim. Acta*, **582** (2), 243–249.
9. Xiong, L., Zhang, R., and Regnier, F.E. (2004) Potential of silica monolithic columns in peptide separations. *J. Chromatogr. A*, **1030** (1–2), 187–194.
10. Shah, A.J., Crespi, F., and Heidbreder, C. (2002) Amino acid neurotransmitters: separation approaches and diagnostic value. *J. Chromatogr. B*, **781** (1–2), 151–163.
11. Tanaka, N., Kimura, H., Tokuda, D., et al. (2004) Simple and comprehensive two-dimensional reversed-phase HPLC using monolithic silica columns. *Anal. Chem.*, **76** (5), 1273–1281.
12. Merck website (2005) Speed and Performance in Monolithic Form, Chromolith® HPLC Columns http://www.merck.de/en/index.html (accessed March 7, 2005).
13. Tzanavaras, P.D., Themelis, D.G., Zotou, A., Stratis, J., et al. (2008) Optimization and validation of a dissolution test for selegiline hydrochloride tablets by a novel rapid HPLC assay using a monolithic stationary phase. *J. Pharm. Biomed. Anal.*, **46** (4), 670–675.
14. Liu, Y., Antonucci, V., Shen, Y., Vailaya, A., et al. (2005) Applications of monolithic columns to pharmaceutical process development. *J. Liq. Chromatogr. Related Technol.*, **28** (3), 341–356.
15. Cabrera, K., Lubda, D., Eggenweiler, H., Minakuchi, H., and Nakanishi, K. (2000) A new monolithic-type HPLC column for fast separations. *J. High Resolut. Chromatogr.*, **23** (1), 93–99.
16. Liu, J., Sun, J., Zhang, W., Gao, K., et al. (2008) HPLC determination of rifampicin and related compounds in pharmaceuticals using monolithic column. *J. Pharm. Biomed. Anal.*, **46** (2), 405–409.
17. Mizuguchi, T., Ogasawara, C., and Shimada, K. (2005) Application of monolithic silica column for HPLC separation of estrogens. *J. Chromatogr.*, **3**, 101–104.
18. Mc Fadden, K., Gillespie, J., Carney, B., and O'Driscoll, D. (2006) Development and application of a high-performance liquid chromatography method using monolithic columns for the analysis of

ecstasy tablets. *J. Chromatogr. A*, **1120** (1–2), 54–60.
19. Ali, I., and Aboul-Enein, H.Y. (2005) Fast determination of haloperidol in pharmaceutical preparations using HPLC with a monolithic silica column. *J. Liq. Chromatogr. Related Technol.*, **28** (20), 3169–3179.
20. Pavli, V., and Kmetec, V. (2004) Fast separation of bacitracin on monolithic silica columns. *J. Pharm. Biomed. Anal.*, **36** (2), 257–264.
21. Aboul-Enein, H.Y., Hoenen, H., Ghanem, A., and Koll, M. (2005) Reversed phase liquid chromatographic method for the high-throughput analysis of clopidogrel in pharmaceutical formulations using a monolithic silica column. *J. Liq. Chromatogr. Related Technol.*, **28** (9), 1357–1365.
22. Aboul-Enein, H.Y., and Ali, I. (2005) Determination of tadalafil in pharmaceutical preparation by HPLC using monolithic silica column. *Talanta*, **65** (1), 276–280.
23. Aboul-Enein, H.Y., and Hefnawy, M.M. (2003) Rapid determination of sildenafil citrate in pharmaceutical preparations using monolithic silica HPLC column. *J. Liq. Chromatogr. Related Technol.*, **26** (17), 2897–2908.
24. Aboul-Enein, H.Y., and Hefnawy, M.M. (2003) High throughput analysis of lamivudine in pharmaceutical preparations using monolithic silica HPLC column. *Anal. Lett.*, **36** (11), 2527–2538.
25. Koelling-Speer, I. (2006) 16-O-Methylcafestol determination in coffee using different HPLC columns and eluents with diode array and MS detection. *J. Computer Optical Disk*, **21**, 204–208.
26. Shanker, K., Gupta, M.M., Srivastava, S.K., Bawankule, D.U., *et al.* (2007) Determination of bioactive nitrile glycosides in drumstick (*Moringa oleifera*) by reverse phase HPLC. *Food Chem.*, **105** (1), 376–382.
27. Gupta, M.M., Singh, D.V., Tripathi, A.K., Pandey, R., *et al.* (2005) Simultaneous determination of vincristine, vinblastine, catharanthine, and vindoline in leaves of *Catharanthus roseus* by high-performance liquid chromatography. *J. Chromatogr. Sci.*, **43** (9), 450–453.
28. Schmidt, A.H. (2005) Fast HPLC for quality control of *Harpagophytum procumbens* by using a monolithic silica column: method transfer from conventional particle-based silica column. *J. Chromatogr. A*, **1073** (1–2), 377–381.
29. Pan, C., Zhang, H., Chen, S., Xu, Y., and Jiang, S. (2006) Determination of chloramphenicol residues in honey by monolithic column liquid chromatography-mass spectrometry after use of quenchers clean-up. *Acta Chromatographica*, **17**, 320–327.
30. Aboul-Enein, H.Y., Ghanem, A., Hoenen, H., *et al.* (2005) Determination of vardenafil in pharmaceutical formulation by HPLC using conventional C18 and monolithic silica columns. *J. Liq. Chromatogr. Related Technol.*, **28** (4), 593–604.
31. El Deeb, S., and Watzig, H. (2006) Performance comparison between monolithic C18 and conventional C18 particle-packed columns in the liquid chromatographic determination of propranolol HCl. *Turk. J. Chem.*, **30**, 543–552.
32. El Deeb, S., Schepers, U., and Waetzig, H. (2006) Evaluation of monolithic C18 HPLC columns for the fast analysis of pilocarpine hydrochloride in the presence of its degradation products. *Pharmazie*, **61** (9), 751–756.
33. Novakova, L., Matysova, L., Solichova, D., Koupparis, M.A., *et al.* (2004) Comparison of performance of C18 monolithic rod columns and conventional C18 particle-packed columns in liquid chromatographic determination of Estrogel and Ketoprofen gel. *J. Chromatogr. B*, **813** (1–2), 191–197.
34. Karin Cabrera (2008) Recent advances in silica-based monolithic HPLC columns, http://chromatographyonline.findanalytichem.com/lcgc/article/articleDetail.jsp?id=512609&sk=&date=&pageID=4 (accessed April 20, 2009).
35. El Deeb, S. (2006) High efficiency separation techniques: Fast HPLC using monolithic silica columns and

chiral separation using capillary zone electrophoresis.
36. Mistry, K., and Grinberg, N. (2005) Application of monolithic columns in high performance liquid chromatography. *J. Liq. Chromatogr. Related Technol.*, **28** (7–8), 1055–1074.
37. Smith, J.H. (2002) Chromatographic properties of silica-based monolithic HPLC columns, p. 83.
38. Altria, K.D., Marsh, A., and Clark, B. (2006) High performance liquid chromatographic analysis of pharmaceuticals using oil-in-water microemulsion eluent and monolithic column. *J. Chromatogr.*, **63** (7–8), 309–314.
39. El Deeb, S., Schepers, U., and Waetzig, H. (2006) Fast HPLC method for the determination of glimepiride, glibenclamide, and related substances using monolithic column and flow program. *J. Sep. Sci.*, **29** (11), 1571–1577.
40. Kaminiski, L., El Deeb, S., and Wätzig, H. (2008) Repeatability of monolithic HPLC columns while using a flow program. *J. Sep. Sci.*, **31** (10), 1745–1749.
41. Tzanavaras, P.D., and Themelis, D.G. (2007) High-throughput HPLC assay of acyclovir and its major impurity guanine using a monolithic column and a flow gradient approach. *J. Pharm. Biomed. Anal.*, **43** (4), 1526–1530.
42. El Deeb, S., Preu, L., and Waetzig, H. (2004) Evaluation of monolithic HPLC columns for various pharmaceutical separations: Method transfer from conventional phases and batch to batch repeatability. *J. Pharm. Biomed. Anal.*, **44** (1), 85–95.
43. El Deeb, S., Preu, L., and Waetzig, H. (2007) A strategy to develop fast RP-HPLC methods using monolithic silica columns. *J. Sep. Sci.*, **30** (13), 1993–2001.
44. Kimura, H., Tanigawa, T., Morisaka, H., Ikegami, T., et al. (2004) Simple 2D-HPLC using a monolithic silica column for peptide separation. *J. Sep. Sci.*, **27** (10–11), 897–904.

45. Li, F., Wang, L., Yang, G., and Zhao, Y. (2008) Synthesis and evaluation of octadecyl ether-bonded monolithic silica column. *Huaxue Tongbao*, **71** (3), 220–223.
46. Guo, N., Zhao, Y., Zhang, Y., et al. (2005) Roles of TiO_2 nanoparticles in the formation of monolithic silica-gels via bulk sol-gel process. *Zh Neorg Khim*, **50** (2), 197–203.
47. Miyazaki, S., Miah, M.Y., Morisato, K., et al. (2005) Titania-coated monolithic silica as separation medium for high performance liquid chromatography of phosphorus-containing compounds. *J. Sep. Sci.*, **28** (1), 39–44.
48. Chankvetadze, B., Kubota, T., Ikai, T., Yamamoto, C., et al. (2006) High-performance liquid chromatographic enantioseparations on capillary columns containing crosslinked polysaccharide phenylcarbamate derivatives attached to monolithic silica. *J. Sep. Sci.*, **29** (13), 1988–1995.
49. Mathias, B., Clarissa, H., and Armin, M. (2006) Enantiomer separation on monolithic silica HPLC columns using chemically bonded methylated and methylated/acetylated 6-O-tert-butyldimethylsilylated beta-cyclodextrin. *J. Sep. Sci.*, **29** (11), 1561–1570.
50. Xiaoli, D., Jing, D., Junjie, O., et al. (2007) Preparation and evaluation of a vancomycin-immobilized silica monolith as chiral stationary phase for CEC. *Electrophoresis*, **28** (15), 2606–2612.
51. Ikegami, T., Fujita, H., Horie, K., et al. (2006) HILIC mode separation of polar compounds by monolithic silica capillary columns coated with polyacrylamide. *Anal. Bioanal. Chem.*, **386** (3), 578–585.
52. Chuanhui, X., Jiwei, H., Xiao, H., et al. (2005) Preparation of monolithic silica column with strong cation-exchange stationary phase for capillary electrochromatography. *J. Sep. Sci.*, **28** (8), 751–756.
53. Jiwei, H., Chuanhui, X., Ruijun, T., et al. (2007) Separation of basic and acidic compounds by capillary electrochromatography using

monolithic silica capillary columns with zwitterionic stationary phase. *J. Sep. Sci.*, **30** (6), 891–899.
54. Obando, M.A., Estela, J.M., and Cerda, V. (2008) Simultaneous determination of hydrochlorothiazide and losartan potassium in tablets by high-performance low-pressure chromatography using a multi-syringe burette coupled to a monolithic column. *Anal. Bioanal. Chem.*, **391** (6), 2349–2356.
55. Gonzalez-San Miguel, H.M., Alpizar-Lorenzo, J.M., and Cerda, V. (2007) Simultaneous determination of β-lactamic antibiotics by a new high-performance low-pressure chromatographic system using a multisyringe burette coupled to a monolithic column (MSC). *Anal. Bioanal. Chem.*, **387** (2), 663–671.
56. Satinsky, D., Chocholous, P., Salabova, M., and Solich, P. (2006) Simple determination of betamethasone and chloramphenicol in a pharmaceutical preparation using a short monolithic column coupled to a sequential injection system. *J. Sep. Sci.*, **29** (16), 2494–2499.
57. Waite, S., McGinley, M. (2006) Phenomenex website for Onyx HPLC columns. www.phenomenex.com (accessed April 7, 2009)
58. Plumb, R., Dear, G., Mallett, D., and Ayrton, J. (2001) Direct analysis of pharmaceutical compounds in human plasma with chromatographic resolution using an alkyl-bonded silica rod column. *Rapid Commun. Mass Spectrom.*, **15** (12), 986–993.
59. Zang, X., Luo, R., Song, N., Chen, T., et al. (2005) A novel online solid-phase extraction approach integrated with a monolithic column and tandem mass spectrometry for direct plasma analysis of multiple drugs and metabolites. *Rapid Commun. Mass Spectrom.*, **19** (22), 3259–3268.
60. Aboul-Enein, H.Y., and Hefnawy, M.M. (2005) Liquid chromatographic high-throughput analysis of ketamine and its metabolites in human plasma using a monolithic silica column and solid phase extraction. *Talanta*, **65** (1), 67–73.
61. Foroutan, S.M., Zarghi, A., Shafaati, A., and Khoddam, A. (2006) Application of monolithic column in quantification of gliclazide in human plasma by liquid chromatography. *J. Pharm. Biomed. Anal.*, **42** (4), 513–516.
62. Zarghi, A., Foroutan, S.M., Shafaati, A., and Khoddam, A. (2006) HPLC determination of omeprazole in human plasma using a monolithic column. *Arzneimittelforschung*, **56** (6), 382–386.
63. Zarghi, A., Foroutan, S.M., Shafaati, A., and Khoddam, A. (2006) Development an ion-pair liquid chromatographic method for determination of sotalol in plasma using a monolithic column. *J. Pharm. Biomed. Anal.*, **41** (4), 1433–1437.
64. Caufield, W.V., and Stewart, J.T. (2002) Rapid determination of selected drugs of abuse in human plasma using a monolithic silica HPLC column and solid phase extraction. *J. Liq. Chromatogr. Related Technol.*, **25** (19), 2977–2998.
65. Souverain, S., Rudaz, S., and Veuthey, J.L. (2003) Use of monolithic supports for fast analysis of drugs and metabolites in plasma by direct injection. *Chromatographia*, **57** (9–10), 569–575.
66. Huang, M.Q., Mao, Y., Jemal, M., and Arnold, M. (2006) Increased productivity in quantitative bioanalysis using a monolithic column coupled with high-flow direct-injection liquid chromatography/tandem mass spectrometry. *Rapid Commun. Mass Spectrom.*, **20** (11), 1709–1714.
67. Borges, V., Yang, E., Dunn, J., and Henion, J. (2004) High-throughput liquid chromatography-tandem mass spectrometry determination of bupropion and its metabolites in human, mouse and rat plasma using a monolithic column. *J. Chromatogr. B*, **804** (2), 277–287.

11
Monolithic Stationary Phases for Fast Ion Chromatography

Pavel N. Nesterenko and Paul R. Haddad

11.1
Fast Ion Chromatography

Ion chromatography (IC) is a high-performance ion-exchange separation technique used predominantly for the separation of inorganic anions and cations. Typical separations performed under conventional conditions require about 20 min, so there is considerable interest in the development of faster separations. It is noteworthy that IC separations frequently involve the determination of a relatively small number of ions and typical IC-certified analytical methods usually provide quantitative information on only one or very few ions in a sample. For example, the EPA 321.8 method concerns the IC determination of bromate in drinking waters, EPA 314 covers IC determination of perchlorate drinking water, EPA method 1636 covers IC determination of chromium (VI), and EPA 300 involved the determination of seven common inorganic anions (bromide, chloride, fluoride, nitrate, nitrite, phosphate and sulfate) in water. This means that there is a strong demand for the development of fast IC methods for relatively small numbers of analytes, with an emphasis on methods that can be easily developed and used effectively in routine analysis.

In this chapter we will concentrate on those IC methods where the major emphasis is on the speed of the separation, and where that speed is generated through the use of a silica-based monolithic stationary phase. The historical development of fast IC separations will be examined briefly, after which monolithic IC stationary phases will be introduced (especially those in which the ion-exchange functionality is formed by coating the stationary phase with a suitable surfactant). Finally, analytical applications of fast IC will be reviewed.

11.2
Historical Development of Fast Ion Chromatography

The first reported ion-exchange separation of 6–10 metal cations by column ion-exchange chromatography was achieved as part of the Manhattan project and

Monolithic Silicas in Separation Science. Edited by K.K. Unger, N. Tanaka, and E. Machtejevas
© 2011 WILEY-VCH Verlag GmbH & Co. KGaA, Weinheim
ISBN: 978-3-527-32575-7

required a run time of about 125 h. Despite being extremely slow, this separation was nevertheless considered to be a breakthrough in separation science [1]. The establishment of IC as an analytical method did not occur until 1975 [2] and one of its most significant advantages was the ability to accomplish the separation of ions in the relatively short time span of 40–50 min. Since that time, the acceleration of IC analysis has been considered as a high priority in the continued development of IC [3]. The term "fast ion chromatography (FIC)" originated some 30 years ago, when Dionex (Sunnyvale, USA) released an ion-exchange column under the product name "Fast Run Anion Separator". This approach was based on the use of a long (250 × 4.0 mm ID) ion-exchange column packed with coarse, but uniform, agglomerated type ion-exchange resin particles of 25 μm diameter. This column could operate at relatively high flow rates up to 3.25 ml/min, which provided the separation of a mixture of common inorganic anions in the then remarkably short time of 9 min [4]. It should be noted that the agglomerated structure of this ion-exchange resin provided good mass-transfer kinetics and reasonable column separation efficiency. This same approach remains in current use, with more advanced chromatographic columns (IonPac Fast Anion III and IonPac Fast Cation II) being available commercially. A similar approach based on pumping eluent through a particle-packed ion-exchange column at high flow rate was also introduced by Alltech, USA as the Rocket® format (53 × 7.0 mm ID) cation-exchange column, which was packed with 3-μm silica particles coated by poly(butadiene-maleic acid) copolymer and used for the simultaneous separation of alkali and alkaline-earth metal cations. The separation took 6 min at a flow rate of 2.5 ml/min.

Since that time, several other technical solutions using particle-packed columns have been suggested to achieve fast separation of ions [5–8]. An obvious decrease in separation time can be achieved by using short (30–50 mm) and ultrashort (<10 mm) columns packed with fine particles of suitable ion-exchange materials. However, the diameter of particles of major commercially available organopolymer ion-exchange materials usually exceeds 5 μm [9], so the application of short columns packed with polymer particles is usually restricted by low column efficiency. Perhaps the best FIC separation and determination of inorganic anions was achieved on a short (30 × 4 mm ID) column packed with high-capacity organopolymer anion-exchanger (Dionex AS20) by Tyrrell *et al.* [10], who separated seven inorganic anions in 2.6 min using a hydroxide gradient and a flow rate of 1.8 ml/min. The reported column efficiency was 43 000 theoretical plates/m. Importantly, the authors used gradient elution to reduce the run time. However, higher separation efficiency has been demonstrated for short chromatographic columns (30 mm × 4.6 mm ID) packed with 3-μm octadecyl silica particles dynamically modified with the long-chain cationic surfactant didodecyldimethylammonium bromide (DDAB) [11]. The isocratic separation of nine anions in 160 s was obtained with 5 mM phthalate eluent (pH 7.5). Pelletier and Lucy [12] demonstrated FIC separation of seven inorganic anions (iodate, chloride, nitrite, bromide, nitrate, phosphate, sulfate) on 13- and 20-mm long × 4.6 mm ID columns packed with pH-resistant Zorbax Extend-C18 1.8-μm particles coated with DDAB. A 40-s separation was achieved at 2 ml/min with a 2.5-mM 4-hydroxybenzoic acid eluent at pH 10. Recently, Kanatyeva et al. [13] reported FIC separations on 50 × 4.0 mm ID columns packed with low-capacity anion-exchanger prepared by modification of nonporous MICRA silica 1.5-µm microbeads. Due to the small particle size, an impressive efficiency of 190 000 theoretical plates/m was achieved. However, as expected, the column had a very high backpressure and the maximum column efficiency could not be exploited with common chromatographic equipment.

A further option in achieving fast IC separation is an increase of the column temperature. This provides a reduction of eluent viscosity and thereby permits the usage of higher flow rates. It should also be noted that van Deemter plots (as H against u) at higher column temperatures exhibit a minimum plate height that shows a small shift to higher flow rates and the C term in the van Deemter equation is also reduced. In practice, this means that there is an improvement of column efficiency at increased flow rates when higher temperatures are used.

In common with other areas of HPLC a remarkable breakthrough in the achievement of rapid separations in IC accompanied the appearance of monolithic columns in the 1990s. The flow-through channel structure of monolithic columns depends on the column matrix and can greatly improve permeability, separation efficiency and applicability of the columns. The use of poly(methacrylate) (PMA), poly(styrene-divinylbenzene) (PS-DVB) and silica-based monolithic columns has been reported in IC. However, the highest efficiencies in the separation of small molecules and ions have been achieved with silica-based monolithic stationary phases [14, 15]. A comparison of the maximum efficiency reported for the separation of inorganic anions on columns of different geometry and monolithic matrix type is presented in Table 11.1. The data show that silica-based monolithic columns have at least three times higher efficiency and therefore provide the greatest possibilities for FIC [25, 26]. For this reason, this chapter will focus on the use of various types of modified silica monoliths, and on the corresponding FIC separations.

Table 11.1 Maximum efficiency (theoretical plates/m) of different types of monolithic ion-exchange columns reported for the separation of inorganic anions.

Matrix	Column shape		
	Rod ($L >$ ID, ID > 2 mm)	Capillary ($L \gg$ ID, ID < 0.25 mm)	Disk ($L \leq$ ID, $L < 5$ mm)
Silica	112 200 [16]	106 600 [17]	101 000 [18]
Methacrylate	40 000 [19]	13 500 [20, 21]	940 [22]
PS-DVB	No data	13 000 [23]	~ 1600 [24]

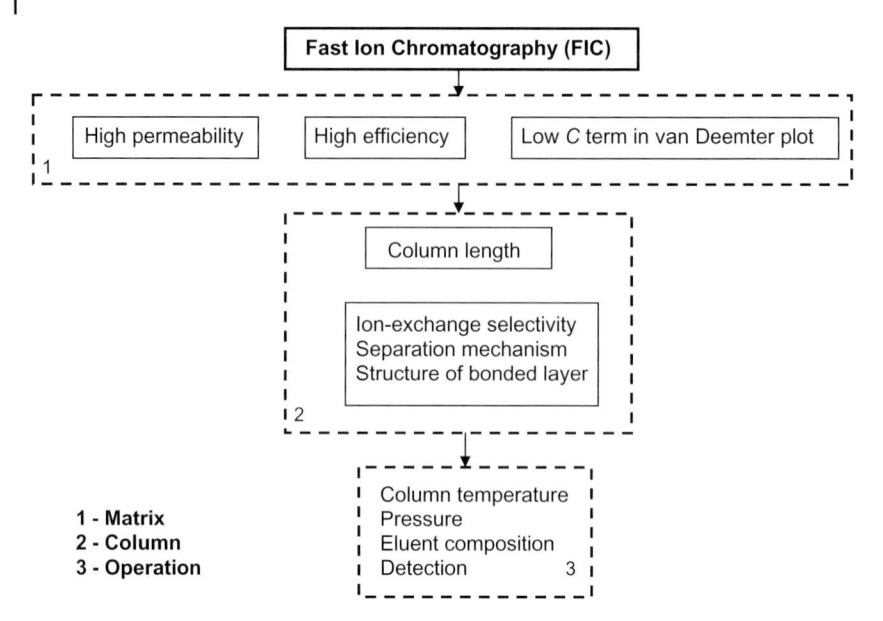

Figure 11.1 Scheme of important conditions and parameters to achieve FIC separations.

The diagram shows:
- Fast Ion Chromatography (FIC)
  - 1 - Matrix: High permeability, High efficiency, Low *C* term in van Deemter plot
  - 2 - Column: Column length; Ion-exchange selectivity, Separation mechanism, Structure of bonded layer
  - 3 - Operation: Column temperature, Pressure, Eluent composition, Detection

The essence of the above-mentioned approaches to FIC can be expressed by the block diagram presented in Figure 11.1. According to this scheme the parameters influencing the speed of an IC analysis can be divided into three groups associated with the properties of the matrix of the ion exchanger, the column length and surface chemistry of the ion exchanger, and the operational conditions. The exceptional properties of porous silica-based reversed-phase monoliths provide strong advantages in the separation of small molecules.

11.3
3 Advantages of the Bimodal Porous Structure of the Silica Monolith Matrix

Monolithic columns have high mechanical stability and permeability, enhanced column efficiency at high flow rates and as a result, can accomplish high resolution chromatography in a short time compared with packed columns [14, 15]. These advantages of monolithic silica-based columns are associated with their bimodal porous structure comprising large flow-through transport channels of average diameter 2 μm and mesopores of diameter 13 nm within a rigid silica skeleton of the monolith. Such a structure is very different from the typical "cauliflower"-type structure of organopolymer monoliths, which is less efficient in IC applications. There are no reported investigations on the relationship between pore size in a silica monolith and separation efficiency of ions.

Silica monolithic ion exchangers have excellent mechanical stability and can operate under pressures very high pressure. Thus, Ikegami et al. [27] reported the fast separation of nucleosides on poly(acrylic acid) (PAA) modified silica monolith capillary column with eluent delivered at 19.6 MPa. However, usually, the operation of monolithic columns does not need such high pressure because of their high permeability, K, expressed as:

$$K = \frac{F\eta L}{\pi r^2 \Delta P} = \frac{u\eta L\varepsilon}{\Delta P} \qquad (11.1)$$

where ΔP is the column backpressure at flow rate F or the linear velocity u of the eluent having viscosity η; and L, r and ε are the column length, the column radius and the porosity of the monolith. The K values $(4.10–5.08) \times 10^{-14}\,m^2$ reported for capillary silica-based monolithic columns coated with ion-exchange polymer are 3–5 times higher than for particle-packed columns [27, 28]. It has been demonstrated that the covalent grafting of a polymer layer to the surface of silica monolithic columns does not significantly change their permeability [29].

The maximum efficiency reported for the separation of inorganic ions on a monolithic column is about 110 000 theoretical plates/m (Table 11.1), which approximately matches the efficiency of 3-μm particle-packed columns. The optimum efficiency for silica monolith columns is achieved at linear velocities of the eluent of about 1 mm/s (Figure 11.2). However, the more significant parameter for the use of these columns in FIC is the value of the C term in the van Deemter plot:

$$H = A + Bu + C/u \qquad (11.2)$$

where H is the height equivalent to a theoretical plate. The C term shows the slope of the decrease of the column efficiency with increases in flow rate above the optimum. Figure 11.2 presents three van Deemter plots recalculated from the literature for inorganic ions separated on different capillary and analytical size columns. The calculated values of the C term are (6.00 ± 0.66), (2.73 ± 0.94) and (1.13 ± 0.09) ms for curves 1 [30], 2 [17] and 3 [31], respectively. The possible impact of extracolumn band broadening in the C term was avoided in [30], for which the sample was injected directly onto the column and capacitively coupled contactless conductivity in-column detection (C^4D) was used. It should be noted that capillary columns prepared in-situ in fused-silica capillaries have a C-term that is 2.4–5.3 times higher than for an analytical column prepared originally as a silica monolithic rod and then clad with thermoshrinking polyetheretherketone (PEEK) tubing. Although capillary columns are less suitable for FIC separations at high flow rates, they still can compete with analytical-sized columns at optimum linear velocity if high-speed separation is achieved by using very short columns. In general, the C term values are in good agreement with data for various organic molecules reviewed recently by Sioufi [32].

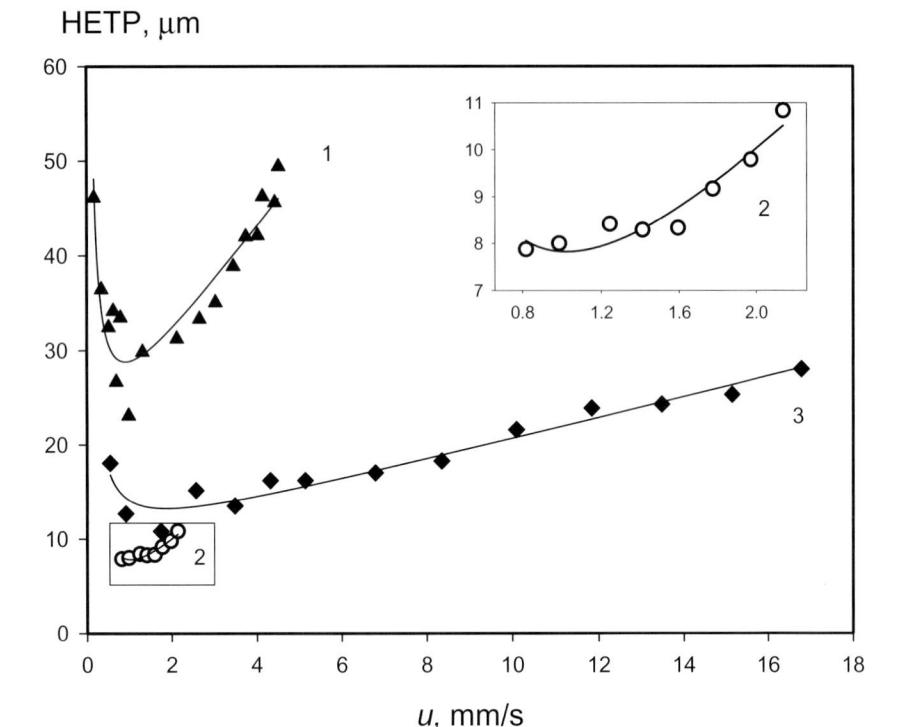

Figure 11.2 Recalculated van Deemter plots for anion-exchange monolithic silica columns. Columns: 1 – Onyx C18, 80 mm × 0.1 mm ID column dynamically coated with DDMAU, eluent: 0.5 mM phthalate [30]; 2 – 330 mm × 0.1 mm ID silica monolith column modified with MOP-copolymer DMAPAA-Q, eluent – 0.05 M sodium phosphate (pH 6.6) [17]; 3 SpeedRod C18 50 × 4.6 mm ID column dynamically coated with DDAB, eluent 6 mM cyanophenol (pH 7.0) [31]. Solutes nitrate (1) bromate (2) and iodate (3).

11.4
Type and Properties of Silica Monolithic Columns Used in IC

The preparation of silica-based monolithic ion exchangers is not a trivial task. There are two types of silica-based monolith columns that have been used in IC. The first class represents analytical-size chromatographic columns, while the second includes capillary columns.

The production of analytical-sized columns is based on initial preparation of calcinated porous silica rods, followed by cladding with PEEK. Since 2000 this type of porous silica monolith has been available commercially under the trade name Chromolith from Merck KGaA and later from Phenomenex Inc under the trade name Onyx. The product range of these companies includes 25-, 50- and

11.4 Type and Properties of Silica Monolithic Columns Used in IC

100-mm long analytical columns of internal diameter 2.0, 3.0 and 4.6 mm, together with 5- or 10-mm long guard columns of internal diameter 4.6 mm. There are some limitations on length (<100 mm) and internal diameter (>2 mm) of the produced columns, which arise chiefly from difficulties in the preparation of columns having internal diameter less than 2 mm, especially at the stages of calcination of long fragile bare silica rods and their cladding with PEEK.

All analytical-size silica monolithic columns have the same general porous structure, including approximately 2.0-μm transport pores and mesopores of diameter approximately 13 nm. The developed mesoporous structure of the monolith provides a specific surface area of $300\,m^2/g$ [14, 15]. The range of columns includes bare silica, octyl- and octadecyl-modified silica rods. The most popular type of C18 silica rod has $3.6\,\mu mol/m^2$ bonded groups, which corresponds to 18% carbon load. The carbon load in the case of octyl-modified columns is about 11%. Thus, silica-based monolithic columns can be used in FIC as bare unmodified silica, dynamically modified bare silica, dynamically modified reversed-phase C18 columns, and chemically modified or covalently bound columns. Examples of separations performed using these modes are given in Table 11.2.

Table 11.2 FIC separations on analytical size silica-based monolithic columns (id of columns is 4.6 mm, unless stated otherwise) with stable surfactant coating.

L, cm	Coating	N, plates/m	Separated ions	Flow rate, ml/min	Time, min	Ref
Bare-silica monolith						
10	Bare silica	44 980	Cl^-, Li^+, Na^+, K^+	5.0	4.6	[33]
		33 300	Cu^{2+}, Cd^{2+}, Mn^{2+}	5.0	4	[34]
10	Coated with latex anion-exchange particles	43 200	IO_3^-, Cl^-, NO_2^-, Br^- NO_3^-, HPO_4^{2-},	1.0	9	[35]
10	Covalently bound iminodiacetic acid	33 000	K^+, Mg^{2+}, Ca^{2+}, Ba^{2+}	2.0	7	[36]
		10 000	Mg^{2+}, Ca^{2+}	5.0	0.6	[37]
10	Covalently bound lysine	24 300	NO_2^-, BrO_3^-, Br^-,NO_3^-,I^-,SCN^-	4.9	1.7	[38]
C18 columns coated with cationic surfactants						
5	DDAB	37 000	IO_3^-,Cl^-, NO_2^-, Br^-, NO_3^-, HPO_4^{2-}, SO_4^{2-}	10 17	0.5	[31]

Table 11.2 Continued

L, cm	Coating	N, plates/m	Separated ions	Flow rate, ml/min	Time, min	Ref
2.5	DDAB	62 000	HPO_4^{2-}, Cl^-, NO_2^-, Br^-, NO_3^-, ClO_3^-, SO_4^{2-}, I^-	2.0	1.7	[39]
1.0	DDAB	106 000	IO_3^-, HPO_4^{2-}, Cl^-, NO_2^-, Br^-, NO_3^-, SO_4^{2-}	2.0	2.2	[18]
0.5	DDAB	101 000	HPO_4^{2-}, Cl^-, NO_2^-, Br^-, NO_3^-, SO_4^{2-}	2.0	1.2	[18]
2.5	CPC	90 880	Acetate, NO_2^-, Br^- NO_3^-, WO_4^{2-}, CrO_4^{2-}	3.0	0.7	[40]
5+10	CTAC	13 000	NO_2^-, Br^- NO_3^-	3.0	2.8	[41]
10	CTAB	112 200[a]	F^-, $H_2PO_4^-$, Cl^-, Br^-, NO_2^-, NO_3^-, I^-, SO_4^{2-}	2.0	5.5	[16]
10	CTAB + poly(oxyethylene)	15 870	OH^-	1.0	3.0	[42]
10	DDMAU	–	IO_3^-, BrO_3^-, NO_2^-, Br^-, NO_3^-, I^-, SCN^-	Dual gradient	4.7	[43]
2.5	DDMAU	28 000	IO_3^-, BrO_3^-, NO_2^-, Br^-, NO_3^-	2.0	3.5	[43]
C18 columns coated with anionic surfactants						
5	DOSS	58 000	Cu^{2+}, Mg^{2+}, Ca^{2+}, Sr^{2+}, Ba^{2+}	8.0	1.7	[39]

[a] Calculated using $N = 5.54 (t_R/w_{0.5})^2$ from the last eluting peak on chromatogram.

The second type is capillary columns of internal diameter less than 0.2 mm. In this case the production of the silica monolith does not include any cladding step since the monolith is synthesized *in situ* within a fused-silica capillary of suitable length. This provides more flexibility not only for the chemical modification of the surface with various functional groups but also for the modulation of the porous structure of silica backbone. Table 11.3 gives some examples of separations of ions performed on capillary silica-based monolithic columns.

Table 11.3 Capillary IC separations on silica-based capillary monolithic columns.

Column size, mm	Coating	N, plates/m	Separated ions	u, mm/s	t, min	Ref
Anions						
200 × 0.1 mm	Bare silica modified with DDAB	16 450[a]	IO_3^-, BrO_3^-, NO_2^-, Br^-, NO_3^-	2.1 µl/min	12	[44]
200 × 0.1 mm	Bare silica modified with CTAC	7500[a]	IO_3^-, BrO_3^-, NO_2^-, Br^-, NO_3^-	11.1 µl/min	0.9	[45]
80 × 0.1 mm	C18 column modified with DDMAU	56 200	BrO_3^-, NO_2^-, Br^-, NO_3^-, I^-, SO_4^{2-}, $S_2O_3^{2-}$, SCN^-, ClO_4^-	0.7	32	[30]
45 × 0.1 mm	C18 column modified with DDMAU	–	IO_3^-, BrO_3^-, NO_2^-, Br^-, NO_3^-	0.4	2.8	[30]
330 × 0.1 mm	Covalently bonded polymer layer of DMAPAA-Q	106 600[a]	IO_3^-, BrO_3^-, NO_3^-, I^-, SCN^-	1.4	17	[17]
300 × 0.2 mm	Covalently bonded polymer layer of DAHMA	24 200[a]	BrO_3^-, Br^-, NO_3^-, I^-,	–	6.5	[46]
300 × 0.2 mm	Covalently bonded polymer layer of DMAEA-Q	29 000[a]	BrO_3^-, Br^-, NO_3^-, I^-,	–	7.0	[46]
Cations						
150 × 0.1 mm	C18 column modified with DOSS	48 400	Mg^{2+}, Ca^{2+}, Sr^{2+}, Ba^{2+}	1.0 µl/min	9	[47]
300 × 0.2 mm	Covalently bonded polymer layer of pSSA	43 500	Uracil, thymine, cytosine, adenine	–	10.5	[46]
300 × 0.2 mm	Covalently bonded polymer layer of AMPS	29 400	Uracil, thymine, cytosine, adenine	–	16.5	[46]

Table 11.3 Continued

Column size, mm	Coating	N, plates/m	Separated ions	u, mm/s	t, min	Ref
200 × 0.2 mm	Covalently bonded poly(acrylic acid)	6300	Uridine, guanosine, adenosine	13.1	1.3	[28]
		147 060	Uracil, thymine, cytosine, adenine	1.02	5.0	[27]
Amphiphilic organic molecules						
150 × 0.1 mm	C18 column modified with DOSS	41 900	Asn, Gly, Pro, Glu, Ala, Val, Met, Ile, Leu, Lys, His, Phe	–	14	[47]
250 × 0.2 mm	Covalently bonded polymer layer of DAHMA	80 000	5'-CMP, 5'-AMP, 5'-UMP, 5'-GMP	–	20	[46]
300 × 0.2 mm	Covalently bonded polymer layer of DMAEA-Q	32 200	5'-CMP, 5'-AMP, 5'-UMP, 5'-GMP	–	9	[46]

a) Calculated using $N = 5.54\,(t_R/w_{0.5})^2$ from the last eluting peak on chromatogram.

11.5
Modification of Silica Monoliths for IC Separations

Of the silica monolithic materials discussed above, only bare silica is directly suited to ion-exchange separations. The other materials need to be modified to impart ion-exchange characteristics and these modifications are considered below.

11.5.1
Bare-Silica Monoliths as Ion Exchangers

Bare-silica gel particles exhibit weak cation-exchange properties arising from the surface silanol groups and can therefore be used in an unmodified form for the separation of alkali and alkaline-earth metal ions using diluted inorganic acids as the eluent [48, 49]. Unmodified silica-based monoliths have similar cation-exchange selectivity, but exhibit a weaker ability to retain these cations. The retention of alkali, alkaline-earth and transition-metal cations on a silica monolith was observed in acetate buffer at pH higher than 3.5 and in the presence of acetonitrile [33, 34, 50]. The separation of three metal cations was obtained in 4.0–4.6 min at a flow rate 5.0 ml/min.

A bare-silica monolith can be converted to an anion exchanger by coating with cationic surfactants, such as DDAB or cetyltrimethylammonium chloride (CTAC) [44, 45]. In the former case, a double-layer surfactant structure is formed at the surface, as shown by Takeuchi and Yeung [51]. The first layer arises from electrostatic interactions between silanols and quaternary ammonium "head" groups on the surfactant molecule, whereas the second layer is formed by hydrophobic interaction of alkyl "tails" of the surfactant molecules. Thus, alkylammonium groups in the second layer are exposed and can interact with analyte anions by ion exchange. In the case of CTAC, the stability of the surfactant coating on the monolith was maintained by addition of 0.1 mM CTAC to the eluent [45]. A remarkable improvement in coating stability was obtained by replacement of CTAC (which has a single long-chain alkyl moiety and is described as a "single-leg" surfactant) by DDAB (which has two long-chain alkyl moieties and is described as a "double-leg" surfactant) [44]. The ion-exchange capacity of the DDAB-coated anion-exchange column was found to be unchanged after 10 h of work, as indicated by a constant retention time for nitrate. With both surfactants, fast separation of 5 anions in 2 min was achieved.

Another possibility of modification of a bare-silica monolith is to coat the material with anion-exchange Dionex AS9-SC latex nanoparticles of diameter 110 nm [35]. A 100×4.6 mm Chromolith monolithic silica column was coated with nanoparticles in 0.01 M HCl solution according to a previously reported procedure [52] and resulted in an anion-exchange capacity of $41 \pm 4\,\mu eq/column$. The ion-exchange selectivities of the prepared column and a commercial Dionex IonPac AS9-SC column (which uses the same nanoparticles) were well correlated, except for very weak retention of fluoride on the monolithic column. The nanoparticle coating was also found to be very stable. The authors did not use flow rates higher than 2 ml/min, so the separation of 6 anions was accomplished in 9 min.

11.5.2
Coated Reversed-Phase Silica Monolithic Ion-Exchange Columns

As mentioned earlier, there is a range of commercially available reversed-phase silica monolith columns of different length and internal diameter and these materials can be converted to ion exchangers by coating with surfactants. In this case, a monolayer of surfactant is adsorbed through hydrophobic interactions with the stationary-phase surface. Such coated columns have been used extensively for the fast separations of ions either by coating the stationary phase with a hydrophobic surfactant (Tables 11.2 and 11.3) or by using a less-hydrophobic surfactant added to the eluent (Table 11.4).

Table 11.4 Fast IC separations of inorganic anions on analytical-sized silica-based C18 monolithic columns (ID of columns is 4.6 mm) using surfactants added to the eluent.

L, cm	Eluent	N, plates/m	Separated ions	P, MPa	Flow rate, ml/min	Time, min	Ref
5	0.5 mM TBAH–0.31 mM phthalic acid pH 5.5	34 250	Cl^-, Br^-, NO_3^-, BF_4^-, SO_4^{2-}	8.7	6.0	1.4	[53]
5	1.0 mM TBAH–0.8 mM phthalic acid pH 6.0,	21 300	Cl^-, NO_3^-, Cr(VI), SO_4^{2-}	7.8	6.0	1.0	[54]
5	0.25 mM TBAH–0.18 mM phthalic acid pH 5.5, 7% acetonitrile	37 600	Cl^-, NO_3^-, ClO_3^-, I^-, ClO_4^-, SO_4^{2-}	8.8	6.0	1.0	[55]
5	1.5 mM TBAH–1.1 mM phthalic acid pH 5.5	–	Cl^-, NO_2^-, Br^-, NO_3^-, ClO_3^-, SO_4^{2-}, I^-		6.0	1.0	[56]
5	1.5 mM TBAH–1.1 mM phthalic acid, pH 5.5	42 900[a]	$H_2PO_4^{2-}$, Cl^-, NO_2^-, Br^-, NO_3^-, ClO_3^-, I^-, SO_4^{2-}	17.2	16.0	0.25	[57]
10	2.5 mM TBAB, 10 mM Na_2HPO_4 (pH 5.6), 1.0% methanol	24 000[a]	As(III), AsB, DMA, MMA, As(V)	6.8	1.0	3.0	[58]
10	10 mM KCl–0.2 mM DDAAA, pH 6	9300	NO_2^-, Br^-, NO_3^-, I^-, SCN^-	–	4.5	3.7	[59]
1	10 mM KCl, 10 mM phosphate buffer, 0.2 mM DDAAA, pH 3,	60 800	NO_2^-, Br^-, NO_3^-, I^-, SCN^-	–	Flow gradient	6.5	[60]
10	60 mM LiCl–0.3 mM Li-DS	24 500	(Na^+, K^+, NH_4^+), H^+,	4.5	2.0	3	[61]
5	2 mM ethylenediamine–0.1 mM Li-DS, pH 6.0	14 700[a]	(Na^+, K^+, NH_4^+), H^+, Mg^{2+}, Ca^{2+}		4.0	4	[62]
5	5 mM K_2EDTA–0.1 mM Li-DS	18 800[a]	(Na^+, K^+, NH_4^+), H^+	–	1.0	1.4	[63]

a) Calculated using $N = 5.54\,(t_R/w_{0.5})^2$ from the last eluting peak on chromatogram.

11.5 Modification of Silica Monoliths for IC Separations

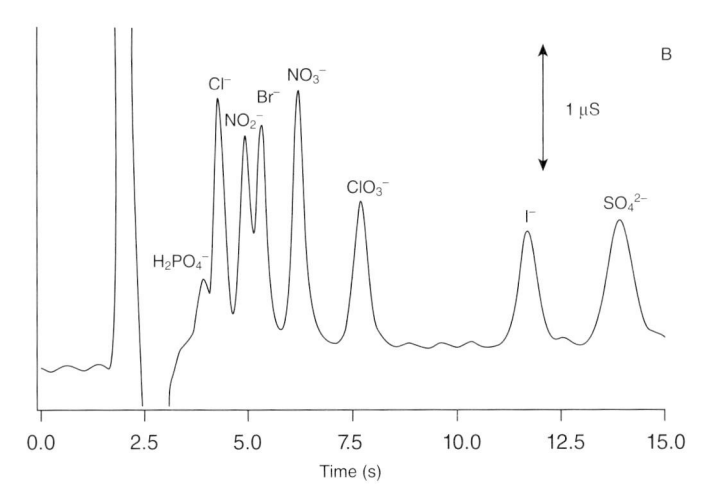

Figure 11.3 Separation of common anions in 15 s at 16 ml/min. Experimental conditions: Speed ROD, 1.5 mM TBA–1.1 mM phthalate with 5% (v/v) acetonitrile, 20 μl injection. Analyte concentration approximately 25 times detection limit. Reproduced with permission from [57].

A very rapid separation of 8 inorganic anions in 15 s at a flow rate 16 ml/min on a 5-cm long Chromolith Speed Rod RP-18e column was obtained by Hatsis and Lucy in 2002 [57] (Figure 11.3). Surprisingly, the backpressure at this flow rate was only 17.2 MPa. Since that time, some publications have appeared on the use of slightly modified elution systems for the quantitative determination of tetrafluoroborate [53], chromate [54] and perchlorate [55, 64]. The fast separation of five common species of arsenic in 3 min with inductively coupled plasma mass spectrometry (ICP-MS) detection has also been reported using the relatively low flow rate of 1 ml/min (Figure 11.4).

Because of its simplicity, the coating of commercially available reversed-phase silica monolithic columns remains the most popular way for preparation of ion-exchange columns. However, the key disadvantage of ion-exchange columns prepared by coating with ionic surfactants is the stability of the adsorbed layer and hence the maintenance of the ion-exchange capacity of the coated column. It has been reported that surfactants having one long *n*-alkyl chain (CTAC, CTAB, CPC, Li-DS, DDAAA) do not always provide appropriate stability of the coating [16, 18, 59, 61] and the ion-exchange capacity of the prepared columns reduces over time. To stabilize the ion-exchange capacity and enhance the reproducibility of retention times, addition of the surfactants to the eluent is often required [59–61, 63]. This problem can be overcome by using "double-leg" surfactants, such as DDAB [11, 39] and DOSS [39], or the zwitterionic surfactant DDMAU that has long methylene chain between oppositely charged groups [30, 43] for coating. Another drawback of coated ion exchangers is nonuniform distribution of surfactant or ion-exchange capacity profile along the column [47] and the possibility of the formation of hemimicelles or similar aggregates of surfactant

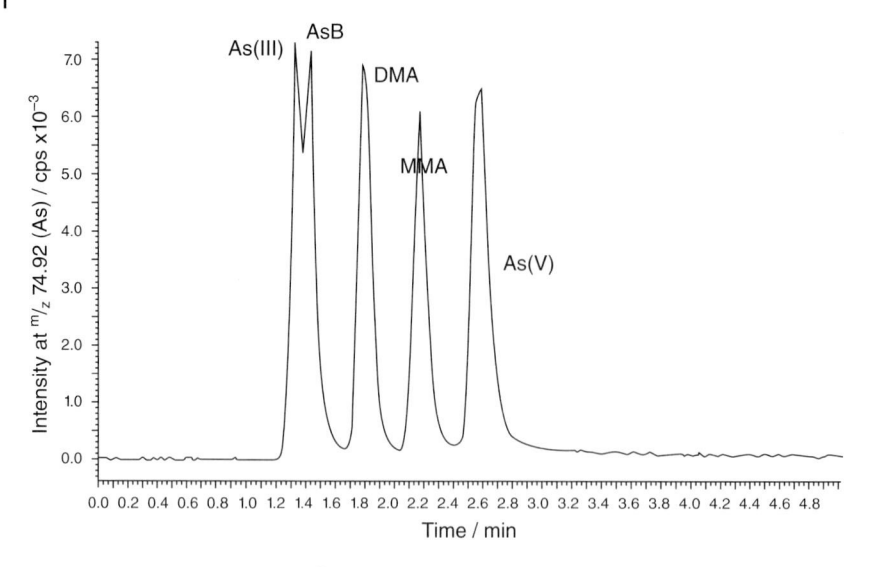

Figure 11.4 Ion-pair separation of 5 arsenic species. Column: Chromolith RP-18e 100 × 4.6 mm ID. Eluent: 2.5 mM TBAB–10 mM sodium phosphate buffer (pH 5.6) and 1.0% (v/v) methanol. Detection ICP-MS. Reproduced with permission from [58].

molecules at the hydrophobic surface [65]. These aggregates transform gradually into a monolayer structure and this may contribute to some instability of the ion-exchange properties of coated columns. This is especially pertinent when the concentration of surfactant in the coating solution is close to the critical micelle concentration (cmc) of the surfactant and prolonged washing is required to obtain an equilibrated coating. It should be noted that the critical aggregation concentration of surfactants at hydrophobic surfaces, such as octadecylsilica, can be 10–50 times less than the corresponding cmc values. Finally, a significant potential advantage of the coating approach is the possibility of varying the column ion-exchange capacity by addition of organic solvents to the coating solution. Thus, Glenn and Lucy varied the ion-exchange capacity of Chromolith RP-18e column of 4.6 mm ID in the range 1–31 µeq/cm by addition of up to 35% acetonitrile to a CTAB coating solution [16].

A number of impressive FIC separations have been obtained with coated columns, especially using C18 monolithic columns coated with DDAB. Pelletier and Lucy obtained separation of 6 inorganic anions in 1.2 min and 7 inorganic anions in 2.2 min on columns of only 5 mm and 10 mm length, respectively, with eluent delivered at 2 ml/min [18]. Connolly *et al.* [39] used a 2.5-cm long column with the same coating and the same flow rate and separated 8 anions in 1.7 min. Hatsis and Lucy [31] applied a high flow rate of 10 ml/min and while some separation efficiency was lost for a 5-cm long column (see Table 11.2), they separated 7 anions in 0.5 min. There is a clear relationship in FIC between the applied flow rate of the eluent and column length, which reflects a special balance between column efficiency and the number of separated components, and the run time.

The type of detection used has become a very important parameter in FIC on ultrashort and capillary columns. For example, Pelletier and Lucy [18] demonstrated that the peak efficiency for separation on a 1-cm long column with non-suppressed conductivity detection was approximately 1.6 times higher than for suppressed conductivity detection due to dispersion in the suppressor. To avoid these problems O'Riordain et al. [30] used C^{4}D on-column detection in the separation of anions on a reversed-phase capillary column (Onyx C18, 150 × 0.1 mm) modified with DDMAU. However, the authors also noted some reduction in detection sensitivity (by a factor of approximately 1.75) when detection was performed on the chromatographic bed, compared to detection in an open-tubular fused-silica capillary of the same internal diameter.

11.5.3
Silica Monoliths with Covalently Bonded Ion-Exchange Groups

The problem of column stability observed with coated ion exchangers can be addressed by covalent attachment of ion-exchange groups to the silica backbone (Tables 11.2 and 11.5). Thus, Sugrue et al. grafted iminodiacetic acid [36, 37] and lysine [38] to the surface of commercial bare-silica analytical-size monolithic columns and successfully used them for the separation of alkaline-earth and transition-metal cations and inorganic anions. Both columns demonstrated excellent efficiency, selectivity and stability. An example of the FIC separation of 5 inorganic anions is shown in Figure 11.5.

A number of capillary columns having a covalently grafted polymeric layer of ion-exchange groups have been prepared by Tanaka et al. (Tables 11.3 and 11.5). An improved sol-gel technology was utilized for the *in situ* preparation of the silica skeleton that resulted in a remarkable increase in efficiency up to 147 000 theoretical plates/m, even with polymer coatings that are usually less efficient in separation of ions because of poor kinetics of mass transfer (Table 11.3). The authors suggested that the main use of the prepared column would be for hydrophilic-interaction liquid chromatography (HILIC) of biomolecules, but most of the column have also been tested for separation of inorganic ions. Capillary IC is not generally considered to be a very useful approach to FIC separations [66] but the possibility of cation-exchange separation of uridine, guanosine, adenosine in 1.3 min has been demonstrated [28].

11.6
Operational Parameters

The operational conditions may significantly increase separation speed. The key parameters are applied pressure for the eluent delivery, column temperature and use of elution gradients. The maximum recommended pressure that can be applied to monolithic silica columns is about 20 MPa and Hatsis and Lucy used 17.2 MPa to pump the eluent through a 50 × 4.6 mm ID at a flow rate of 16 ml/min to obtain the separation of 8 anions in 15 s [57]. This flow rate could be even

Table 11.5 Structures of chemically modified silica-based monolithic columns for IC.

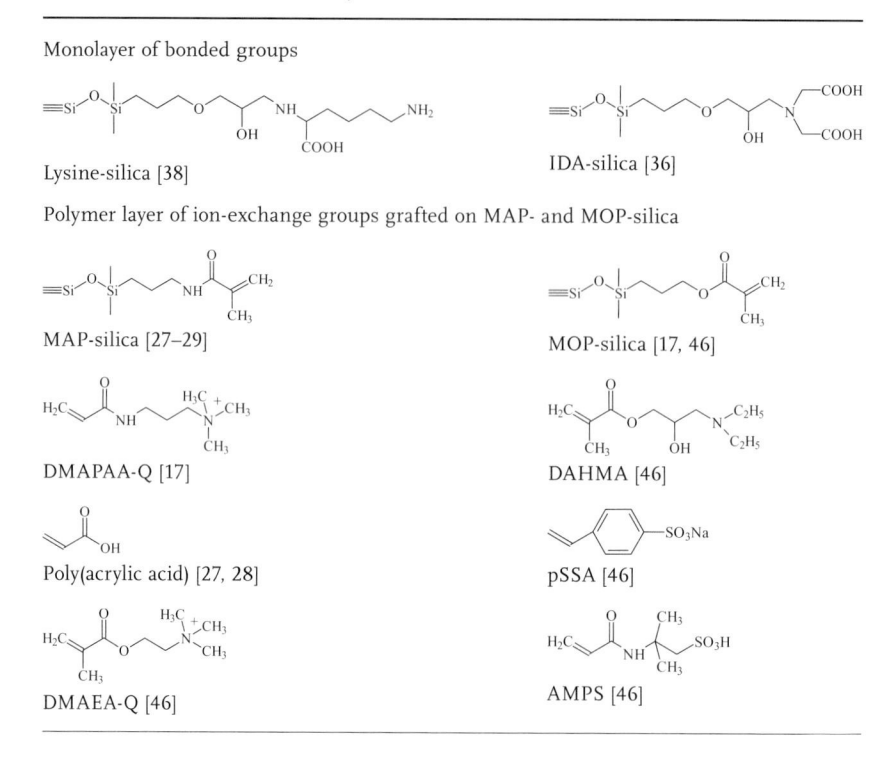

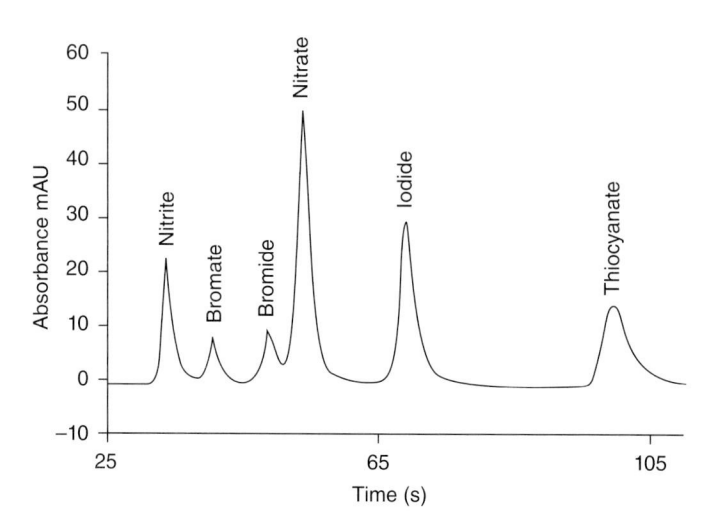

Monolayer of bonded groups

Lysine-silica [38]

IDA-silica [36]

Polymer layer of ion-exchange groups grafted on MAP- and MOP-silica

MAP-silica [27–29]

MOP-silica [17, 46]

DMAPAA-Q [17]

DAHMA [46]

Poly(acrylic acid) [27, 28]

pSSA [46]

DMAEA-Q [46]

AMPS [46]

Figure 11.5 Fast separation of anions using a 100 × 4.6 mm lysine-bonded silica monolith. Eluent: 50 mM phosphate buffer (pH 3.0); flow rate = 4.9 ml/min. UV 214 nm. Reproduced with permission from [38].

higher if elevated column temperatures were used to reduce the eluent viscosity. Some attempts to implement this approach were made by Victory *et al.* [67], who used an increased column temperature to minimize eluent viscosity and to obtain an optimal flow rate in low-pressure IC using a peristaltic pump and a short monolithic column. However, the technology of the column production restricts the maximum possible column temperature to 60 °C. It is pertinent to note that the tubing used in IC equipment is constructed from polymers, which also have limitations on maximum operational pressure and temperature.

There is great potential for the application of flow-rate gradients in FIC. The small change of the van Deemter *C* term for monolithic columns means that loss of column efficiency at high flow rates will be minimal. This property was used in FIC by implementing a gradually increasing flow rate of an eluent of constant composition [43, 60, 67]. Such an eluent flow gradient can be combined with an eluent concentration gradient (i.e., "double-gradient elution") to obtain an even faster separation [43, 60]. A double gradient of eluent flow rate and pH allowed the separation time of a model mixture of anions on a short monolithic column coated with the carboxybetaine surfactant DDAAA to be reduced from 16 min under isocratic/constant flow elution to 3 min under double-gradient conditions (Figure 11.6). The conditional peak efficiency for thiocyanate under double-gradient elution was 385 000 plates/m.

11.7
Analytical Applications

The ideal application of FIC is the determination of a small number of analytes in a large number of samples, such as for the monitoring of an industrial process or for environmental control. Some applications of FIC are listed in Table 11.6, with special emphasis on short separation times (<3 min.) and low limits of detection (LOD). It should be noted that FIC has direct applicability for the rapid determination of bromide, nitrite and nitrate in very complex samples, such as seawater, or for the determination of alkaline-earth metal cations in brines. Another useful application of FIC for complex samples is the determination of five arsenic species in food and urine.

A more complex FIC system has been designed for the simultaneous determination of 3 anions and 3 cations comprising one pump, one injection valve and one detector, but two parallel separation lines with short reversed-phase silica monolithic columns. The first column was 25 × 4.6 mm and was coated with DDAB, while the other was 50 × 4.6 mm and was coated with DOSS. An eluent composition of 2.5 mM phthalate–1.5 mM ethylenediamine (pH 4.5) allowed simultaneous direct conductivity detection of anions and indirect conductivity detection of cations. At a flow rate of 2.0 ml/min, the separation and determination of chloride, nitrate, copper, calcium and magnesium was accomplished in 5 min.

At the present time, the only monolithic silica-based column designed specifically for use in IC is the Metrosep Dual 4 column, which is based on a C18

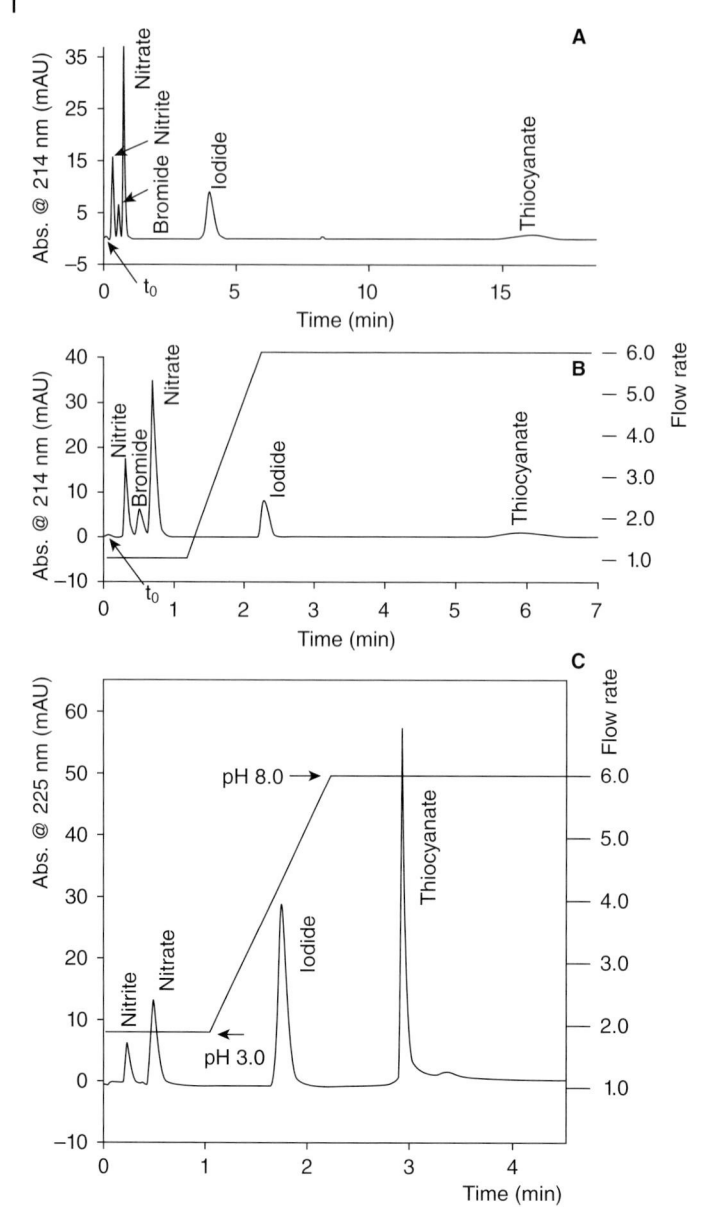

Figure 11.6 FIC separations of model mixture of anions obtained using flow-gradient (B) and double (flow and pH)-gradient elution (C). Column: 10 × 4.6 mm DDAAA modified silica monolith. (A) Isocratic isofluentic separation in 18 min at a constant flow rate of 1 ml/min. (B) Separation in 6.5 min obtained using an eluent flow gradient. (C) Eluent delivered at 1 ml/min for first 1 min, increased to 6 ml/min between 1 and 2 min. Eluent: 10 mM KCl, 10 mM phosphate buffer (pH 3), 0.2 mM carboxybetaine. Reproduced with permission from [60].

Table 11.6 Practical applications of FIC using silica-based monolithic columns.

Sample	Ions	t, min	Detection	LOD	Ref.
Seawater	Br^-	1.3	UV 210 nm	4.5 mg/l	[45]
	Br^-	1.5	UV 210 nm	1.6 mg/l	[44]
Estuarine water and seawater	Cl^-, Br^-, NO_3^-	2.0	C^4D	–	[59]
	NO_2^-, NO_3^-	3.0	UV 214 nm	0.8 and 1.6 µg/l	[41]
Drinking water	Cl^-, NO_3^-, SO_4^{2-}	3.0	C^4D	0.014 and 0.062 mM	[30]
Rainwater	H^+	1.0	Conductivity	0.37 µg/l	[63]
	H^+, Mg^{2+}, Ca^{2+}	4.0	Conductivity	1.0, 48.6 and 80 µg/l	[62]
	H^+, Na^+, NH_4^+, K^+	3.0	Conductivity	6.2 µg/l, 0.46, 0.22 and 0.38 mg/l	[61]
Ham	NO_2^-	0.7	UV 210 nm	21.3 µg/l	[40]
Pool water	NO_3^-, SO_4^{2-}	0.7	UV 279 nm	0.50 and 0.22 mg/l	[40]
Waste water	CrO_4^{2-}	0.8	UV 210 nm	4.86 µg/l	[40]
Brines	Mg^{2+}, Ca^{2+}	0.7	PCR	–	[37]
Food, urine	As(III), AsB, DMA, MMA, As(V)	3.0	ICP-MS	0.107, 0.084, 0.120, 0.121 and 0.101 µg/l As	[58]
Ground water[a]	Cr(VI)	1.0	Conductivity	1.02 mg/l	[54]
Ground water, juice[a]	ClO_4^-	1.0	Conductivity	1.56 mg/l	[55]
Ionic liquids	BF_4^-	1.4	Conductivity	1.4 mg/l	[53]

a) As spiked with analyte ion.

covalently modified silica monolith and is coated with strongly retained cationic surfactant. This column is recommended for the determination of perchlorate in accordance with the EPA 314 standard method.

11.8 Future Work

Instability of the coating and a limited variety of chemically modified surfaces are the key limitations in using modified silica monolithic columns in IC. Future development of FIC will require the introduction of new surface functionalities on the silica surface in order to obtain optimum selectivity. For example, there are no carboxylated monolithic silica-based columns, whereas particulate carboxylic silica-based cation exchangers are used frequently in IC determinations of alkali and alkaline-earth metal cations.

FIC can be successfully performed with low-pressure chromatography equipment, so the development of corresponding equipment is expected. The combination of optimum ion-exchange selectivity, column size and operational conditions may result in new analytical approaches similar to simultaneous multicomponent flow-injection analysis.

Abbreviations

AMPS	– 2-acrylamido-2-methylpropanesulfonic acid
AsB	– arsenobetaine
CPC	– cetylpyridinium chloride
CTAC	– cetyltrimethylammonium chloride
DAHMA	– 3-diethylamino-2-hydroxylpropylmethacrylate
DDAAA	– (dodecyldimethylamino)acetic acid
DDAB	– didodecyldimethylammonium bromide
DDMAU	– N-dodecyl-N,N-(dimethylammonio)undecanoate
DMA	– dimethylarsenic acid
DMAEA-Q	– 2-(trimethylammonium)ethyl methacrylate chloride salt
DMAPAA-Q	– N-[3-(dimethylamino)propyl]acrylamide methyl chloride salt
DOSS	– dioctylsulphosuccinate sodium
IC	– ion chromatography
MAP	– (3-methacrylamidopropyl)trimethoxysilane
MOP	– (3-methacryloxypropyl)trimethoxysilane
MMA	– monomethylarsonic acid
pSSA	– p-styrenesulfonic acid sodium salt
TBAB	– tetrabutylammonium bromide
TBAH	– tetrabutyl ammonium hydroxide

References

1 Kettele, B.H., Boyd, G.E. (1947) The exchange adsorption of ions from aqueous solutions by organic zeolites. IV. The separation of the yttrium group rare earths. *J. Am. Chem. Soc.*, **69**, 2800–2812.

2 Small, H., Stevens, T.S., Bauman, W.C. (1975) Novel ion-exchange chromatographic method using conductimetric detection. *Anal. Chem.*, **47** (11), 1801–1809.

3 Haddad, P.R., Nesterenko, P.N., Buchberger, W. (2008) Recent developments and emerging directions in ion chromatography. *J. Chromatogr. A*, **1184** (1–2), 456–473.

4 Itoh, H., Shinbori, Y. (1982) Ion chromatography of anions on a fast run column. *Bunseki Kagaku*, **31** (4), T39–T43.

5 Paull, B., Nesterenko, P.N. (2005) New possibilities in ion chromatography

using porous monolithic stationary-phase media. *Trends Analyt. Chem.*, **24** (4), 295–303.

6 Schaller, D., Hilder, E.F., Haddad, P.R. (2006) Monolithic stationary phases for fast ion chromatography and capillary electrochromatography of inorganic ions. *J. Sep. Sci.*, **29** (12), 1705–1719.

7 Ai, S.Y., Xu, Q. (2006) Separation of inorganic anions and cations by high-speed ion chromatography. *Curr. Anal. Chem.*, **2** (4), 389–396.

8 Chambers, S.D., Glenn, K.M., Lucy, C.A. (2007) Developments in ion chromatography using monolithic columns. *J. Sep. Sci.*, **30** (11), 1628–1645.

9 Paull, B., Nesterenko, P.N. (2005) Novel ion chromatographic stationary phases for the analysis of complex matrices. *Analyst*, **130** (2), 134–146.

10 Tyrrell, E., Hilder, E.F., Shalliker, R.A., Dicinoski, G.W., Shellie, R.A., Breadmore, M.C., Pohl, C.A., Haddad, P.R. (2008) Packing procedures for high efficiency, short ion-exchange columns for rapid separation of inorganic anions. *J. Chromatogr. A*, **1208** (1–2), 95–100.

11 Connolly, D., Paull, B. (2002) Fast ion chromatography of common inorganic anions on a short ODS column permanently coated with didodecyldimethylammonium bromide. *J. Chromatogr. A*, **953** (1–2), 299–303.

12 Pelletier, S., Lucy, C.A. (2006) Fast and high-resolution ion chromatography at high pH on short columns packed with 1.8 µm surfactant coated silica reverse-phase particles. *J. Chromatogr. A*, **1125** (2), 189–194.

13 Kanatyeva, A.Y., Viktorova, E.N., Korolev, A.A., Kurganov, A.A. (2007) Comparison of nonporous silica-based ion exchangers and monolithic ion exchangers in separations of inorganic anions. *J. Sep. Sci.*, **30** (17), 2836–2842.

14 Cabrera, K., Wieland, G., Lubda, D., Nakanishi, K., Soga, N., Minakuchi, H., Unger, K.K. (1998) SilicaROD (TM) – A new challenge in fast high-performance liquid chromatography separations. *Trend. Anal. Chem.*, **17** (1), 50–53.

15 Cabrera, K. (2004) Applications of silica based monolithic HPLC columns. *J. Sep. Sci.*, **27** (10–11), 843–852.

16 Glenn, K.M., Lucy, C.A. (2008) Stability of surfactant coated columns for ion chromatography. *Analyst*, **133** (11), 1581–1586.

17 Jaafar, J., Watanabe, Y., Ikegami, T., Miyamoto, K., Tanaka, N. (2008) Anion exchange silica monolith for capillary liquid chromatography. *Anal. Bioanal. Chem.*, **391** (7), 2551–2556.

18 Pelletier, S., Lucy, C.A. (2006) Achieving rapid low-pressure ion chromatography separations on short silica-based monolithic columns. *J. Chromatogr. A*, **1118** (1), 12–18.

19 Evenhuis, C.J., Buchberger, W., Hilder, E.F., Flook, K.J., Pohl, C.A., Nesterenko, P.N., Haddad, P.R. (2008) Separation of inorganic anions on a high capacity porous polymeric monolithic column and application to direct determination of anions in seawater. *J. Sep. Sci.*, **31** (14), 2598–2604.

20 Kurganov, A.A., Korolev, A.A., Viktorova, E.N., Kanat'eva, A.Y. (2009) The synthesis and study of monolithic capillary columns for ion chromatography of anions. *Russ. J. Phys. Chem. A*, **83** (2), 303–307.

21 Hutchinson, J.P., Hilder, E.F., Shellie, R.A., Smith, J.A., Haddad, P.R. (2006) Towards high capacity latex-coated porous polymer monoliths as ion-exchange stationary phases. *Analyst*, **131** (2), 215–221.

22 Nesterenko, P.N., Rybalko, M.A. (2004) The use of a continuous flow gradient for the separation of inorganic anions on a monolithic disk. *Mendeleev Comm.*, **14** (3), 121–123.

23 Zakaria, P., Hutchinson, J.P., Avdalovic, N., Liu, Y., Haddad, P.R. (2005) Latex-coated polymeric monolithic ion-exchange stationary phases. 2. Micro-ion chromatography. *Anal. Chem.*, **77** (2), 417–423.

24 Tyrrell, E., Nesterenko, P.N., Paull, B. (2006) Flow analysis method using chelating CIM monolithic disks for monitoring dissolved labile copper in environmental water samples. *J. Liq. Chromatogr. Relat. Technol.*, **29** (15), 2201–2216.

25 Paull, B., Nesterenko, P.N. (2005) New porous monolithic phases for rapid ion analysis. *Eur. Pharm. Rev.*, **10** (1), 47–53.

26 Lucy, C.A., Pelletier, S. (2006) Advances in fast ion chromatography. *Eur. Pharm. Rev.*, **11** (2), 79–84.

27 Ikegami, T., Horie, K., Jaafar, J., Hosoya, K., Tanaka, N. (2007) Preparation of highly efficient monolithic silica capillary columns for the separations in weak cation-exchange and HILIC modes. *J. Biochem. Biophys. Methods*, **70** (1), 31–37.

28 Horie, K., Ikegami, T., Hosoya, K., Saad, N., Fiehn, O., Tanaka, N. (2007) Highly efficient monolithic silica capillary columns modified with poly(acrylic acid) for hydrophilic interaction chromatography. *J. Chromatogr. A*, **1164** (1–2), 198–205.

29 Ikegami, T., Fujita, H., Horie, K., Hosoya, K., Tanaka, N. (2006) HILIC mode separation of polar compounds by monolithic silica capillary columns coated with polyacrylamide. *Anal. Bioanal. Chem.*, **386** (3), 578–585.

30 Riordain, C.O., Gillespie, E., Connolly, D., Nesterenko, P.N., Paull, B. (2007) Capillary ion chromatography of inorganic anions on octadecyl silica monolith modified with an amphoteric surfactant. *J. Chromatogr. A*, **1142** (2), 185–193.

31 Hatsis, P., Lucy, C.A. (2003) Improved sensitivity characterization of high-speed ion chromatography of inorganic anions. *Anal. Chem.*, **75** (4), 995–1001.

32 Siouffi, A.M. (2006) About the *C* term in the van Deemter's equation of plate height in monoliths. *J. Chromatogr. A*, **1126** (1-2), 86–94.

33 Pack, B.W., Risley, D.S. (2005) Evaluation of a monolithic silica column operated in the hydrophilic interaction chromatography mode with evaporative light scattering detection for the separation and detection of counter-ions. *J. Chromatogr. A*, **1073** (1–2), 269–275.

34 Sugrue, E., Nesterenko, P.N., Paull, B. (2005) Solvent enhanced ion chromatography of alkaline earth and transition metal ions on porous monolithic silica. *Anal. Chim. Acta*, **553** (1–2), 27–35.

35 Glenn, K.M., Lucy, C.A., Haddad, P.R. (2007) Ion chromatography on a latex-coated silica monolith column. *J. Chromatogr. A*, **1155** (1), 8–14.

36 Sugrue, E., Nesterenko, P., Paull, B. (2004) Ion exchange properties of monolithic and particle type iminodiacetic acid modified silica. *J. Sep. Sci.*, **27** (10–11), 921–930.

37 Sugrue, E., Nesterenko, P., Paull, B. (2003) Iminodiacetic acid functionalised monolithic silica chelating ion exchanger for rapid determination of alkaline earth metal ions in high ionic strength samples. *Analyst*, **128** (5), 417–420.

38 Sugrue, E., Nesterenko, P.N., Paull, B. (2005) Fast ion chromatography of inorganic anions and cations on a lysine bonded porous silica monolith. *J. Chromatogr. A*, **1075** (1–2), 162–175.

39 Connolly, D., Victory, D., Paull, B. (2004) Rapid, low pressure, and simultaneous ion chromatography of common inorganic anions and cations on short permanently coated monolithic columns. *J. Sep. Sci.*, **27** (10–11), 912–920.

40 Li, J., Zhu, Y., Guo, Y.Y. (2006) Fast determination of anions on a short coated column. *J. Chromatogr. A*, **1118** (1), 46–50.

41 Ito, K., Takayama, Y., Makabe, N., Mitsui, R., Hirokawa, T. (2005) Ion chromatography for determination of nitrite and nitrate in seawater using monolithic ODS columns. *J. Chromatogr. A*, **1083** (1–2), 63–67.

42 Xu, Q., Mori, M., Tanaka, K., Ikedo, M., Hu, W., Haddad, P.R. (2004) Ion chromatographic determination of hydroxide ion on monolithic reversed-phase silica gel columns coated with

43 Riordain, C.O., Barron, L., Nesterenko, E., Nesterenko, P.N., Paull, B. (2006) Double gradient ion chromatography using short monolithic columns modified with a long chained zwitterionic carboxybetaine surfactant. *J. Chromatogr. A*, **1109** (1), 111–119.

44 Suzuki, A., Lim, L.W., Takeuchi, T. (2007) Rapid separation of inorganic anions by capillary ion chromatography using monolithic silica columns modified with dilauryldimethylammonium ion. *Anal. Sci.*, **23** (9), 1081–1084.

45 Suzuki, A., Lim, L.W., Hiroi, T., Takeuchi, T. (2006) Rapid determination of bromide in seawater samples by capillary ion chromatography using monolithic silica columns modified with cetyltrimethylammonium ion. *Talanta*, **70** (1), 190–193.

46 Ikegami, T., Ichimaru, J., Kajiwara, W., Nagasawa, N., Hosoya, K., Tanaka, N. (2007) Anion- and cation-exchange microHPLC utilizing poly(methacrylates)-coated monolithic silica capillary columns. *Anal. Sci.*, **23** (1), 109–113.

47 Gillespie, E., Connolly, D., Macka, M., Nesterenko, P.N., Paull, B. (2007) Use of contactless conductivity detection for non-invasive characterisation of monolithic stationary-phase coatings for application in capillary ion chromatography. *Analyst*, **132** (12), 1238–1245.

48 Smith, R.L., Pietrzyk, D.J. (1984) Liquid-chromatographic separation of metal-ions on a silica column. *Anal. Chem.*, **56** (4), 610–614.

49 Kolpachnikova, M.G., Penner, N.A., Nesterenko, P.N. (1998) Effect of temperature on retention of alkali and alkaline-earth metal ions on some aminocarboxylic acid functionalised silica based ion exchangers. *J. Chromatogr. A*, **826** (1), 15–23.

50 Sutton, P.A., Nesterenko, P.N. (2007) Retention characteristics of aromatic hydrocarbons on silica and aminopropyl-modified monolithic columns in normal-phase HPLC. *J. Sep. Sci.*, **30** (17), 2900–2909.

51 Takeuchi, T., Yeung, E.S. (1986) High-performance liquid-chromatographic separation of inorganic anions on a silica-gel column modified with a quaternary ammonium salt. *J. Chromatogr.*, **370** (1), 83–92.

52 Hutchinson, J.P., Hilder, E.F., Macka, M., Avdalovic, N., Haddad, P.R. (2006) Preparation and characterisation of anion-exchange latex-coated silica monoliths for capillary electrochromatography. *J. Chromatogr. A*, **1109** (1), 10–18.

53 Zhou, S., Yu, H., Yang, L., Ai, H.J. (2008) Fast determination of tetrafluoroborate by high-performance liquid chromatography using a monolithic column. *J. Chromatogr. A*, **1206** (2), 200–203.

54 Ai, H.J., Yu, H., Zhou, S., Li, R.S. (2008) Fast determination of Cr(VI) by ion-pair chromatography using a monolithic column. *Fenxi Ceshi Xuebao*, **27** (9), 991–993.

55 Yu, H., Li, R.S. (2008) Fast analysis of perchlorate by ion-pair chromatography using a monolithic column. *Fenxi Huaxue*, **36** (6), 835–838.

56 Li, R.S., Yu, H., Ai, H.J. (2008) Fast analysis of inorganic anions by ion-interaction chromatography using a monolithic column. *Fenxi Kexue Xuebao*, **24** (1), 6–10.

57 Hatsis, P., Lucy, C.A. (2002) Ultra-fast HPLC separation of common anions using a monolithic stationary phase. *Analyst*, **127** (4), 451–454.

58 Pearson, G.F., Greenway, G.M., Brima, E.I., Haris, P.I. (2007) Rapid arsenic speciation using ion pair LC-ICPMS with a monolithic silica column reveals increased urinary DMA excretion after ingestion of rice. *J. Anal. Atom. Spectrom.*, **22** (4), 361–369.

59 Riordain, C.O., Nesterenko, P., Paull, B. (2005) Zwitterionic ion chromatography with carboxybetaine surfactant-coated particle packed and monolithic type columns. *J. Chromatogr. A*, **1070**, 71–78.

60 Paull, B., O'Riordain, C., Nesterenko, P.N. (2005) Double gradient ion chromatography on a short carboxybetaine coated monolithic anion exchanger. *Chem. Commun.*, **2**, 215–217.

61 Xu, Q., Tanaka, K., Mori, M., Helaleh, M.I., Toada, H., Hu, W., Hasebe, K. (2003) Monolithic ODS-silica gel column for determination of hydrogen, sodium, ammonium and potassium in acid rain by ion chromatography. *Chromatographia*, **57** (1–2), 19–22.

62 Xu, Q., Mori, M., Tanaka, K., Ikedo, M., Hu, W. (2004) Dodecylsulfate-coated monolithic octadecyl-bonded silica stationary phase for high-speed separation of hydrogen, magnesium and calcium in rainwater. *J. Chromatogr. A*, **1026** (1–2), 191–194.

63 Xu, Q., Tanaka, K., Mori, M., Helaleh, M.I., Hu, W., Hasebe, K., Toada, H. (2003) Monolithic octadecylsilyl-silica gel column for the high-speed ion chromatographic determination of acidity. *J. Chromatogr. A.*, **997** (1–2), 183–190.

64 Hong, Y., Rui-Shu, L. (2008) Fast analysis of perchlorate by ion-pair chromatography using a monolithic column. *Chin. J. Anal. Chem.*, **36** (6), 835–838.

65 Nesterenko, P.N., Haddad, P.R., Hu, W.Z. (2003) Studies on the separation of hydronium ion using surfactant-modified reversed-phase stationary phases and eluents containing an acidified electrolyte. *J. Chromatogr. A*, **986** (2), 263–273.

66 Kuban, P., Dasgupta, P.K. (2004) Capillary ion chromatography. *J. Sep. Sci.*, **27** (17–18), 1441–1457.

67 Victory, D., Nesterenko, P., Paull, B. (2004) Low-pressure gradient micro-ion chromatography with ultra-short monolithic anion exchange column. *Analyst*, **129** (8), 700–701.

12
Monolithic Chiral Stationary Phases for Liquid-Phase Enantioseparation Techniques

Bezhan Chankvetadze

12.1
Introduction

In spite of significant developments over the last 3 decades, resolution of enantiomers still remains one of the hot topics of separation science. With multiple technologies available, such as gas chromatography (GC), liquid chromatography (LC), super/subcritical fluid chromatography (SFC), capillary electrophoresis (CE) and capillary electrochromatography (CEC), the separation of two enantiomers of a given chiral analyte no longer represents a problem for separation scientists. In addition to the above-mentioned methods, some techniques are available in various formats in order to deal with the given separation problem in an optimal way. For instance, high-performance liquid chromatography (HPLC) is best suited for classical nonsophisticated separations of nonvolatile and thermolabile compounds if no strict limits exist for the available sample size, mobile phase and packing materials, while capillary liquid chromatography (CLC) or nano-liquid chromatography (nano-LC) can be employed when the available sample size, mobile phase and packing materials are limited. On the opposite end, preparative or product-scale LC can be performed in various formats (such as recycling, simulating moving-bed [SMB], etc.) when the goal is a preparative isolation or purification of enantiomers. The above-mentioned techniques can be employed in a single-column mode, parallel-column mode or multidimensional mode depending on the separation problem. In addition, technology variations (exemplified just for the case of LC), the particle-based chiral stationary phases (CSPs) with various chemistry and complementary chiral recognition ability are available for almost all liquid-phase enantioseparation techniques. As this very concise summary shows, there are powerful sets of the state-of-the-art technologies and materials at the disposal of separation scientists for solving almost any analytical or preparative enantiomer resolution problem. With that in mind, what drives our research for new technologies and materials in this field? The number of chiral fine chemicals (pharmaceuticals, agrochemicals, food additives, materials with desired optical, electric and magnetic properties, etc.) entering diverse fields of our life in enantiomerically pure forms increases steadily. For instance, among

Monolithic Silicas in Separation Science. Edited by K.K. Unger, N. Tanaka, and E. Machtejevas
© 2011 WILEY-VCH Verlag GmbH & Co. KGaA, Weinheim
ISBN: 978-3-527-32575-7

15 FDA-approved drugs in the period of January–August 2003, 64% were single enantiomers, 14% racemates and 22% achiral compounds [1]. Enantiomerically pure chiral compounds need to be prepared and their purity has to be examined with reliable and fast analytical techniques. In addition, chromatographic techniques can be used for production of enantiomerically pure chiral compounds [2, 3]. However, even if the choice of the production step falls in favor of chiral-pool, diastereomeric crystallization or asymmetric synthesis/catalysis, chiral separations are still needed to optimize chiral catalysts [4], to analyze the enantiomeric purity of involved materials, intermediates and final products [5]. Fast development of combinatorial synthetic strategies facilitates the accumulation of a large set of potential drug candidates in a rather short time. Fast analytical techniques are required in order to analyze these compounds with regard to their (enantiomeric) purity as well as in various biological matrices (after reaching the stage of clinical trials) [6]. Short analysis times are attractive not only from the viewpoint of time saving but sometimes also allow saving mobile phases. The latter is very important considering the current financial pressure, shortages of the acetonitrile supply and adverse impact of synthetic chemicals on the environment.

Quite often, the compounds of interest or their characteristic response must be recovered from very complex matrices. This implies very high requirements to the efficiency (peak capacity) of a separation system. Thus, reliable, highly efficient, high-throughput and highly productive separation materials and technologies are in demand. How do monolithic materials enable these current demands to be addressed and is the monolithic strategy the only way in order to address the above-mentioned requirements?

Based on the well-known fundamental Equation 12.1 in chromatography it is obvious that the column efficiency increases with decreasing particle diameter (d_p). This is related to a smaller

$$H = \frac{1}{\left(\frac{1}{C_e d_p}\right) + \frac{D_m}{C_m d_p^2 u}} + \frac{C_d D_m}{u} + \frac{C_{sm} d_p^2 u}{D_m} = Au^{\frac{1}{3}} + \frac{B}{u} + Cu \qquad (12.1)$$

A term (while the voids between particles are reducing with decreasing particle size, d_p) and a smaller C term due the shorter diffusion path length inside the particle. In the case of particle-packed column, the plate height H is roughly proportional to d_p whereas the backpressure, ΔP, is inversely proportional to d_p^2 [7]. Thus, a linear increase in column efficiency due to the use of small particles is accompanied by a quadratic increase in column backpressure. This limits the application of small particles in HPLC to 1.8–2.0 μm at present. What are the principal advantages of monoliths from this point of view? The flow-through monolithic channel is laminar. This means that monoliths have nearly no void volume and do not develop eddies. The transport of an analyte through the monolithic bed is basically perfussive and in the monolithic media with a low portion of mesopores, the analyte diffusion into and out of the pores does not significantly contribute to the band broadening. Thus, monolithic materials enable higher peak efficiencies to be achieved by alternative mechanisms (different flow path) compared to the approach used in particle-packed columns. Due to reducing the role of diffusion (which is slow) in solute transfer through the monolithic bed, the peak efficiency does not decrease as sharply as in particle-packed columns when increasing the linear flow rate of the mobile phase. This allows the application of higher flow rates and thus shortens the analysis time without a significant sacrifice of plate numbers.

Due to the flow characteristics through monolithic beds mentioned above this material may exhibit a higher dynamic binding capacity, especially for large molecules. This may provide a higher loading capacity in preparative and product-scale separations (in spite of lower surface area and overall porosity).

Thus, the major advantages associated with the use of monolithic separation media are a higher peak efficiency and shorter analysis time in analytical-scale separations and higher loadability (and consequently higher productivity) in preparative purifications.

As for any kind of material, the monoliths as separation media have not only advantages but also some disadvantages. To the latter belong the problems associated with the *in situ* characterization and modification of monoliths and with column to column reproducibility of separation characteristics. Therefore, the advantages and disadvantages of monoliths and alternative materials needs to be very carefully compared before taking a decision in favor of one or another material for solving a given separation and /or purification problem.

The goal of this chapter is to summarize the achievements in the field of enantioselectively modified monolithic chromatographic media, to stress the currently existing problems and provide the author's vision for their solution. Although this book is about monolithic silica, organic monoliths are also briefly overviewed for the sake of completeness and fair comparison. The particle-based silica monoliths are not included in this chapter.

12.2
Organic Monolithic Materials for the Separation of Enantiomers

There are certain advantages and disadvantages for the use of organic- and silica-based materials for separation science in general. These advantages and disadvantages basically relate to the chemistry of the materials, their mechanical and pH stability, complexity and reproducibility of their preparation, availability in various morphology (pore structure and diameter, particle size and uniformity and so on) and are relevant also for materials used for the separation of enantiomers. In addition to the above-mentioned features, the possibility of enantioselective modification has to be considered in the latter case.

Monoliths based on organic polymers are typically prepared from acrylamide, methacrylate or styrene monomers or are prepared by ring-opening polymerization [8, 9].

Organic monoliths for the separation of enantiomers have been applied only in the capillary format and basically for CEC enantioseparations [9–16]. Actually, the first monolithic columns applied for enantioseparations were organic monoliths [10]. The organic monolithic chiral column was prepared by the direct copolymerization of the chiral monomer, 2-hydroxyethyl methacrylate (N-L-valine-3,5-dimethylanilide) carbamate with ethylene dimethacrylate, 2-acrylamido-2-methyl-1-propanesulfonic acid and butyl or glycidyl methacrylate in the presence of a porogenic solvent. The hydrophilicity of the stationary phase, which may be enhanced further by the hydrolysis of the epoxide functionalities of the glycidyl methacrylate moieties within the monolith, was found to have a pronounced effect on the enantioseparation. Using the most hydrophilic monolithic capillary column and optimized elution conditions, the separation of N-(3,5-dinitrobenzoyl) leucine diallylamide enantiomers with an efficiency of 61 000 plates m^{-1} and a resolution of 2.0 was achieved in this pioneering study.

Monolithic columns with the chiral anion-exchange-type selectors incorporated into the polymer matrix obtained by *in situ* copolymerization of a chiral monomer [11–13] or attached to the surface of a reactive monolith in the subsequent derivatization step [14, 15] both provided useful chiral columns for CEC separations.

In the initial study by Lämmerhofer *et al.* [11] the effect of chromatographic conditions on the performance of chiral monolithic poly(O-[2-(methacryloyloxy)-ethylcarbamoyl]-10,11-dihydroquinidine-co-ethylene dimethacrylate-co-2-hydroxyethyl methacrylate) columns were investigated for the CEC separation of enantiomers. The flow velocity was found to be dependent on the pore size of the monolith and both the pH and the composition of the mobile phase. The length of both open and monolithic segments of the capillary column was found to exert a substantial effect on the run times. The use of monoliths as short as 8.5 cm and the "short-end" injection technique enabled the separations in less than 5 min despite the high retentivity of the quinidine selector. High column efficiencies of close to 250 000 plates/m and good selectivities were achieved for the separations of numerous enantiomers using the chiral monolithic capillaries with the optimized chromatographic conditions. Later, in a series of papers the same group reported the optimization of the chiral selector, as well as the morphology of monolithic matrix for a better understanding of the mechanisms of flow generation and peak dispersion in monolithic capillary columns [12–15]. Thus, a new cinchona alkaloid-based chiral selector was proposed in 2003 and incorporated into the 2-hydroxyethylmethacrylate-based monolithic polymeric matrix. Various reagent compositions and two initiating strategy (UV and thermal) were applied in order to prepare the capillary columns with distinct morphology [12]. Van Deemter coefficients were determined for different columns and an elegant explanation of band-broadening mechanisms depending on the morphology of monoliths has been proposed [13].

One major disadvantage of monolithic materials for separation science in general is that these materials have to be prepared *in situ*. This makes the optimization of their morphology as well as their characterization difficult. Therefore, Lämmerhofer's group proposed a strategy of in-capillary chiral selector attachment to preoptimized monolithic supports [14, 15]. With the example of a test racemic analyte, N-(3,5-dinitrobenzoyl)-leucine, the authors could demonstrate the advantage of this approach compared with an *in situ* copolymerization approach since higher enantioselectivity and resolution could be achieved with the novel postmodification technique [14].

A similar postmodification strategy of polymethacrylate-based monolith with terguride as a chiral selector was later applied by Messina *et al.* [16] for the preparation of chiral capillary columns useful for CEC separation of enantiomers of arylpropionic acids.

Schmid *et al.* prepared a chiral continuous bed for ligand-exchange separations by copolymerization of methacrylamide, piperazine diacrylamide as a crosslinker, vinylsulfonic acid as a comonomer for EOF generation, and N-(2-hydroxy-3-allyloxypropyl)-l-4-hydroxyproline as a chiral selector [17]. The polymer was used for CEC separations of amino acids [17] and hydroxyl acids [18].

Kornysova *et al.* prepared an organic monolith containing vancomycin as a chiral selector by *in situ* copolymerization of N-(hydroxymethyl) acrylamide, piperazine diacrylamide, and allyl glycidyl ether [19]. After conversion of the epoxy groups to aldehyde functionality, vancomycin was covalently attached to the polymer by reductive amination.

Machtejevas and Maruska [20] prepared continuous beds with immobilized human serum albumin as a chiral selector. The interaction sites of the chiral selector were noncovalently protected by salicylic acid or L-tryptophan added during the protein allylation and polymerization steps, thus, avoiding their blocking during the preparation process of a chiral stationary phase.

Molecularly imprinted organic polymer monoliths have also been synthesized and used for enantioseparations in CEC [21].

Some other kinds of chiral selectors have been also attached to organic monoliths used for separation of enantiomers basically in CEC. However, the limited space and limited number of references does not allow mentioning all of these publications in this short chapter. The interested reader can find additional information on the topic in review papers by Preinerstorfer and Lämmerhofer [9] and Gübitz and Schmid [22].

12.3
Silica-Based Monolithic Materials for the Separation of Enantiomers

Most of the silica-based monolithic materials are bimodally structured and contain besides the 1.7–2.0-μm diameter macropores additional mesopores of about 12–25 nm diameter. Macropores are responsible for a convective, bulk flow through the monolith, while mesopores provide a large surface area for the interaction of the analyte molecules with the stationary phase. The technology and chemistry for synthesis of silica-based monoliths, as well as the cladding technology of monoliths have been continuously optimized during the last decade. Thus, it is currently possible to prepare monolithic silica columns with acceptable reproducibility in common-size, as well as in capillary columns from 75 up to 500 µm ID . Several review papers have been devoted to the methodologies of fabrication of silica monolithic phases in both common size and capillary columns [23, 24], as well as to chromatographic performance and fundamental aspects of this kind of separation media for liquid phase separations [25]. This short chapter does not deal with the above-mentioned general aspects of preparation of silica monoliths and discusses only the immobilization strategies of chiral selectors onto the surface of monolithic silica and the separation ability of enantiomers by these materials.

12.3.1
Monolithic Silica Columns with Physically Adsorbed Chiral Selector

A chiral selector can be immobilized onto the surface of monolithic silica by physical adsorption as well as by covalent attachment to it. Both of these strategies have their advantages and disadvantages. Physical adsorption is a rather fast and simple process. In addition, the amount of a chiral selector loaded onto silica can be easily varied and a chiral selector attached to a monolith by physical adsorption can be removed from the surface completely and if desirable to be replaced by another chiral selector.

The strategy of physical adsorption for the preparation of monolithic silica-based CSPs has been used only in a few studies in common-size [26] and capillary columns [27–29].

Enantioselective modification of commercially available Chromolith ® Silica C18 (15 × 4.6 mm) column *in situ* with cellulose tris(3,5-dimethylphenylcarbamate) dissolved in acetone was reported in 2003 [26]. The mobile phase flow rate in the range of 20 ml/min could be achieved with a common HPLC instrument in this column leading to enantioseparations within very short analysis times (Figure 12.1).

The same strategy has been extended to monolithic fused-silica capillary columns with 100-µm ID and baseline enantioseparations of several chiral chemicals (Figure 12.2) and pharmaceuticals has been achieved within a few minutes [27].

The advantage of using monolithic fused silica capillaries over particle-packed capillary columns is that the former does not need a frit. Frits are technically not always easy to fabricate and, what is more important, they may cause additional band broadening.

Later, a different polysaccharide-based chiral selector, namely amylose tris(3,5-dimethylphenylcarbamate), was used to modify monolithic silica *in situ* inside fused-silica capillaries [28]. Again, highly efficient separations with very short (subminute) separation times were reported (Figure 12.3).

Furthermore, Zou's group used the above-mentioned technology for coating cellulose tris(3,5-dimethylphenylcarbamate) onto monolithic silica and used the capillary columns for CEC separation of neutral and basic analytes in both aqueous and nonaqueous CEC [29].

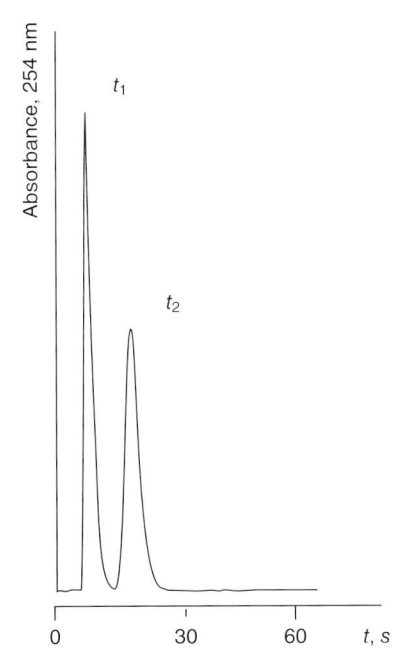

Figure 12.1 Fast separation of enantiomers of 2,2,2-trifluoro-1-(9-anthryl)ethanol on Chromolith ® RP18 silica column modified by coating with cellulose tris(3,5-dimethylphenylcarbamate) (Reproduced with permission from ref. [26]).

As was shown in several publications [26–29] the immobilization strategy by *in situ* physical adsorption of a chiral selector onto the surface of monolithic silica is, in principle, a useful strategy for a rather easy and fast preparation of monolithic CSPs for liquid-phase enantioseparations. The same studies showed that further optimization of the chemistry and morphology of the available monoliths is necessary in order to employ all potential advantages of silica-based monolithic chiral columns.

12.3.2
Monolithic Silica Columns with Covalently Attached Chiral Selector

Coated-type monolithic chiral columns have some disadvantages such as the limited selection of mobile phases and lower stability. Preparation of monolithic silica columns for liquid-phase enantioseparations by covalent attachment of a chiral selector has been studied over the last decade quite intensively. Various low and high molecular weight chiral selectors have been attached to monolithic silica (Table 12.1) [30–48] *in situ* inside common-size [39, 42, 43, 48] and capillary columns [30–38, 40, 41, 44, 45].

The covalent attachment of the chiral selector to monolithic silica is a more time- and technology-demanding multistep process. The surface of monolithic

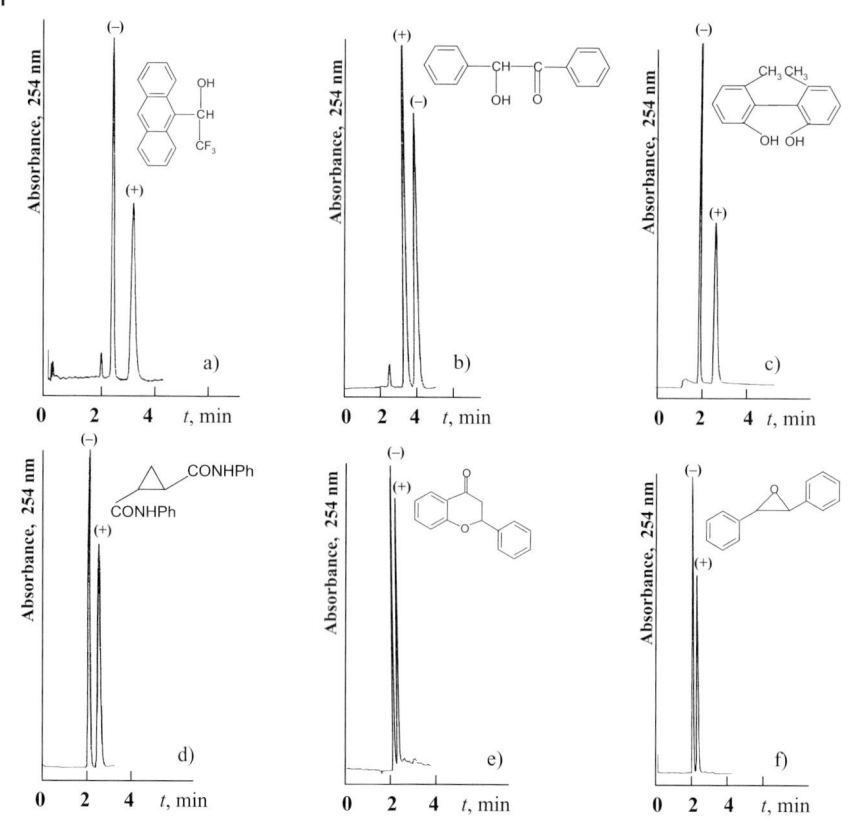

Figure 12.2 Separation of enantiomers of chiral chemicals on monolithic fused silica capillary column modified by coating with cellulose tris(3,5-dimethylphenylcarbamate) (Reproduced with permission from ref. [27]).

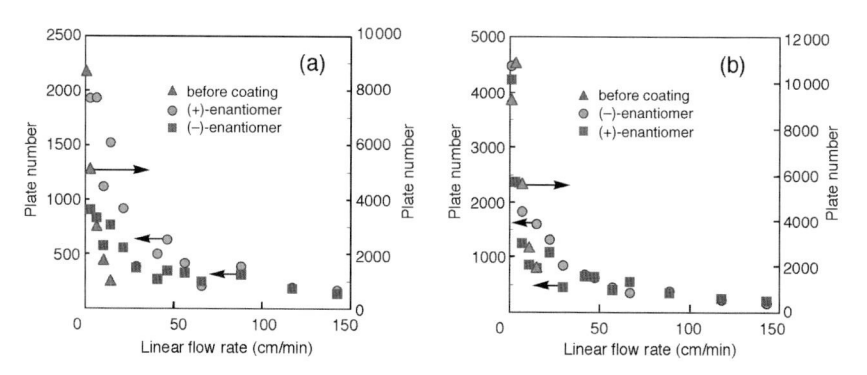

Figure 12.3 The dependence of the plate numbers on the linear flow rate of the mobile phase for 1,2,2,2-tetraphenylethanol (a) and 2,2′-dihydroxy-6,6′-dimethylbiphenyl (b) in unmodified monolithic fused silica capillary column and in the same capillary column modified with a 50-mg/ml amylase tris(3,5-dimethylphenylcarbamate) solution in chloroform (Reproduced with permission from ref. [28]).

Table 12.1 Monolithic silica columns with covalently attached chiral selector

Chiral selector	Immobilization method	Chiral Analytes	Separation mode	Ref.
β- and γ-Cyclodextrin	Activation of monolithic silica with (3-isocynatooropyl)triethoxysilane	Benzoin, DNS-amino acids	CEC	[30]
L-Amino acid amides	Activation of monolithic silica with γ-glycidoxypropyltrimethoxysilane	DNS-amino acids	Nano-LC, CEC	[31, 32]
L-Hydroxyproline	Activation of monolithic silica with γ-glycidoxypropyltrimethoxysilane	DNS-amino acids, free amino acids, hydroxyl acids, dipeptides	Nano-LC, CEC	[33]
(S)-N-(4-Allyloxy-3,5-dichlorobenzoyl)-2-amino-3,3-dimethylbutane phosphonic acid (cation exchanger)	Activation of monolithic silica with 3-mercaptopropyltrimethoxysilane and radical addition of vinyl-group containing monomer	Basic chiral drugs (Atenolol, benzetimide, celiprolol, clenbuterol, ephedrine, isoxsuprine, mefloquin-t-butylcarbamate, metoprolol, pronethalol, propranolol, sotalol, talinolol)	CEC	34, 46
O-9-(tert-Butylcarbamoyl) quinidine (anion exchanger)	Activation of monolithic silica with 3-mercaptopropyltrimethoxysilane and radical addition of vinyl-group containing monomer	Phospinic acid pseudodipeptide derivatives	CEC, Nano-LC	[35]
Bovine serum albumin (BSA)	Encapsulation of BSA on the sol-gel formation step	D,L-tryptophane	CEC	36–38
Cellulose-3,5-dimethylphenylcarbamate	Activation of monolithic silica with γ-glycidoxypropyltrimethoxysilane and polyasaccharide attachment by using BF_3 as a catalyst.	Benzoin, cobalt(III) tris(acetylacetonate), 2,2′-dihydroxy-6,6′-dimethylbiphenyl, flavanone, 2-phenylcyclohexanone, 2,2,2-trifluoro-1-(9-anthryl)ethanol, 1,2,2,2-tetraphenylethanol, Tröger's base.	HPLC	[39]

Table 12.1 Continued

Chiral selector	Immobilization method	Chiral Analytes	Separation mode	Ref.
Cellulose-3,5-dimethylphenylcarbamate, Amylose-3,5-dimethylphenylcarbamate, Cellulose-3,5-dichlorophenylcarbamate	AIBN-initiated radical copolymerization/crosslinking of preactivated polysaccharide derivatives	Tröger's base, trans-stilbene oxide; benzoin, 1,2,2,2-tetraphenylethanol, 2-phenylcyclohexanone; 2,2,2-trifluoro-1-(9-anthryl)ethanol, cobalt(III) tris(acetylacetonate), flavanone, trans-cyclopropandicarboxylic acid dianilide, 2,2'-dihydroxy-6,6'-dimethylbiphenyl.	Nano-LC	[40]
Cellulose-3,5-dimethylphenylcarbamate	Activation of monolithic silica with γ-glycidoxypropyltrimethoxysilane and polysaccharide attachment by using BF_3 as a catalyst	Benzoin, indapamide, pindolol, praziquantel, tetrahydropalmatine, Tröger's base	CEC	[41]
β-cyclodextrin	Activation of monolithic silica with 3-mercaptopropyltrimethoxysilane and AIBN initiated radical addition of the chiral selector	Chromakalin, Prominal, Oxazepam, Methadone and some other drugs	HPLC	[42]
tert-Butylcarbamoylquinine	Activation of monolithic silica with 3-mercaptopropyltrimethoxysilane and AIBN initiated radical addition of the chiral selector	N-derivatized amino acids (DNB- Ac, DNZ-, Bz-, Z-amino acids), Suprofen	HPLC	[43]
Aminosulfonic acid derived strong cation exchanger trans-(1S,2S)-2-(N-4-allyloxy-3,5-dichlorobenzoyl)amino cyclohexanesulfonic acid		Mefloquine, Mefloquine-O-t-butylcarbamate, pronethalol	CEC, nano-LC	44, 45
Human serum albumin		Triptophan, warfarin	CEC	[47]
Brush-type chiral selector	Copper-catalyzed azide-alkyne cycloaddition	N-(3,5-dinitrobenzoyl)amino acid dialkyl amide derivatives	HPLC	[48]

silica must be preactivated before the attachment of a chiral selector that also has to contain a reactive functional group. In a few cases difunctional chiral selectors can be crosslinked onto the surface of (monolithic) silica [40] or a monofunctional chiral selector can be crosslinked with auxiliary crosslinking reagents.

Although covalent attachment of polysaccharide derivative to silica surface involves additional steps compared to the modification by simple coating by physical adsorption, this approach may result in columns with higher peak efficiency as well as a wider coverage of chiral analytes due to an extended mobile-phase choice. Thus, during *in situ* coating of monolithic silica columns with a chiral selector, inhomogeneous films can be created on the silica surface. This problem becomes especially severe when the modification is performed with concentrated solutions of high molecular weight chiral selectors possessing high viscosity. The inhomogeneous distribution of a chiral selector inside the column may adversely affect both the peak symmetry and efficiency. In contrast to this, the silanizing bifunctional reagents are attached to the surface of silica from rather diluted solutions that typically results in a homogeneous distribution of activated functional groups on the surface of silica monoliths, where a chiral selector is attached during the following step. This may appear advantageous from the viewpoint of peak symmetry and efficiency.

The scheme of the covalent immobilization of cellulose 3,5-dimethylphenylcarbamate derivative *in situ* on the surface of monolithic silica in a 50 mm × 4.6 mm polyether ether ketone high-performance liquid chromatographic column is shown in Figure 12.4. The column obtained by this technology exhibited comparable enantiomer coverage but lower enantioseparation factors compared with the column of the same dimension but packed with wide pore silica (nominal pore size 100 nm) with the nominal particle diameter of 5 µm.

The lower separation factors observed with the monolithic column could be attributed to the lover content of the chiral selector in the case of the silica monolith. Additional contributions that must be considered in this particular case are the following: In the polysaccharide esters and phenylcarbamate derivatives used as chiral selectors in commercially available coated-type columns, hydroxyl groups in the structure of the polysaccharides are completely converted into ester or phenylcarbamate moieties. Such chiral selectors do not contain reactive functionalities required for their covalent attachment to the surface. Thus, in the chiral selectors in all covalently immobilized polysaccharide-based CSPs (particulate and monoliths) not all hydroxyl groups are converted to the ester or carbamate moieties but at least a small portion of them are used for the covalent linkage between the chiral selector and the silica. Thus, chiral selectors in coated-type and covalently immobilized polysaccharide-based CSPs are not absolutely identical. One additional point that may affect the column efficiency rather than the separation factor is the fact that commercially available monolithic silica contains mesopores in the range of 10–15 nm, while polysaccharide derivatives in particle-based CSPs are coated on a wide-pore silica (with the nominal pore diameter commonly in the range of 100 nm). The difference in the pore size between

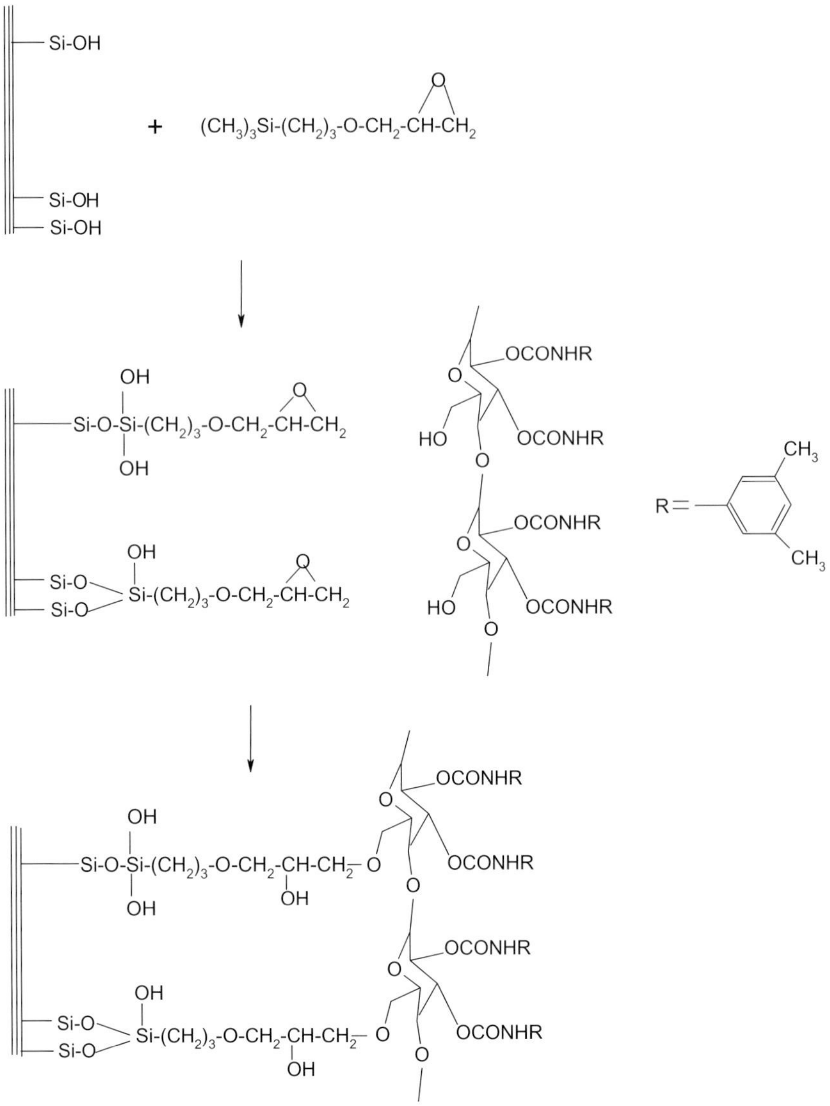

Figure 12.4 Immobilization schema of cellulose 3,5-dimethylphenylcarbamate on monolithic silica. (Reproduced with permission from ref. [39]).

particle-based and monolithic CSPs may not be as critical for low molecular weight chiral selectors as it is for high molecular weight CSPs, like polysaccharide derivatives.

As can be seen from the separation of the enantiomers of 2,2′-dihydroxy-6,6′-dimethylbiphenyl shown in Figure 12.5, monolithic silica columns with covalently immobilized polysaccharide derivatives show promising enantiomer-

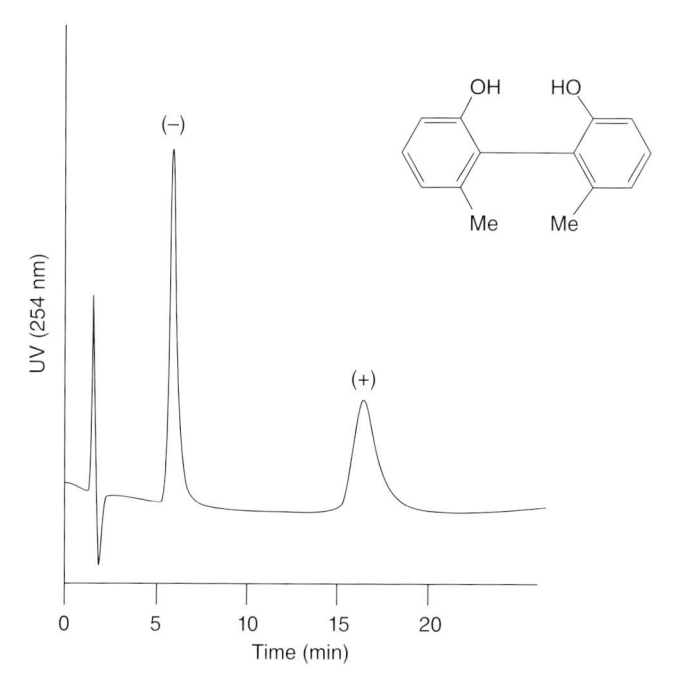

Figure 12.5 Enantioseparation of 2,2′-dihydroxy-6,6′-dimethylbiphenyl on monolithic Chromolith ® silica after covalent immobilization of 16.7% (w/w) of cellulose 3,5-dimethylphenylcarbamate. (Reproduced with permission from ref. [39]).

resolving ability and may be very useful for fast analytical, as well as for highly productive preparative-scale separations.

There are rather few examples published on the *in situ* modification of 4.6 ID monolithic columns [26, 39, 42, 43, 48].

In studies by Lubda *et al.* systematic comparisons between monolithic and particle-packed chiral columns have been performed using β-CD [42] and *tert*-butylcarbamoyl quinine [43] as chiral selectors. At the optimal flow rate, monolithic columns showed reduced plate height comparable to that of the column packed with 5-μm particles, but the advantage of a monolithic column became evident at the higher flow rates.

Svec and coworkers reported recently *in situ* activation of monolithic silica inside the commercially available 4.6 mm ID column with 3-(azydopropyl)trimethoxysilane and subsequently coupled an alkyne-containing brush-type chiral selector to it by using copper as a catalyst under mild reaction conditions [48]. Good separation factors and acceptable performance was observed towards π-acidic amino acid amide derivatives in the case of both monolithic silica (after endcapping) and 10-μm particulate silica used for comparison. However, before final endcapping 10-μm particulate silica performed better.

In all studies mentioned above in this subsection, monolithic silica columns commercially available from Merck [39, 42, 43] or Phenomenex [48] have been used as a starting columns for the modification. This is not the case for the studies performed on monolithic silica capillary columns. In most of these studies home-made silica monoliths have been used, but in a few cases the prototypes of the monolithic capillary columns later commercialized by Merck under the name Chromolith Si CapRod have been used [34, 35, 44, 45].

The studies performed in order to obtain enantioselective modification of monolithic silica columns illustrate interesting chemistry of immobilization [40] (Table 12.1) or clarify some fundamental aspects [45] of microscale separation technologies. However, the same studies clearly indicate serious limitations of the current technology, especially with regard to some chiral selectors. For instance, due to the required *in situ* activation of the monolith and attachment of the chiral selector it may not always be easy to immobilize a desired amount of the chiral selector by a single immobilization step. A multiple *in situ* coating and covalent immobilization of a chiral selector onto monolithic silica make it possible to adjust acceptable retention and separation factors for chiral analytes. However, a very critical issue is the significant decrease of separation efficiency with increasing chiral selector content in the packing material (Figure 12.6).

The most likely origin of this phenomenon seems to be the nonideal mesopore size, as well as other effects related to the morphology and chemistry of available monolithic silica. Solutions to this problem must be found in further studies. Otherwise, monolithic silica capillary columns at least those containing polysaccharide-based or other high molecular weight chiral selectors may not compete with packed columns due to the lower performance in enantiomer separation.

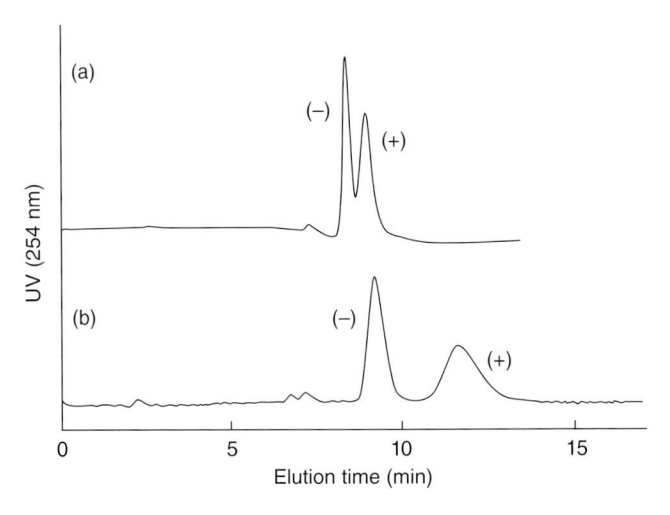

Figure 12.6 Enantioseparation of 2,2,2-trifluoro-1-(9-anthryl)ethanol after the first (a) and the second (b) covalent immobilization of CMDMPC onto monolithic silica. (Reproduced with permission from ref. [40]).

12.4
Summary of the Present State-of-the-Art and Problems to be Solved in the Future

As has been shown in numerous studies and summarized in recent review papers [9, 22, 49] on this and related topics, the preparation of organic and silica-based monolithic chiral columns for liquid-phase enantioseparation techniques is possible. It has also been shown that due to the lower flow resistance it is possible to perform faster enantioseparations with monolithic columns [26, 27]. Comparisons between particle-packed and monolithic columns (performed in the case of silica monoliths) indicate a comparable peak efficiency between monolithic silica and 5-μm [42, 43] or 10-μm [48] particles at the optimal flow rate, but some advantages of the monolithic materials at higher flow rates [42, 43].

The major problems when working with monolithic materials for enantioseparations are the following: (a) Chemical transformations (sometimes difficult and multistep) must be performed *in situ* inside the common diameter or even capillary chromatographic columns. These are not ideal vessels for mixing mass and heat transfer. (b) Materials from the viewpoint of morphology and chemistry need to be characterized also *in situ*. Both of these hindrances might be counterbalanced by the perfect performance of chiral monolithic columns for analytical and preparative separation of enantiomers. However, this state has not yet been reached. Thus, it is still a long time before one may see a competitive monolithic chiral column on the market. The researchers involved in the field and industry can contribute to this development by making achiral starting monolithic columns with various morphology (domain and mesopore size) and chemistries available for a lower price.

References

1 Agranat, I. (2004) Trends in the development of chiral drugs. *Drug Discov. Today*, **9**, 105–110.

2 Zhang, Y., Wu, D.-R., Wang-Iverson, D.B., and Tymiak, A.A. (2005) Enantioselective chromatography in drug discovery. *Drug Discov. Today*, **10**, 571–577.

3 Miller, L., and Potter, M. (2008) Preparative chromatographic resolution of racemates using HPLC and SFC in a pharmaceutical discovery environment. *J. Chromatogr. B*, **875**, 230–236.

4 Belder, D., Ludwig, M., Wang, L.-W., and Reetz, M.T. (2006) Enantioselective catalysis and analysis on a chip. *Angew. Chem. Int. Ed.*, **45**, 2463–2466.

5 He, B.L., Shi, Y., Kleintop, B., and Raglione, T. (2008) Direct and indirect separation of five isomers of Brivanib Alaninate using chiral high-performance liquid chromatography. *J. Chromatogr. B*, **875**, 122–135.

6 Zhang, Y., Caporuscio, C., Dai, J., Witkus, M., Rose, A., Santella, J., D'Arenzio, C., Wang-Iverson, D.B., and Tymiak, A.A. (2008) Development and implementation of a stereoselective normal-phase liquid chromatography-tandem mass spectrometry method for the determination of intrinsic metabolic clearance in human liver microsomes. *J. Chromatogr. B*, **875**, 154–160.

7 Tanaka, N., Nagayama, H., Kobayashi, H., Ikegami, T., Hosoya, K., Ishizuka, N., Minakuchi, H., Nakanishi, K., Cabrera, K., and Lubda, D. (2000) Monolithic silica columns for HPLC, micro-HPLC, and CEC. *J. High Resolut. Chromatogr.*, **23**, 111–116.

8 Svec, F., Tennikova, T.B., and Deyl, Z. (eds) (2003) *Monolithic Materials: Preparation, Properties and Applications, J. Chromatogr. Library, V. 67*, Elsevier, Amsterdam.

9 Preinerstorfer, B., and Lämmerhofer, M. (2007) Recent accomplishments in the field of enantiomer separation by CEC. *Electrophoresis*, **28**, 2527–2565.

10 Peters, E.C., Lewandowski, K., Petro, M., Svec, F., and Fréchet, J.M. (1998) Chiral electrochromatography with a "moulded" rigid monolithic capillary column. *J. Anal. Commun.*, **35**, 83–86.

11 Lämmerhofer, M., Svec, F., Fréchet, J.M.J., and Lindner, W. (2000) Chiral monolithic columns for enantioselective capillary electrochromatography prepared by copolymerization of a monomer with quinidine functionality. 2. Effect of chromatographic conditions on the chiral separations. *Anal. Chem.*, **72**, 4623–4628.

12 Lämmerhofer, M., Tobler, E., Zarbl, E., and Lindner, W. (2003) Macroporous monolithic chiral stationary phases for capillary electrochromatography: New chiral monomer derived from cinchona alkaloid with enhanced enantioselectivity. *Electrophoresis*, **24**, 2986–2999.

13 Lämmerhofer, M. (2005) Chirally-functionalized monolithic materials for stereoselective capillary electrochromatography. *Anal. Bioanal. Chem.*, **382**, 873–877.

14 Preinerstorfer, B., Bicker, W., Lindner, W., and Lämmerhofer, M. (2004) Development of reactive thiol-modified monolithic capillaries and in-column surface functionalization by radical addition of a chromatographic ligand for capillary electrochromatography. *J. Chromatogr. A*, **1044**, 187–199.

15 Preinerstorfer, B., Lindner, W., and Lämmerhofer, M. (2005) Polymethacrylate-type monoliths functionalized with chiral amino phosphonic acid-derived strong cation exchange moieties for enantioselective nonaqueous capillary electrochromatography and investigation of the chemical composition of the monolithic polymers. *Electrophoresis*, **26**, 2005–2018.

16 Messina, A., Flieger, M., Bachechi, F., and Sinibaldi, M. (2006) Enantioseparation of 2-aryloxypropionic acids on chiral porous monolithic columns by capillary electrochromatography. Evaluation of column performance and enantioselectivity. *J. Chromatogr. A*, **1120**, 69–74.

17 Schmid, M.G., Grobuschek, N., Tuscher, C., Gübitz, G., Végvári, A., Machtejevas, E., Maruka, A., and Hjertén, S. (2000) Chiral separation of amino acids by ligand-exchange capillary electrochromatography using continuous beds. *Electrophoresis*, **21**, 3141–3144.

18 Schmid, M.G., Grobuschek, N., Lecnik, O., Gübitz, G., Végvári, Á., and Hjertén, S. (2001) Enantioseparation of hydroxy acids on easy-to-prepare continuous beds for capillary electrochromatography. *Electrophoresis*, **22**, 2616–2619.

19 Kornysova, O., Owens, P.K., and Maruska, A. (2001) Continuous beds with vancomycin as chiral stationary phase for capillary electrochromatography. *Electrophoresis*, **22**, 3335–3338.

20 Machtejevas, E., and Maruška, A. (2002) A new approach to human serum albumin chiral stationary phase synthesis and its use in capillary liquid chromatography and capillary electrochromatography. *J. Sep. Sci.*, **25**, 1303–1309.

21 Ansell, R.J. (2005) Molecularly imprinted polymers for the enantioseparation of chiral drugs. *Adv. Drug Deliv. Rev.*, **57**, 1809–1835.

22 Gübitz, G., and Schmid, M.G. (2004) Recent advances in chiral separation principles in capillary electrophoresis and capillary electrochromatography. *Electrophoresis*, **25**, 3981–3996.

23 Siouffi, A.M. (2003) Silica gel-based monoliths prepared by the sol-gel method: Facts and figures. *J. Chromatogr. A*, **1000**, 801–818.

24 Kobayashi, H., Ikegami, T., Kimura, H., Hara, T., Tokuda, D., and Tanaka, N. (2006) Properties of monolithic silica columns for HPLC. *Anal. Sci.*, **22**, 491–501.

25 Guiochon, G. (2007) Monolithic columns in high-performance liquid chromatography. *J. Chromatogr. A*, **1168**, 101–168.

26 Chankvetadze, B., Yamamoto, C., and Okamoto, Y. (2003) Very fast enantioseparations in HPLC using cellulose tris(3,5-dimethylphenylcarbamate) as chiral stationary phase. *Chem. Lett.*, **32**, 850–851.

27 Chankvetadze, B., Yamamoto, C., Tanaka, N., Nakanishi, K., and Okamoto, Y. (2004) Enantioseparations on monolithic silica capillary column modified with cellulose tris(3,5-dimethylphenylcarbamate). *J. Sep. Sci.*, **27**, 905–911.

28 Chankvetadze, B., Yamamoto, C., Kamigaito, M., Tanaka, N., Nakanishi, K., and Okamoto, Y. (2006) High-performance liquid chromatographic enantioseparations on capillary columns containing monolithic silica modified with amylose tris(3,5-dimethylphenylcarbamate). *J. Chromatogr. A*, **1110**, 46–52.

29 Qin, F., Xie, C., Feng, S., Qu, J., Kong, L., Ye, M., and Zou, H. (2006) Monolithic silica capillary column with coated cellulose tris(3,5-dimethylphenyl-carbamate)for capillary electrochoromatographic separation of enantiomers. *Electrophoresis*, **27**, 1050–1059.

30 Chen, Z., Ozawa, H., Uchiyama, K., and Hobo, T. (2003) Cyclodextrin-modified monolithic columns for resolving dansyl amino acid enantiomers and positional isomers by capillary electrochromatography. *Electrophoresis*, **24**, 2550–2558.

31 Chen, Z., Niitsuma, M., Uchiyama, K., and Hobo, T. (2003) Comparison of enantioseparations using Cu(II) complexes with L-amino acid amides as chiral selectors or chiral stationary phases by capillary electrophoresis, capillary electrochromatography and micro liquid chromatography. *J. Chromatogr. A*, **990**, 75–82.

32 Chen, Z., and Hobo, T. (2001) Chemically L-phenylalaninamide-modified monolithic silica column prepared by a sol–gel process for enantioseparation of dansyl amino acids by ligand exchange-capillary electrochromatography. *Anal. Chem.*, **73**, 3348–3357.

33 Chen, Z., Nishiyama, T., Uchiyama, K., and Hobo, T. (2004) Electrochromatographic enantioseparation using chiral ligand exchange monolithic sol-gel column. *Anal. Chim. Acta*, **501**, 17–23.

34 Preinerstorfer, B., Lubda, D., Lindner, W., and Lämmerhofer, M. (2006) Monolithic silica-based capillary column with strong chiral cation-exchange type surface modification for enantioselective non-aqueous capillary electrochromatography. *J. Chromatogr. A*, **1106**, 94–105.

35 Preinerstorfer, B., Lubda, D., Mucha, A., Kafarski, P., Lindner, W., and Lämmerhofer, M. (2006) Stereoselective separations of chiral phosphinic acid pseudodipeptides by CEC using silica monoliths modified with an anion-exchange-type chiral selector. *Electrophoresis*, **27**, 4312–4320.

36 Kato, M., Matsumoto, N., Sakai-Kato, K., and Toyo'oka, T. (2003) Investigation of chromatographic performances and binding characteristics of BSA-encapsulated capillary column prepared by the sol-gel method. *J. Pharm. Biomed. Anal.*, **30**, 1845–1850.

37 Sakai-Kato, K., Kato, M., Nakakuki, H., and Toyo'oka, T. (2003) Investigation of structure and enantioselectivity of BSA-encapsulated sol-gel columns prepared for capillary electrochromatography. *J. Pharm. Biomed. Anal.*, **31**, 299–309.

38 Kato, M., Saruwatari, H., Sakai-Kato, K., and Toyo'oka, T. (2004) Silica sol-gel/organic hybrid material for protein encapsulated column of capillary electrochromatography. *J. Chromatogr. A*, **1044**, 267–270.

39 Chankvetadze, B., Ikai, T., Yamamoto, C., and Okamoto, Y. (2004) High-performance liquid chromatographic enantioseparations on monolithic silica column containing covalently attached 3,5-dimethylphenylcarbamate derivative of cellulose. *J. Chromatogr. A*, **1042** (1–2), 55–60.

40 Chankvetadze, B., Kubota, T., Ikai, T., Yamamoto, C., Tanaka, N., Nakanishi, K., and Okamoto, Y. (2006) High-performance liquid chromatographic enantioseparations on capillary columns containing crosslinked polysaccharide phenylcarbamate derivatives attached to monolithic silica. *J. Sep. Sci.*, **29**, 1988–1995.

41 Dong, X., Wu, R., Dong, J., Wu, M., Zhu, Y., and Zou, H. (2008) The covalently bonded cellulose tris(3,5-dimethylphenylcarbamate) on a silica monolithic capillary column for enantioseparation in capillary electrochromatography. *J. Chromatogr. B*, **875**, 317–322.

42 Lubda, D., Cabrera, K., Nakanishi, K., and Lindner, W. (2003) Monolithic silica columns with chemically bonded β-cyclodextrin as a stationary phase for enantiomer separations of chiral pharmaceuticals. *Anal. Bioanal. Chem.*, **377**, 892–901.

43 Lubda, D., and Lindner, W. (2004) Monolithic silica columns with chemically bonded *tert*-butylcarbamoylquinine chiral anion-exchanger selector as a stationary phase for enantiomer separations. *J. Chromatogr. A*, **1036**, 135–143.

44 Calleri, E., Massolini, G., Lubda, D., Temporini, C., Loiodice, F., and Caccialanza, G. (2004) Evaluation of a monolithic epoxy silica support for penicillin G acylase immobilization. *J. Chromatogr. A*, **1031**, 93–100.

45 Preinerstorfer, B., Hoffmann, C., Lubda, D., Lämmerhofer, M., and Lindner, W. (2008) Enantioselective silica-based monoliths modified with a novel aminosulfonic acid-derived strong cation exchanger for electrically driven and pressure-driven capillary chromatography. *Electrophoresis*, **29**, 1626–1637.

46 Preinerstorfer, B., Lämmerhofer, M., Hoffmann, C., Lubda, D., and Lindner, W. (2008) Deconvolution of electrokinetic and chromatographic contributions to solute migration in stereoselective ion-exchange capillary electrochromatography on monolithic silica capillary columns. *J. Sep. Sci.*, **31**, 3065–3078.

47 Mallik, R., and Hage, D.S. (2009) Development of an affinity silica monolith containing human serum albumin for chiral separations. *J. Pharm. Biomed. Anal.*, **46**, 820–830.

48 Slater, M.D., Fréchet, J.M.J., and Svec, F. (2009) In-column preparation of a brush-type chiral stationary phase using click chemistry and a silica monolith. *J. Sep. Sci.*, **32**, 21–28.

49 Qin, F., Xie, C., Yu, Z., Kong, L., Ye, M., and Zou, H. (2006) Monolithic enantioselective stationary phases for capillary electrochromatography. *J. Sep. Sci.*, **29**, 1332–1343.

13
High-Speed and High-Efficiency Separations by Utilizing Monolithic Silica Capillary Columns

Takeshi Hara, Kosuke Miyamoto, Satoshi Makino, Shohei Miwa, Tohru Ikegami, Masayoshi Ohira, and Nobuo Tanaka

13.1
Introduction

In the development of HPLC equipment and the columns, an increase in the number of theoretical plates and a decrease in the separation time were desired, and achieved simultaneously by using a column packed with smaller particles. Starting from 50–60-cm columns packed with 35–50-μm particles in the 1970s, 25–30-cm columns of 10-μm particles, 15–25-cm columns of 5-μm particles, 10–15-cm columns of 3-μm particles, then 5–10-cm columns of about 2-μm particles have been provided. However, the increase in the number of theoretical plates that can be utilized for actual separations has been rather modest compared to the performance limit expected [1], because of the use of a shorter columns packed with the smaller particles. The trend was dictated by the principle of economy (the considerations of time and solvent consumption) or by the technological requirement associated with such columns (the need for higher pressure for smaller particles) as well as the issue of detection sensitivity. Increased consumption of solvent with a long column will not be a problem for capillary HPLC, although long analysis time could be a problem, if it is excessive.

It seems that the particle size of the next-generation columns has not become clear yet, although it seems that it would be smaller than 3 μm. The production of commercial columns packed with particles of various sizes around 2 μm typically shows the issue of instrumental requirement for fast separations. Contrary to the previous cases for columns of 10-, 5-, or 3-μm particles, recent high-efficiency commercial columns are packed with particles of a variety of size specification, depending on the policy of each manufacturer as to which pressure range to aim at between conventional HPLC and ultrahigh-pressure liquid chromatography. Actually, columns packed with 1.5, 1.7, 1.8, 2.0, 2.1, 2.2, 2.3, 2.5, 2.6, 2.7, or 2.8 μm are commercially available. There have also been a few recent reports showing the interest in generating a large number of theoretical plates [2, 3], but the study was limited to about one hundred thousand theoretical plates.

Monolithic Silicas in Separation Science. Edited by K.K. Unger, N. Tanaka, and E. Machtejevas
© 2011 WILEY-VCH Verlag GmbH & Co. KGaA, Weinheim
ISBN: 978-3-527-32575-7

The instrumental requirement is more favorable for monolithic materials [4]. The development of high-performance columns having controlled domain structures has been successful with capillary-type monoliths [5–10]. Monolithic columns can be in competition for high-speed separations, because column efficiency equivalent to 2–2.5-µm particles was reported with a pressure drop equivalent to 5-µm particles [8]. The capillary column preparation (below 530 µm) is more straightforward than the preparation of a column of 1 mm ID or larger, especially for long ones. This is because the preparation in a capillary can form monolithic silica structures covalently attached to the tube wall for the column that can be used without cladding.

The advantages of monolithic silica capillary columns include the high permeability associated with large-sized through-pores and high-efficiency provided by small-sized skeletons. Another feature of monolithic silica capillary column is facile modification of silica surfaces resulting in various types of stationary phases without much loss of column efficiency [11–13]. Cocontinuous structures producing a large number of theoretical plates can be maintained through modification reactions, unless the formed stationary phase, often polymer chains, hinders solute mass transfer. There have been many reports on modifications of silica surfaces for the use in HPLC and capillary electrochromatography [11] including reversed-phase [14–21], ion-exchange [22, 23], HILIC [24–26], and chiral separations [27–29]. Chapters 2 and 3 describe the preparation and modification of a monolithic silica column. This chapter provides an account for the current status of monolithic silica support as a form of a reversed-phase capillary column for high efficiency, then that of a column for high-speed separations in HPLC.

13.2
Preparation of Monolithic Silica Capillary Columns

Two kinds of monolithic silica capillary columns have been prepared; (i) a hybrid structure (abbreviated as MS-H) from a mixture of tetramethoxysilane (TMOS) and methyltrimethoxysilane (MTMS) (v/v 3:1) [7, 9], and (ii) a common silica structure from TMOS only (MS) [5, 6, 8]. Columns of 50–200 µm ID were prepared from a mixture of TMOS and MTMS. Successful preparation of a 530-µm column was also reported [30]. The use of TMOS as a starting material allowed the preparation of 50–100 µm I.D columns, while a 200-µm column could be prepared under certain conditions [8]. Generally speaking, it is easier to prepare the MS-H (hybrid) columns, although the product is slightly less efficient than MS columns.

Table 13.1 shows typical feed compositions for the preparation of long MS-H columns (the column number is followed by the initials of the investigator for the preparation of long capillary columns in Section 13.4) [9]. Depending on a batch of poly(ethylene glycol) (PEG), slight adjustment of feed composition is required. The fused-silica capillary tube was first treated with 1 M NaOH at 40 °C for 3 h, followed by a flush with water, and then kept with 1 M HCl at 40 °C for

13.2 Preparation of Monolithic Silica Capillary Columns

Table 13.1 Feed composition for the preparation of long monolithic silica capillary columns.

Column No.	PEG (g)	TMOS (ml)	TMOS+MTMS (ml)	Urea (g)	AcOH[f] (ml)	Temp. (°C)
MS-50-(1HK)[a]	12.6[c]	40		9.0	100	30
MS-200H-(2DT)[b]	2.05[c]		18	4.05	40	40
MS-200H-(3HK)[b]	1.85[d]		18	4.05	40	40
MS-100H-(4TH,5KK,7ON)[b]	1.80[d]		18	4.05	40	40
MS-100H-(6SM)[b]	1.84[d]		18	4.05	40	40
MS-100H-(8KM)[b]	1.85[e]		18	4.05	40	40
MS-100H-(9KM,10KM, 11KM, 12KM)[b]	1.80[e]		18	4.05	40	40

a) Prepared from TMOS.
b) Prepared from a TMOS-MTMS (3:1 v/v) mixture.
c–e) Indicates different batches of poly(ethylene glycol), MW = 10 000.
f) 0.01 N Acetic acid aqueous solution.
Reproduced from Ref. [9] with permission. Copyright 2008 American Chemical Society.

2 h. After a flush with water, and then with acetone, the capillary tube was air dried at 40 °C.

A TMOS/MTMS mixture (18 ml) was added to a solution of PEG (MW 10 000, 1.80 g) and urea (4.05 g) in 0.01 M acetic acid (40 ml) at 0 °C and stirred for 30 min. The homogeneous solution was then stirred for 10 min at 40 °C, filtered with a 0.45-μm PTFE filter, charged into a fused-silica capillary tube, and allowed to react at 40 °C. The resultant gel was subsequently aged in the capillary overnight at the same temperature. Then, the temperature was raised slowly (over 10–20 h for long capillary columns), and the monolithic silica columns were treated for 4 h at 120 °C to form mesopores with the ammonia generated by the hydrolysis of urea, then cooled and washed with methanol. After air drying, the column was heat treated at 330 °C for 25 h to cause the pyrolysis of organic components in the column. The surface modification (octadecylsilylation) of the silica monoliths can be carried out by reacting with N,N-diethylaminodimethyloctadecylsilane (ODS-DEA) [31]. Endcapping can be carried out with HMDS or trimethylsilylimidazole [32].

Capillary HPLC was easily performed by employing a split flow/injection HPLC system consisting of a conventional pump and an injector with a splitting T-joint [5]. A UV detector with an optical unit including a capillary flow cell (volume: 2–18 nl) connected to an electronic unit with optical fibers, provides a convenient and sensitive means of detection. A microvalve injector (injection volume 10–50 nl) can also be used with a micropump without flow splitting. When two or three capillary columns were connected to form a long-column system, either a stainless steel or PEEK union using graphite Vespel ferrules or a clear connector can be used. The latter resulted in better efficiency in most cases based on easy butt connection. (See reference [9] for experimental details.)

13.3
Properties of Monolithic Silica Capillary Columns

The preparation method of monolithic silicas by a sol-gel reaction accompanied by phase separation (Chapter 2) typically results in cocontinuous structures (Figure 13.1a) consisting of 1–5-μm skeletons and through-pores [33, 34]. The combined size of a skeleton and a through-pore is called a domain size. The through-pores can be maintained at a relatively large size even with small-sized skeletons, so that the column can be operated under relatively low pressure. This is a unique feature associated with a monolithic material. Thus, monolithic silica columns can simultaneously provide high column efficiency and low pressure compared to a particle-packed column [35, 36].

Monolithic silica columns first commercialized in 2000, possessing ca. 2-μm through-pores and 1.3-μm skeletons [37], show column efficiency (a plate height

Figure 13.1 Scanning electron micrographs of monolithic silica columns. (a) Monolithic silica prepared from TMOS in a test tube, and those prepared from a mixture of TMOS and MTMS, (b) in 50 μm, (c) in 100 μm, and (d) in 200 μm fused silica capillary tube. The arrows indicate the through-pore size and the skeleton size (reproduced from Ref. [7] with permission. Copyright 2002 Elsevier).

13.3 Properties of Monolithic Silica Capillary Columns

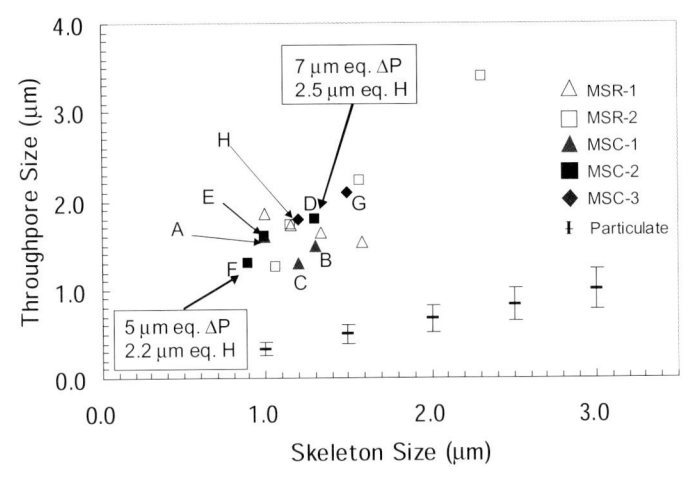

Figure 13.2 Plots of through-pore size versus skeleton size for monolithic silica columns prepared in a test tube (MSR-1 and MSR-2 reported in Refs. [35, 36]), monolithic silica columns (MSC reported in Ref. [8]) prepared in a capillary, and for particle-packed columns. Through-pore-size/skeleton-size ratios in a range 0.25–0.4 are indicated as bars for the particle-packed columns. Reproduced from reference [34] with permission. Copyright 2007 American Chemical Society.

H, or the number of theoretical plates N) similar to a column packed with 3–4-μm particles with pressure drop equivalent to 8–10-μm particles [4, 38, 39]. Such performance is commonly obtained for the capillary columns prepared as above (Figures 13.1b–d). A slight change in the PEG concentration can cause significant differences in the column properties for some batches of PEG. In general, an increase in the PEG concentration of the feed resulted in smaller domain structures, or a column with lower permeability and higher column efficiency.

Figure 13.2 shows the plot of the through-pore size against the size of silica skeleton for monolithic silica columns prepared from TMOS in a test tube (MSR) [35, 36] or in a capillary (MSC) [6–8]. Through-pore-size/skeleton-size ratios in a range 1–2 have been obtained for all the monolithic silica columns that are much larger than those of particle-packed columns at 0.25–0.4 (the range indicated as bars in Figure 13.2) [40]. The particle size dictates the size of interstitial voids for a particle-packed column, while the size ratios can be varied for monolithic columns. Superficially porous silica particles possess somewhat similar features as monolithic materials with respect to the ratio of the size of through-pores to that of a porous layer [41, 42]. Smaller skeletons will produce higher column performance both for a fast and a high-efficiency separation owing to the shorter diffusion path length. Optimum domain size should be found according to the number of theoretical plates desired, just like an optimum particle size for a particulate column [43, 44].

In contrast to the current strategy of manufacturers of particulate columns aiming at tens of thousands theoretical plates at high speed, monolithic silica columns may prove that it is possible to develop two types of advanced materials. (1) Columns for high-speed separations aiming at the performance of sub-2-µm particles (large numbers of theoretical plates, N, with a short elution time, t_0, that is large N/t_0), and (2) those for high-efficiency separations (large N) leading to the generation of hundreds of thousand to one million theoretical plates with the capillary format of monolithic silica columns.

13.4
Monolithic Silica Capillary Columns for High-Efficiency Separations

13.4.1
Performance of Long Monolithic Silica Capillary Columns

Table 13.2 shows a summary of chromatographic properties of the columns shown in Table 13.1 [9]. Typically, a 130–150-cm column can produce 150 000–180 000 theoretical plates in 80% acetonitrile mobile phase for alkylbenzenes with a pressure drop of 7–8 MPa and a t_0 of 15–20 min. Optimum performance was observed at a linear velocity (u) of about 1.5 mm/s in 80% acetonitrile to yield a plate height (H) of about 8–10 µm, and at about 1 mm/s in 80% methanol to give a $H_{min} = 10-12$ µm. As shown in Table 13.2, the columns showed separation impedance (E values) of 500–600, while shorter columns showed smaller E values, or slightly better performance. It has been difficult to prepare a long capillary column with a homogeneous monolithic silica structure along its entire

Table 13.2 Chromatographic properties of long monolithic silica capillary columns prepared from a TMOS-MTMS (3:1 v/v) mixture.

Column No.	L cm	u mm/s	N^a ×10^4	H µm	P MPa	K ×10^{-14} m^2	E
MS-50-C$_{18}$ (1HK)	254[b](260)[c]	2.0	22.6	11.2	20	11.7	1070
MS-200H-C$_{18}$ (2DT)	253[b](260)[c]	1.1	31.3	8.1	12	10.7	610
MS-200H-C$_{18}$ (3HK)	270[b](276)[c]	1.0	30.0	9.0	7.8	11.8	740
MS-100H-C$_{18}$ (4TH)	88[b](94)[c]	1.0	12.5	7.0	5.9	7.2	680
MS-100H-C$_{18}$ (5KK)	133[b](143)[c]	1.0	16.3	8.2	4.8	13.1	510
MS-100H-C$_{18}$ (6SM)	120[b](128)[c]	1.1	15.9	7.5	4.3	14.5	390
MS-100H-C$_{18}$ (7ON)	130[b](138)[c]	1.1	15.9	8.2	4.9	13.7	490
MS-100H-C$_{18}$ (8KM)	442[b](448)[c]	1.1	54.7	8.1	16.1	13.8	480
MS-100H-C$_{18}$ (9KM)	442[b](448)[c]	1.1	45.4	9.7	13.5	16.5	570
MS-100H-C$_{18}$ (10KM)	342[b](348)[c]	1.1	30.3	11.3	7.6	22.8	560

a) Number of theoretical plates measured for hexylbenzene as a solute in 80% acetonitrile.
b) Effective length (cm).
c) Total length (cm).

Reproduced from Ref. [9] with permission. Copyright 2008 American Chemical Society.

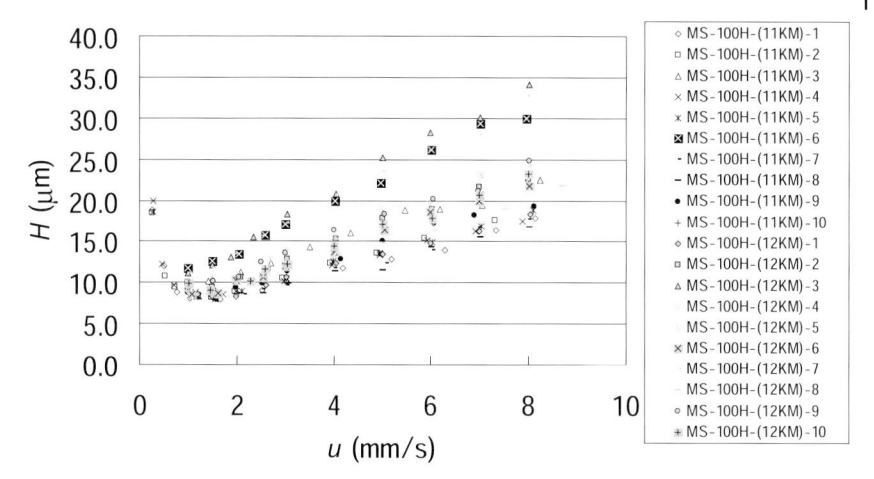

Figure 13.3 Plots of plate height (H) values against linear velocity of mobile phase (u) obtained for twenty columns prepared with the same feed composition. Mobile phase: acetonitrile–water (80/20). Temperature: 30°C. Sample: hexylbenzene. Detection: 210 nm. Column: MS(100)-H–C_{18} (11KM) and Column: MS(100)-H–C_{18} (12KM) reported in Ref. [9]. Reproduced from Ref. [9] with permission. Copyright 2008 American Chemical Society.

length. Sometimes, the column efficiency can be improved by cutting off a 10–20-cm piece at the end.

Figure 13.3 shows van Deemter plots for two batches of ten 150-cm MS(100)-H columns prepared with the same feed composition. Fifteen out of twenty columns gave H values of 8.3 ±0.5 µm at a linear velocity of ca. 1.5 mm/s and separation impedance 500 ±65, in 80% acetonitrile, while all columns showed permeability $K = (1.45 \pm 0.09) \times 10^{-13}\,m^2$. Only nine columns out of the twenty, however, showed H below 20 µm at $u = 8$ mm/s. The results indicated that the reproducibility of monolithic silica capillary columns prepared in a laboratory is poorer than that of commercially available monolithic silica-rod columns [45]. The retention factors were found to be more reproducible with a relative standard deviation of about 2%.

The plots of $\log(t_0/N^2)$ values against $\log(N)$ are shown in Figure 13.4, for the twenty columns evaluated in Figure 13.3 at a 40-MPa pressure limit [43]. (The dashed lines indicate t_0 values required for the generation of a specified number of theoretical plates.) Nine monolithic columns were found to give similar performance over the range, $5 < \log(N) < 6$, presumably indicating the limit of performance for the columns prepared under the conditions described in Section 13.2. The results indicated that 100 000 theoretical plates can be generated with a t_0 of about 250–300 s, 300 000 plates with a t_0 of about 1000 s, and that the columns can give optimum performance at around 1 000 000 theoretical plates with a t_0 of 5000–7500 s. The results suggested the possibility of utilizing the column efficiency over a range of $N = 100\,000$–$1\,000\,000$ for practical separations

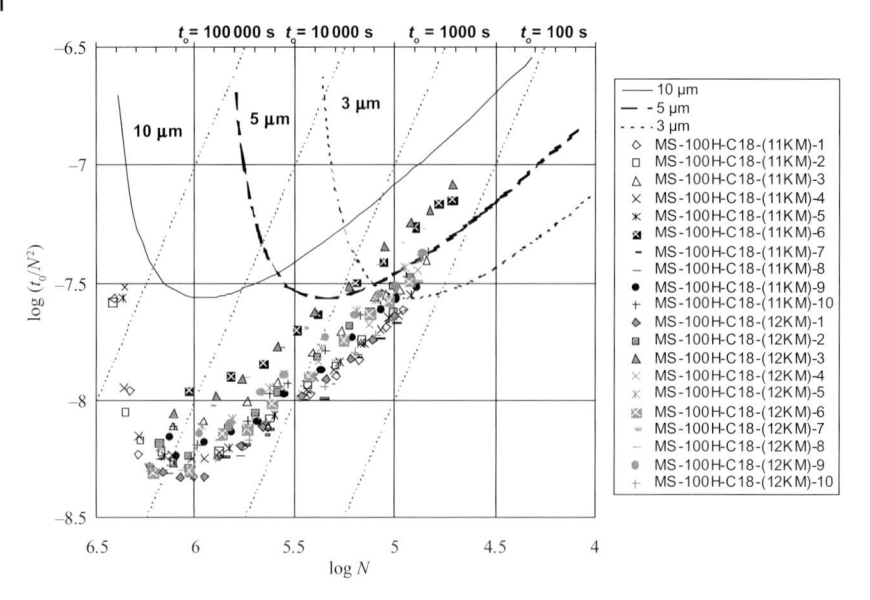

Figure 13.4 Plots of $\log(t_0/N^2)$ against $\log(N)$ for the monolithic silica C_{18} capillary columns evaluated in Figure 13.3. The plots were obtained at a pressure drop of 40 MPa, and by assuming a flow resistance parameter $\Phi = 700$, the viscosity of the mobile phase $\eta = 0.00046$ Pa s, a diffusion coefficient of the solute $D_m = 2.22 \times 10^{-9}$ m^2/s, and a Knox equation $h = 0.65 v^{1/3} + 2/v + 0.08\, v$ for particle-packed columns. The particle diameters (d_p) for the particle-packed columns are 3, 5, and 10 µm. The dotted lines indicate the required t_0 values in seconds. Reproduced from reference [9] with permission. Copyright 2008 American Chemical Society.

using a conventional HPLC instrument. Figure 13.5 shows an example of the performance.

13.4.2
Examples of High-Efficiency Separations

13.4.2.1 Isocratic Mode

It is possible to connect capillary columns in series with a union to make a 10–15-m column system without significant loss of separation efficiency. Figure 13.6a shows the separation of alkylbenzenes on a 1140-cm column system consisting of three columns. At 35 MPa, the column system produced 1 350 000 theoretical plates for the t_0 peak and 1 000 000 theoretical plates for alkylbenzenes with k values of up to 2.1. Smaller numbers of theoretical plates were obtained for alkylbenzenes retained longer, about 800 000 theoretical plates for decylbenzene of $k = 5.6$.

Large numbers of theoretical plates were also observed for polynuclear aromatic hydrocarbons (PAHs). A 1244-cm column system showed good performance for

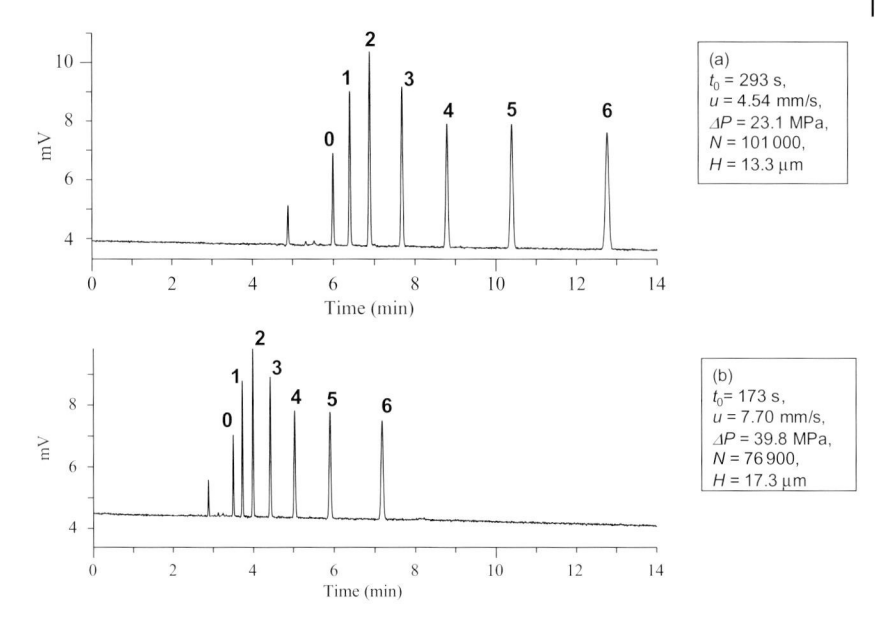

Figure 13.5 Performance of long monolithic silica capillary columns prepared from a TMOS-MTMS (3:1 v/v) mixture. Column: Silica C_{18} column MS(100)-H-C_{18} (5KK). Effective length 133 cm (total length 143 cm). Mobile phase: acetonitrile–water (80/20). Detection: 210 nm. Temperature: 30 °C. Sample: uracil, alkylbenzenes ($C_6H_5C_nH_{2n+1}$, $n=0$–6). Reproduced from Ref. [9] with permission. Copyright 2008 American Chemical Society.

(a) $t_0 = 293$ s, $u = 4.54$ mm/s, $\Delta P = 23.1$ MPa, $N = 101\,000$, $H = 13.3$ µm

(b) $t_0 = 173$ s, $u = 7.70$ mm/s, $\Delta P = 39.8$ MPa, $N = 76\,900$, $H = 17.3$ µm

the separation of 16 PAHs (Figure 13.6b). The column system generated 1 000 000 theoretical plates for PAHs with retention factors (k values) of up to 2.4. The decrease in column efficiencies for solutes with larger k values was shown to be much less than for open tubular columns [46–48], that previously provided successful examples of achieving 10^6 theoretical plates. This is presumably due to much smaller through-pore size of about 2 µm than the diameter of the open tubular columns 11–50 µm.

Another example is an isotopic separation of the benzene isotopologues, benzene, benzene-d, benzene-1,3,5-d_3, benzene-d_6, with a 850-cm capillary column system (350 cm + 500 cm) prepared from a mixture of MTMS and TMOS at a 9/2 ratio, shown in Figure 13.6c. This chromatogram seems to show the resolution of all the H/D isotopologues of benzene contained in the samples purchased as the isotopologues indicated. The use of a high-efficiency column is an easy way to resolve a mixture with small separation factors, 1.0070–1.0072 in this case, and will be effective for resolution of closely related compounds including isotopologues that always resulted in small separation factors [49, 50].

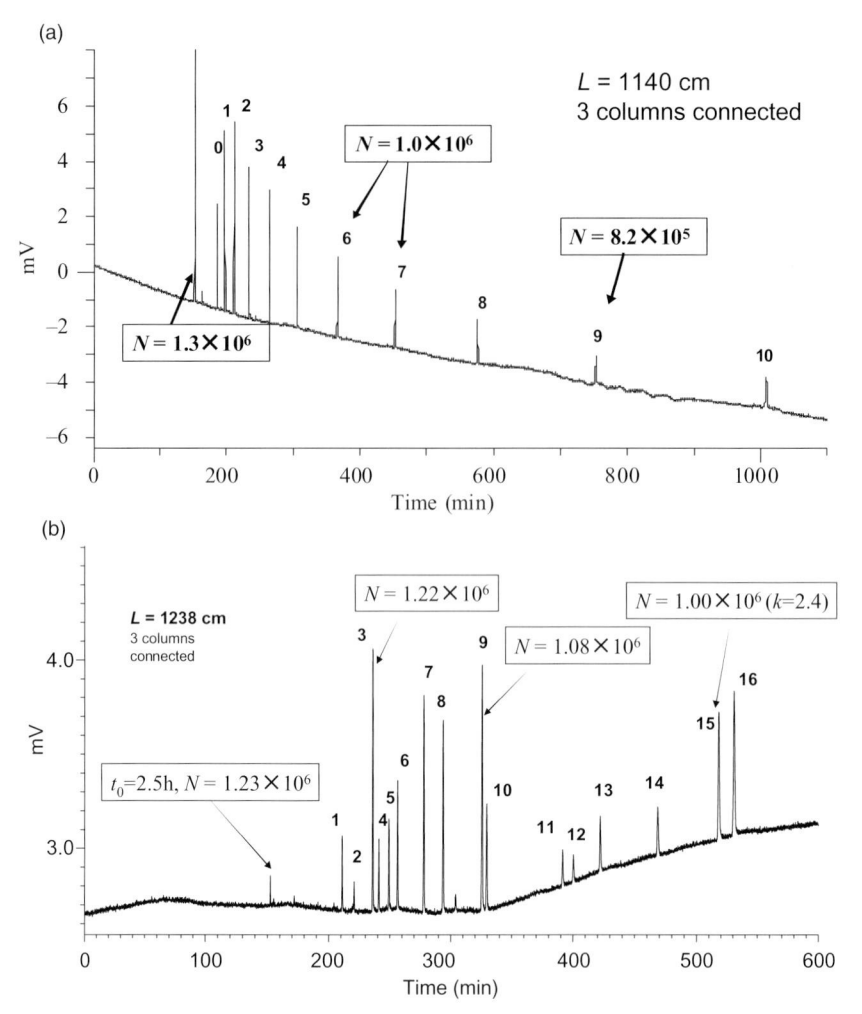

Figure 13.6 Performance of long monolithic silica capillary columns connected in series. (a) Separation of alkylbenzenes. Column: Three monolithic silica C_{18} columns connected, MS(100)-H-C_{18} (8KM and two columns of 9KM), effective length: 1140 cm. Mobile phase: acetonitrile–water (80/20). Detection: 210 nm. Temperature: 30 °C. Sample: thiourea and alkylbenzenes ($C_6H_5C_nH_{2n+1}$, $n=0$–10). $\Delta P=35.4$ MPa. $u=1.24$ mm/s. (b) Separation of polynuclear aromatic hydrocarbons. Column: three monolithic silica C_{18} columns connected, MS(100)-H-C_{18} (8KM, 9KM, and 10KM), effective length 1238 cm (total length 1244 cm). Mobile phase: acetonitrile-water (80/20). Detection: 210 nm. Temperature: 30 °C. Sample: 16-PAHs primary pollutants designated by the EPA. Peak numbers: 1, naphthalene, 2, acenaphthylene, 3, fluorene, 4, acenaphthene, 5, phenanthrene, 6, anthracene, 7, fluoranthene, 8, pyrene, 9, chrysene, 10, benz(a)anthracene, 11, benzo(b)fluoranthene, 12, benzo(k)fluoranthene, 13, benzo(a)pyrene, 14, dibenz(a,h)anthracene, 15, indeno(1,2,3-cd)pyrene, and 16, benzo(g,h,i)perylene. $\Delta P=46.6$ MPa. $u=1.31$ mm/s. (c) Separation of an isotopic mixture of benzene. Column: Two monolithic silica C18 columns connected in series, MS(100)-H-C_{18} (Prepared from a mixture, TMOS/MTMS=9/2), effective length 850 cm. Mobile phase: acetonitrile–methanol–water (10/5/85). Detection: 210 nm. Temperature: 30 °C. Sample:1, benzene-d_6, 2, 1,3,5-benzene-d_3, 3, benzene-d, 4, benzene. $\Delta P=34$ MPa. $u=1.02$ mm/s. Reproduced from Ref. [9] with permission. Copyright 2008 American Chemical Society.

(c)

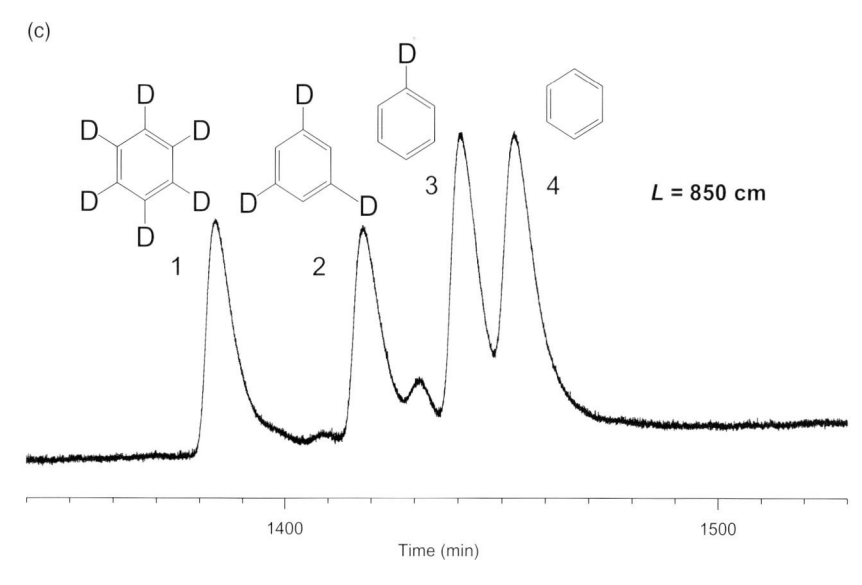

[Chromatogram showing peaks labeled 1 (hexadeuterobenzene), 2 (1,3,5-trideuterobenzene), 3 (monodeuterobenzene), 4 (benzene), with L = 850 cm, time axis from ~1400 to 1500 min]

Figure 13.6 *Continued*

13.4.3
Performance of Long Capillary Columns for Peptides in Gradient Mode

Figure 13.7 compares chromatograms of gradient elution of a bovine serum albumin (BSA) digest sample on a 28.4-cm capillary column (MS(100)-H–C_{18}-(6SM)) and on a 300-cm column, with acetonitrile gradient in the presence of 0.1% formic acid [9]. Gradient times proportional to the column length were employed. Figure 13.7a shows the result of 30-min gradient elution on a 28.4-cm column, showing a peak capacity of ca. 100 in 23 min for the elution range of the peptides. The 300-cm column resulted in a peak capacity of about 380 in 215 min for the elution range of the peptides. The peak capacity obtainable per unit time is more than two times greater with the shorter column than with the longer one, but the absolute peak capacity observed with the 300-cm column was more than three times greater than that obtainable with the 28.4-cm column for the elution range of the peptides in the protein digest [9, 51]. Gradient elution on a short column may provide a large peak capacity per unit time, but it is accomplished by compressing the peaks that are easily separated. The extremely high separation capability of a monolithic silica capillary column was shown very nicely in a proteomic study or in a metabolomic study that handles numerous components in a sample [10, 52].

A 100-µm diameter, 196-cm hybrid monolithic silica column showed a sample-loading capacity of 1–2 ng for peptides with a molecular weight range of 500–1200 in a 20/80 acetonitrile–water mobile phase containing 0.1% trifluoroacetic acid (TFA) at 110 000–150 000 theoretical plates. Early eluting peptides (k of up to 3)

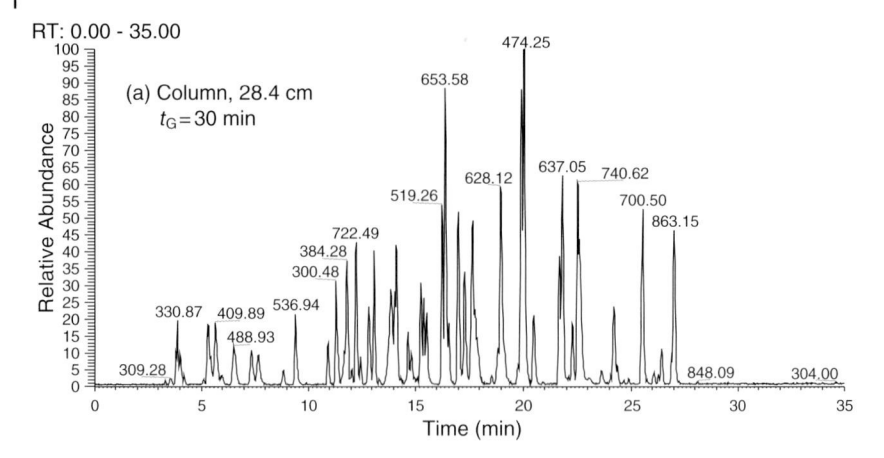

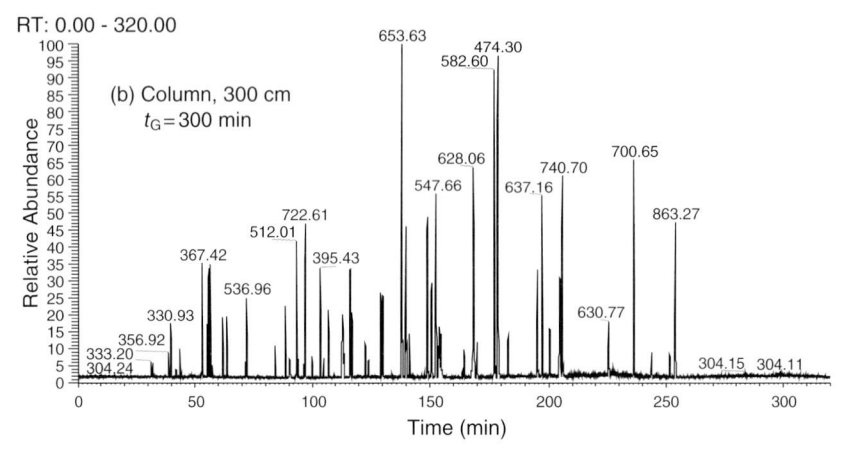

Figure 13.7 Gradient separation of a BSA digest. Column: MS(100)-H–C_{18} (6SM), (a) 28.4 cm, and (b) 300 cm. The mobile phases consisted of (A) 0.1% formic acid in water and (B) 0.1% formic acid in acetonitrile. A linear gradient of 5% to 40% B in 30 min (a) and in 300 min (b) was employed. A BSA digest sample dissolved in water (100 nl, 1 nmol/200 μl) was injected at the start of the gradient with a flow rate 1.2 μl/min (a) and 0.67 μl/min (b). The peak capacity values were obtained from the base peak chromatograms after smoothing. Reproduced from Ref. [9] with permission. Copyright 2008 American Chemical Society.

showed a 20–25% decrease in their N value at 2 ng injection, while longer retained solutes showed a similar decrease with a 0.5–1.0 ng injection, and 50–65% decrease in the N value at a 2-ng injection [9].

The injection of a 1-ng sample into a 100-μm ID column is equivalent to a ca. 2-μg injection into a 4.6-mm ID column based on the cross-section of the columns. The loading capacity found with the high-efficiency monolithic silica capillary column seems to be less than that reported for a conventional particle-packed column [53, 54].The phase ratio of the monolithic silica capillary column, which is smaller than that of a particle-packed column by a factor of about 5, is presumably responsible for the smaller loading capacity compared to a particle-packed column [5]. The amount of stationary phase available for solute binding seems to be the dominant factor. As a matter of fact, the amount of silica in one theoretical plate can be a measure of sample loading capacity [54], ca. $8 \times 10^{-15}\,m^3$ for a 100-μm ID monolithic silica capillary column (a 100-cm column producing 100 000 theoretical plates) with a total porosity of 90%, and is smaller than that of a 4.6-mm ID column packed with particles (a 15-cm column producing 10 000 theoretical plates) having 75% total porosity by a factor of ca. 8000.

Detection of increased numbers of peaks was reported in a gradient elution of peptides and plant metabolites by using long monolithic silica capillary columns [10, 52, 55]. Shallow gradient elution is often effective for increasing resolution. Such conditions will make the elution somewhat similar to an isocratic mode, where high-efficiency columns can provide an effective separation. An increase in peak capacity, however, may not be the primary advantage of using a long column, because the peak-capacity increase is not steep with the increase in column length and separation time [55]. Long monolithic silica capillary columns may be useful for the separation of very complex mixtures over a relatively narrow range of a chromatogram, with little effort required for method development. It is of great interest to show the practical utility that long gradient elution on a long capillary column is effective to identify more proteins in a proteomic study [56].

13.5
Monolithic Silica Capillary Columns for High-Speed Separations

13.5.1
Monolithic Silica Columns Having Increased Phase Ratios

In a series of experiment aiming at high-speed monolithic silica columns, the columns listed in Table 13.3 were prepared from TMOS [8]. Columns MS(100)-A–C (a group of columns, MSC-1 in Figure 13.2) were prepared to examine the effect of the porosity, or the effect of the TMOS concentration in the feed, on the properties with adjusted PEG concentrations to synchronize phase separation and gelation. The lower reaction temperature employed for the preparation compared to previous studies slowed down the polymerization resulting in the slower phase separation and gelation. The higher TMOS concentrations caused faster gelation, while the increased PEG concentrations resulted in the slower phase separation.

Similarly MS(100)-D–F (MSC-2 in Figure 13.2) were prepared with the varied PEG concentrations to examine the effect of domain size on column efficiency at constant porosity. Along with each series, columns **G** and **H** were prepared with lower TMOS concentrations by the previous method [7] that resulted in

Table 13.3 Composition of the feed mixtures for the preparation of monolithic silica columns for high-speed separations.

Column	TMOS (ml)	PEG (g)	Urea (g)	CH_3COOH (ml)
MS(100)-T1.0-G[a]	40	12.4	9.0	100
MS(100)-T1.4-A[b]	56	11.8	9.0	100
MS(100)-T1.6-B[b]	64	10.4	9.0	100
MS(100)-T1.8-C[b]	72	8.4	9.0	100
MS(100)-T1.0-H[a]	40	12.8	9.0	100
MS(100)-T1.4-D[b]	56	11.7	9.0	100
MS(100)-T1.4-E[b]	56	11.8	9.0	100
MS(100)-T1.4-F[b]	56	11.9	9.0	100

a) Prepared at 30 °C.
b) At 25 °C.

Reproduced from Ref. [8] with permission. Copyright 2006 American Chemical Society.

inhomogeneous aggregated skeletons, as shown in Figure 13.8a. Figure 13.2 indicates that an increase in the TMOS concentration resulted in the larger skeleton size and smaller through-pore size, as expected (MSC-1, MSR-1). At a fixed concentration of TMOS, an increase in PEG concentration resulted in the decrease in domain size with almost constant through-pore-size/skeleton-size ratios (MSC-2, MSR-2). The total porosity was found to be about 95% for MS(100)-T1.0-G by SEC, and about 92% for MS(100)-T1.6-B. The difference presumably reflects the difference in the phase ratio. Through-pores accounted for about 65–75% of the column volume for the monolithic silica capillary columns without ODS modifications. With the increase in TMOS concentration in the feed, smaller through-pore volumes and greater mesopore volumes were observed, indicating an increase in the phase ratio

As shown in Figure 13.8, MS(100)-A–C and MS(100)-D–F prepared at 25 °C with the reduced porosity possess higher homogeneity of cocontinuous structures (Figure 13.8), and resulted in higher efficiency (Figure 13.9) than MS(100)-G and MS(100)-H, respectively, prepared at 30 °C. Since the structural inhomogeneity is expected to result from the disturbed coarsening process of the phase-separating domains, the lower temperature contributed to suppress the coarsening by increased viscosity of the whole system. It has been predicted that a decrease in porosity and an increase in homogeneity would improve the performance of monolithic columns [44]. Actually, all columns **A–F** prepared at a lower temperature and with the lower porosity were shown to possess higher homogeneity than columns **G** or **H** (Figure 13.8). For columns **D–F**, the increase in PEG concentration in feed resulted in the higher column efficiency presumably based on the smaller domain size. The increase in a skeleton volume, or the increase in TMOS content in the feed, is reflected in the greater solute retention observed for columns **D–F** than **H**, as shown in Figure 13.9.

(a) MS(100)-T1.0-G

(b) MS(100)-T1.4-A

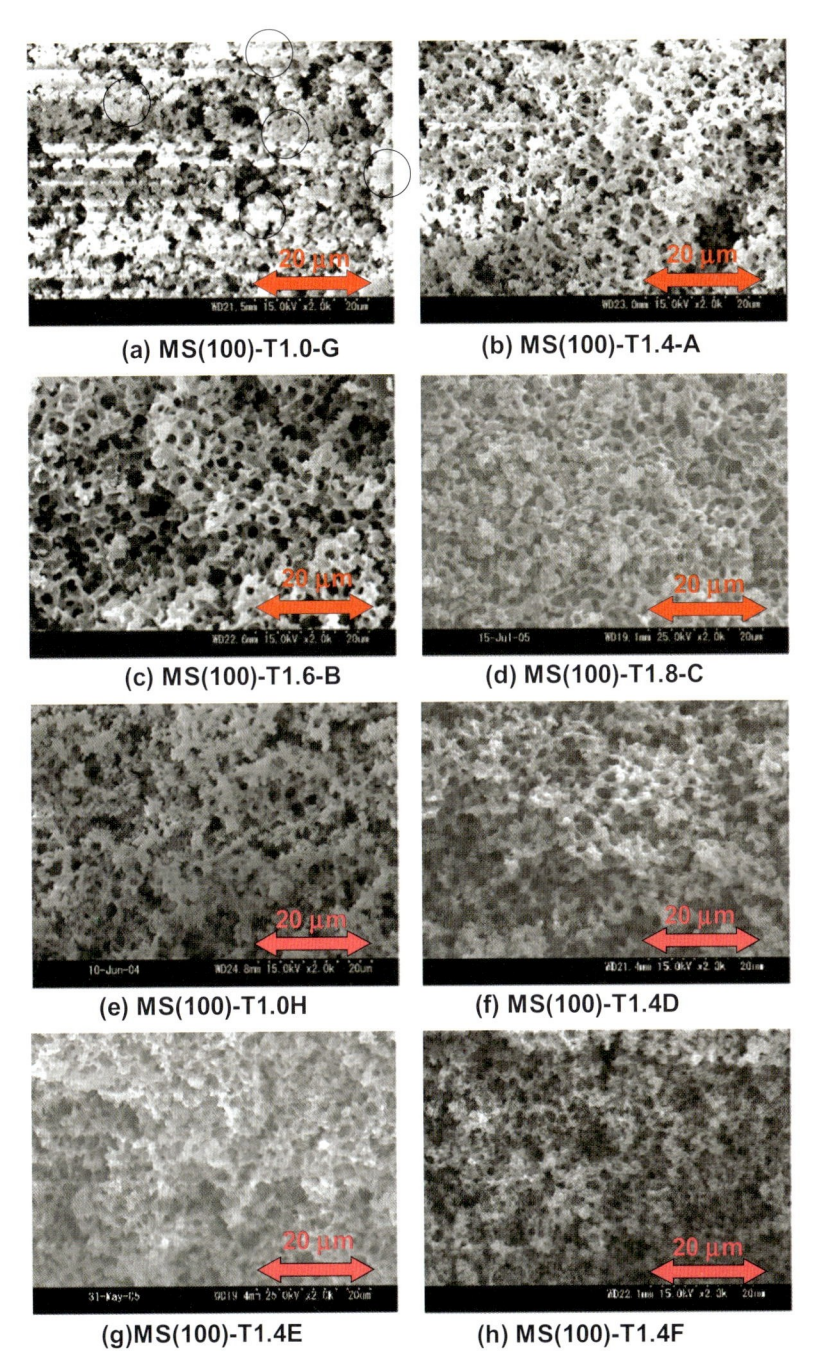

(c) MS(100)-T1.6-B

(d) MS(100)-T1.8-C

(e) MS(100)-T1.0H

(f) MS(100)-T1.4D

(g) MS(100)-T1.4E

(h) MS(100)-T1.4F

Figure 13.8 Scanning electron micrographs of monolithic silica capillary columns. MS(100)-T-1.0G and MS(100)-T-A–C and MS(100)-T-1.0H and MS(100)-T-D–F (Table 13.3) prepared from TMOS, in a 100-μmID capillary tube. Reproduced from Ref. [8] with permission. Copyright 2006 American Chemical Society.

**(a) MS(100)-T1.0-H
(20 cm)**

ΔP = 3.7 MPa
u = 2.02 mm/s
N = 34 000

**(b) MS(100)-T1.4-D
(14.5 cm)**

ΔP = 2.6 MPa
u = 2.06 mm/s
N = 27 000

**(c) MS(100)-T1.4-E
(15 cm)**

ΔP = 3.7 MPa
u = 2.06 mm/s
N = 30 000

**(d) MS(100)-T1.4-F
(15 cm)**

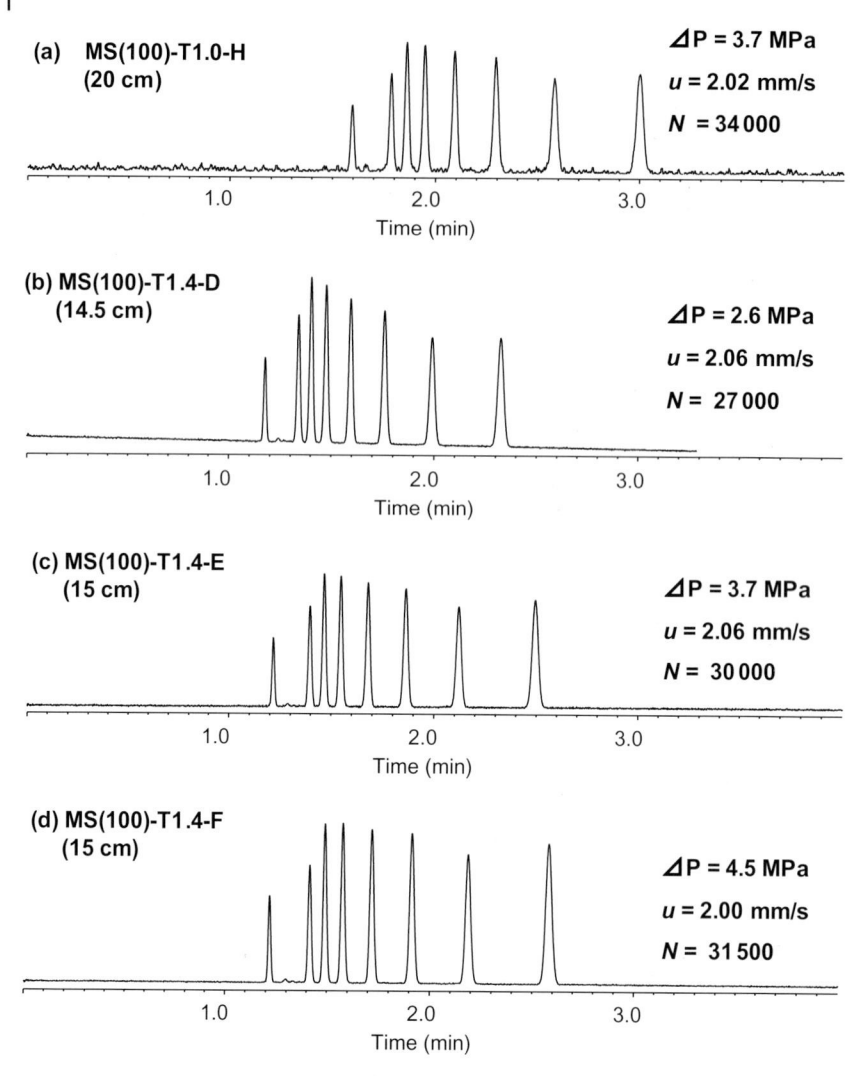

ΔP = 4.5 MPa
u = 2.00 mm/s
N = 31 500

Figure 13.9 Chromatograms obtained for uracil (the first peak) and alkylbenzenes ($C_6H_5(CH_2)_nH$, $n=0$–6). Column: (a) MS(100)-T1.0-H, 25 cm (effective length 20 cm), (b) MS(100)-T1.4-D, 19.5 cm (effective length 14.5 cm), (c) MS(100)-T1.4-E, 20 cm (effective length 15 cm), and (d) MS(100)-T1.4-F, 20 cm (effective length 15 cm). Column diameter: 100 μm. Mobile phase: Acetonitrile/water = 80/20. Temperature: 30 °C. The pressure drop, linear velocity, and number of theoretical plates for hexylbenzene (the last peak) are indicated. Reproduced from Ref. [8] with permission. Copyright 2006 American Chemical Society.

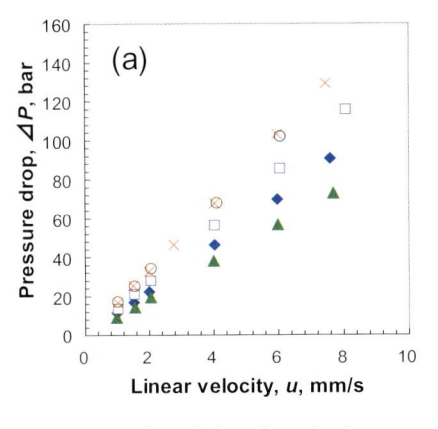

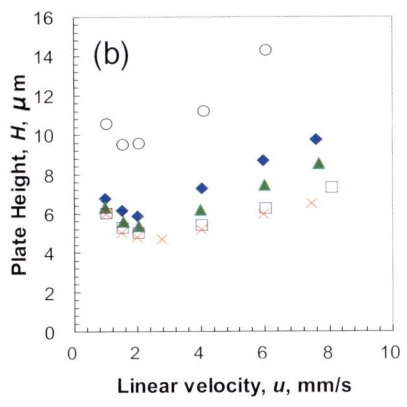

Figure 13.10 Plots of the column backpressure (a, normalized to a column length of 15 cm) and the plate height (b, hexylbenzene as a solute) against the linear velocity of the mobile phase observed for ODS-modified monolithic silica columns. Columns: Mightysil RP18 (○) packed with 5-μm particles, MS(100)-T1.0-H (♦), and MS(100)-T1.4-D (▲), MS(100)-T1.4-E (□), and MS(100)-T1.4-F (X). Mobile phase: acetonitrile/water=80/20. Temperature: 30 °C. Reproduced from Ref. [8] with permission. Copyright 2006 American Chemical Society.

13.5.2
Performance of High-Speed Monolithic Silica Columns

Column **F** having the smallest domain size showed higher column efficiency (N), than the column **D** or **E** by 5–10%, which in turn showed a greater N than column **H** by about 20%. The total performance of column **F** in terms of separation impedance, however, was found to be lower than that of column **D** by ca. 30%, implying that the structure is not optimal, or optimum reaction conditions for each TMOS concentration are very narrow. The adverse effects of inhomogeneous structures are reported to be pronounced at the smaller domain sizes [44].

Figure 13.10a shows the plots of the column pressure drop against the linear velocity of a mobile phase, 80% acetonitrile. A decrease in permeability was observed with a decrease in the domain size. The permeability of MS(100)-T1.4-F was comparable with a column packed with 5-μm particles. The fact that greater permeability and higher column efficiency were observed simultaneously for MS(100)-T1.4-D than MS(100)-T1.0-H suggests that MS(100)-T1.4-D possessed greater structural homogeneity than the product of the previous preparations. MS(100)-T1.4-E and MS(100)-T1.4-F showed a slight increase in column efficiency and pressure drop.

Figure 13.10b shows the plots of plate height (H) against linear velocity of a mobile phase for the elution of hexylbenzene. The smaller minimum plate height and the shift of the optimum linear velocity towards a higher value were observed with the decrease in domain size. The plots for MS(100)-T1.4-F showed a plate height of 4.1 µm for ethylbenzene and 4.7 µm for hexylbenzene at each optimum linear velocity. Such plate-height values can be expected for a column packed with ca. 2–2.5-µm silica particles. The domain size of the monolithic silica column was similar to the size of particles that would have been expected to show a similar column efficiency [35, 39]. Although the column efficiency was somewhat lower than that of a column packed with 2-µm particles or smaller, the results showed that it is possible to prepare monolithic silica columns of increased column efficiencies at high speed by increasing the phase ratio and/or structural homogeneity, as predicted [43].

Figure 13.10 suggests that the gain in column efficiency compared to a column packed with 5-µm particles was slightly greater than two-fold at a similar pressure drop and with a higher linear velocity. In spite of the high column efficiency at the relatively low pressure drop of the monolithic silica capillary columns, the results from Figure 13.10 suggest that the performance of MS(100)-T1.4-E and MS(100)-T1.4-F columns was still not as high as expected based on their domain sizes compared to the MS(100)-T1.4-D column.

13.5.3
Comparison of Performance with a Particle-Packed Column

Figure 13.11a shows a comparison of kinetic plots between the monolithic silica columns and particle-packed columns at a 40-MPa pressure limit. In Figure 13.11a, the curves for monolithic silica columns with smaller domains are located below the curves representing the performance of particle-packed columns in a region of the number of theoretical plates (N) greater than ca. 25 000, thus indicating superior performance in the range. Although the optimum performance of these monolithic silica columns, which is much higher than that of a particle-packed column, is seen in a region of N greater than 100 000, they can compete with the performance of a column packed with 2-µm particles at ca. 25 000 at 40 MPa. Figure 13.11 suggests that the older batches of monolithic silica columns prepared by the previous method are equivalent to a column packed with 3–4 µm particles, whereas the present monolithic silica columns are closer to those packed with 2–2.5 µm particles in performance at 15 000–30 000 theoretical plates under common pressures equivalent to 5-µm particles used in HPLC.

The comparison between the plots for the monolithic silica columns MS(100)-T1.4-D-F and those for 1.4, 2, and 3 µm particles at 100 MPa (shown in Figure 13.11b) suggests that particle-packed columns perform better at $N=80 000$ theoretical plates or lower. At higher pressure drops, the comparison becomes more favorable for a particle-packed column. At 10 MPa, the monolithic silica column MS(100)-T1.4-F performs better than a particle-packed column at greater than 8000 theoretical plates, as shown in Figure 13.11c. The advantage with the monolithic silica columns is that performance similar to that of a column packed with 2–2.5 µm particles can be obtained at a pressure drop similar to a column packed with 5-µm particles. Recently similar performance was achieved with rod-type monolithic silica columns [57].

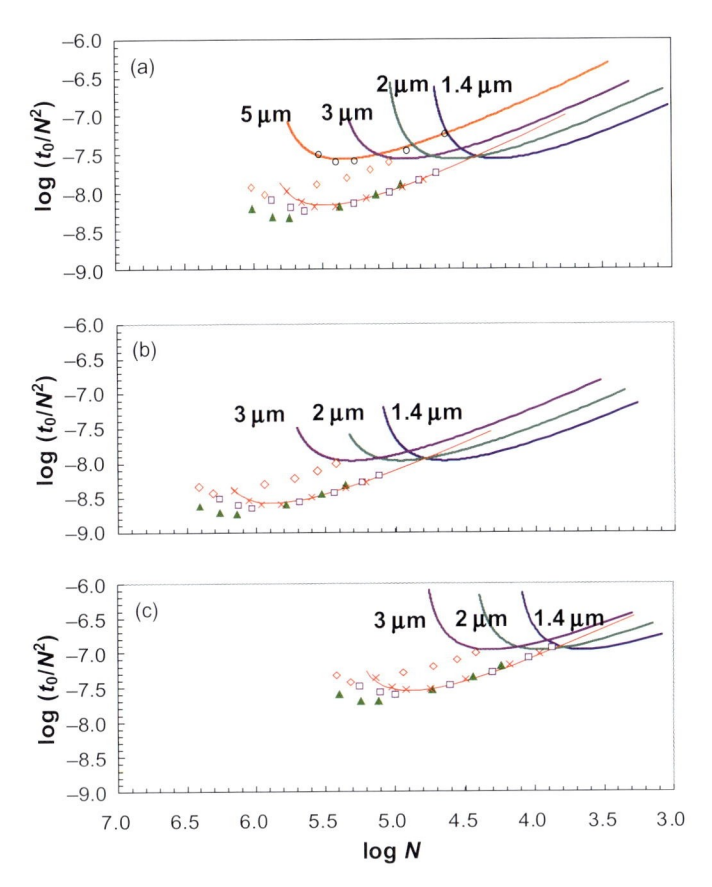

Figure 13.11 Plots of $\log(t_0/N^2)$ against $\log(N)$ for the columns evaluated. The curves for particle-packed columns were obtained by assuming the following parameters: $\eta=0.00046$ Pa s, $\varphi=700$, $D_m=2.22 \times 10^{-9}$ m^2/s, and Knox equation, $h=0.65v^{1/3}+2/v+0.08v$. Pressure: (a) 40 MPa, (b) 100 MPa, and (c) 10 MPa. The particle diameters for the particle-packed columns were 1.4 μm, 2 μm, 3 μm, and 5 μm. The symbols are as in Figure 13.10. Reproduced from Ref. [8] with permission. Copyright 2006 American Chemical Society.

13.6 Future Considerations

Facile preparation of polar-modified surfaces, and stability against high flow rate required for multidimensional applications make monolithic silica even more attractive, in spite of the disadvantages related to the silica preparation and the chemical modification in individual columns. Further studies are in progress for improving monolithic silica columns. Similar results showing increased homogeneity and column efficiency were obtained for hybrid columns with increased phase ratios, as in the case of a column prepared from TMOS [58]. Optimization of reaction conditions of long hybrid capillary columns lead to the generation of 2×10^6 theoretical plates [59].

The generation of ultrahigh efficiency by long capillary format will need high extraskeleton porosity (ca. 0.7 or above), while high-speed separations with N values of 10 000–30 000 need small-sized domains and the lower porosity (ca. 0.4–0.6) [44]. The development of monolithic silica columns having controlled domain size and porosity, and most importantly high homogeneity, will be crucial for high-speed and high-efficiency LC separations in the future, if one recognizes the advantages of monolithic columns over conventional particulate columns. The preparation of the materials for LC requires more rigorously controlled preparation conditions than common sol-gel products. Optimum reaction conditions can be found in a narrow range of concentration of each component.

13.7
Conclusion

The study on the effect of phase ratios on the performance of monolithic silica capillary columns indicated that the two types of columns can be prepared either from TMOS or from a TMOS/MTMS mixture. Those with a greater phase ratio showed better performance at high speed, while highly porous materials were able to provide a higher permeability, leading to large numbers of theoretical plates with a long capillary column. In both cases, preparation at a lower temperature and with an increased phase ratio resulted in higher column efficiency presumably due to the increased homogeneity of the domain structure.

Currently available performance of an efficiency of 10^5–10^6 theoretical plates with t_0 of 5–100 min using a long capillary column of up to 500 cm, or 10^4 theoretical plates with t_0 less than 10 s using a column having a high phase ratio, can be obtained with considerably lower pressure than with particulate counterparts. In other words, most columns of advanced features can be practically used in a conventional pressure range, or below 40 MPa, although the use of higher pressure will lead to even higher performance. Actually, higher performance than any particle-packed column has been shown at $N=25\,000$ or greater with 40 MPa pressure drop using a kinetic plot analysis, while the generation of 1 000 000 theoretical plates was demonstrated for retained solutes that was not possible with particle-packed or open tubular columns.

It seems to be possible to cover the entire range of performance provided by 10-μm to 2-μm particles adequately by the two types of monolithic silica capillary columns for high-speed separations and for high-efficiency separations. For those using conventional HPLC equipment, it would be practical to use a 100–200-cm monolithic silica capillary column at a pressure drop of 20 MPa or lower to generate 100 000–200 000 theoretical plates with a t_0 of 5–20 min. If one can use HPLC equipment with a pressure range of 40–50 MPa, 300 000 theoretical plates can be obtained with a t_0 of 15–20 min, and 1 000 000 theoretical plates with a t_0 around 2.5 h. A long monolithic silica capillary column may have practical utility, especially for resolving very closely related solutes with a separation factor of less than 1.01 that would otherwise be impossible. It would be of great interest whether one can prepare monolithic silica columns with a domain size of ca. 1.5–2 µm or 2.5–3 µm with adequate structural homogeneity. Desmet and coworkers suggested that increased phase ratios and increased homogeneity of the skeletons and through-pores might increase the total performance of monolithic silica columns by several-fold [43]. It should be noted that monolithic silica capillary columns can be used for capillary electrochromatography with much higher efficiency, indicating the contribution of a large A term in pressure-driven operation [5].

References

1 Guiochon, G. (2006) The limits of the separation power of unidimensional column liquid chromatography. *J. Chromatogr. A*, **1126**, 6–49.
2 Cabooter, D., Lestremau, F., Lynen, F., Sandra, P., and Desmet, G. (2008) Kinetic plot method as a tool to design coupled column systems producing 100 000 theoretical plates in the shortest possible time. *J. Chromatogr. A*, **1212**, 23–34.
3 Cabooter, D., Lestremau, F., de Villiers, A., Broeckhoven, K., Lynen, F., Sandra, P., and Desmet, G. (2009) Investigation of the validity of the kinetic plot method to predict the performance of coupled column systems operated at very high pressures under different thermal conditions. *J. Chromatogr. A*, **1216**, 3895–3903.
4 Guiochon, G. (2007) Monolithic columns in high-performance liquid chromatography. *J. Chromatogr. A*, **1168**, 101–168.
5 Ishizuka, N., Minakuchi, H., Nakanishi, K., Soga, N., Nagayama, H., Hosoya, K., and Tanaka, N. (2000) Performance of a silica rod column in a capillary under pressure-driven and electrodriven conditions. *Anal. Chem.*, **72**, 1275–1280.
6 Ishizuka, N., Kobayashi, H., Minakuchi, H., Nakanishi, K., Hirao, K., Hosoya, K., Ikegami, T., and Tanaka, N. (2002) Monolithic silica columns for high-efficiency separations by high-performance liquid chromatography. *J. Chromatogr. A*, **960**, 85–96.
7 Motokawa, M., Kobayashi, H., Ishizuka, N., Minakuchi, H., Nakanishi, K., Jinnai, H., Hosoya, K., Ikegami, T., and Tanaka, N. (2002) Monolithic silica columns with various skeleton sizes and through-pore sizes for capillary liquid chromatography. *J. Chromatogr. A*, **961**, 53–63.
8 Hara, T., Kobayashi, H., Ikegami, T., Nakanishi, K., and Tanaka, N. (2006) Performance of monolithic silica capillary columns with increased phase ratios and small-sized domains. *Anal. Chem.*, **78**, 7632–7642.
9 Miyamoto, K., Hara, T., Kobayashi, H., Morisaka, H., Tokuda, D., Horie, K., Koduki, K., Makino, S., Núñez, O., Yang, C., Kawabe, T., Ikegami, T., Takubo, H., Ishihama, Y., and Tanaka, N. (2008) High-efficiency liquid chromatographic separation utilizing long monolithic silica capillary columns. *Anal. Chem.*, **80**, 8741–8750.
10 Luo, Q.Z., Shen, Y.F., Hixson, K.K., Zhao, R., Yang, F., Moore, R.J., Mottaz, H.M., and Smith, R.D. (2005) Preparation of 20 µm ID silica-based monolithic columns and their performance for proteomics analyses. *Anal Chem.*, **77**, 5028–5035.
11 Li, W., Fries, D.P., and Malik, A. (2004) Sol-gel stationary phases for capillary electrochromatography. *J. Chromatogr. A*, **1044**, 23–52.
12 Wu, R., Hu, L., Wang, F., Ye, M., and Zou, H. (2008) Recent development of monolithic stationary phases with emphasis on microscale chromatographic separation. *J. Chromatogr. A*, **1184**, 369–392.

13 Núñez, O., Nakanishi, K., and Tanaka, N. (2008) Preparation of monolithic silica columns for high-performance liquid chromatography. *J. Chromatogr. A*, **1191**, 231–252.

14 Fujimoto, C. (2000) Preparation of fritless packed silica columns for capillary electrochromatography. *J. High Resolut. Chromatogr.*, **23**, 89–92.

15 Takeuchi, T., Tatsumi, S., Masuoka, S., Hirose, K., Uzu, H., Jin, J.-Y., Fujimoto, C., Ohta, K., Lee, K.-P., Ryoo, J.-J., and Choi, S.-H. (2003) Split flow and bypass flow systems for monolithic capillary columns in liquid chromatography. *J. Chromatogr. A*, **1021**, 55–59.

16 Ye, F., Xie, Z., Wu, X., Lin, X., and Chen, G. (2006) Phenylaminopropyl silica monolithic column for pressure assisted capillary electrochromatography. *J. Chromatogr. A*, **1117**, 170–175.

17 Yan, L.-J., Zhang, Q.-H., Feng, Y.-Q., Zhang, W.-B., Li, T., Zhang, L.-H., and Zhang, Y.K. (2006) Octyl-functionalized hybrid silica monolithic column for reversed-phase capillary electrochromatography. *J. Chromatogr. A*, **1121**, 92–98.

18 Puy, G., Roux, R., Demesmay, C., Rocca, J.-L., Iapichella, J., Galarneau, A., and Brunel, D. (2007) Influence of the hydrothermal treatment on the chromatographic properties of monolithic silica capillaries for nano-liquid chromatography or capillary electrochromatography. *J. Chromatogr. A*, **1160**, 150–159.

19 Núñez, O., Ikegami, T., Miyamoto, K., and Tanaka, N. (2007) Study of a monolithic silica capillary column coated with poly(octadecyl methacrylate) for the reversed-phase liquid chromatographic separation of some polar and non-polar compounds. *J. Chromatogr. A*, **1175**, 7–15.

20 Chen, Y., Chen, J., and Jia, L. (2009) Study of triacontyl-functionalized monolithic silica capillary column for reversed-phase capillary liquid chromatography. *J. Chromatogr. A*, **1216**, 2597–2600.

21 Soonthorntantikul, W., Leepipatpiboon, N., Ikegami, T., Tanaka, N., and Nhujak, T. (2009) Selectivity comparisons of monolithic silica capillary columns modified with poly(octadecyl methacrylate) and octadecyl moieties for halogenated compounds in reversed-phase liquid chromatography. *J. Chromatogr. A*, **1216**, 5868–5874.

22 Hutchinson, J.P., Hilder, E.F., Macka, M., Avdalovic, N., and Haddad, P.R. (2006) Preparation and characterisation of anion-exchange latex-coated silica monoliths for capillary electrochromatography. *J. Chromatogr. A*, **1109**, 10–18.

23 Watanabe, Y., Ikegami, T., Horie, K., Hara, T., Jaafar, J., and Tanaka, N. (2009) Improvement of separation efficiencies of anion-exchange chromatography using monolithic silica capillary columns modified with polyacrylates and polymethacrylates containing tertiary amino or quaternary ammonium groups. *J. Chromatogr. A*, **1216**, 7394–7401.

24 Allen, D., and El Rassi, Z. (2004) Capillary electrochromatography with monolithic silica columns: III. Preparation of hydrophilic silica monoliths having surface-bound cyano groups: chromatographic characterization and application to the separation of carbohydrates, nucleosides, nucleic acid bases and other neutral polar species. *J. Chromatogr. A*, **1029**, 239–247.

25 Horie, K., Ikegami, T., Hosoya, K., Saad, N., Fiehn, O., and Tanaka, N. (2007) Highly efficient monolithic silica capillary columns modified with poly(acrylic acid) for hydrophilic interaction chromatography. *J. Chromatogr. A*, **1164**, 198–205.

26 Ikegami, T., Tomomatsu, K., Takubo, H., Horie, K., and Tanaka, N. (2008) Separation efficiencies in hydrophilic interaction chromatography. *J. Chromatogr. A*, **1184**, 474–503.

27 Chen, Z., Uchiyama, K., and Hobo, T. (2002) Chemically modified chiral monolithic silica column prepared by a sol–gel process for enantiomeric separation by micro high-performance

liquid chromatography. *J. Chromatogr. A*, **942**, 83–91.

28 Liu, Z., Otsuka, K., Terabe, S., Motokawa, M., and Tanaka, N. (2002) Physically adsorbed chiral stationary phase of avidin on monolithic silica column for capillary electrochromatography and capillary liquid chromatography. *Electrophoresis*, **23**, 2973–2981.

29 Chankvetadze, B., Yamamoto, C., Kamigaito, M., Tanaka, N., Nakanishi, K., and Okamoto, Y. (2006) High-performance liquid chromatographic enantioseparations on capillary columns containing monolithic silica modified with amylose tris(3,5-dimethylphenylcarbamate). *J. Chromatogr. A*, **1110**, 46–52.

30 Motokawa, M., Ohira, M., Minakuchi, H., Nakanishi, K., and Tanaka, N. (2006) Performance of octadecylsilylated monolithic silica capillary columns of 530 μm inner diameter in HPLC. *J. Sep. Sci.*, **29**, 2471–2477.

31 Tanaka, N., Kinoshita, H., Araki, M., and Tsuda, T. (1985) On-column preparation of chemically bonded stationary phase with maximum surface coverage and high reproducibility, and its application to packed microcapillary columns. *J. Chromatogr.*, **332**, 57–69.

32 Yang, C., Ikegami, T., Hara, T., and Tanaka, N. (2006) Improved endcapping method of monolithic silica columns. *J. Chromatogr. A*, **1130**, 175–181.

33 Nakanishi, K. (1997) Pore structure control of silica gels based on phase separation. *J. Porous Mater.*, **4**, 67–112.

34 Nakanishi, K., and Tanaka, N. (2007) Sol-gel with phase separation. *Acc. Chem. Res.*, **40**, 863–873.

35 Minakuchi, H., Nakanishi, K., Soga, N., Ishizuka, N., and Tanaka, N. (1997) Effect of skeleton size on the performance of octadecylsilylated continuous porous silica columns in reversed-phase liquid chromatography. *J. Chromatogr. A*, **762**, 135–146.

36 Minakuchi, H., Nakanishi, K., Soga, N., Ishizuka, N., and Tanaka, N. (1998) Effect of domain size on the performance of octadecylsilylated continuous porous silica columns in reversed-phase liquid chromatography. *J. Chromatogr. A*, **797**, 121–131.

37 Cabrera, K., Lubda, D., Eggenweiler, H.-M., Minakuchi, H., and Nakanishi, K. (2000) A new monolithic-type HPLC column for fast separations. *J. High Resolut. Chromatogr.*, **23**, 93–99.

38 Leinweber, F.C., and Tallarek, U. (2003) Chromatographic performance of monolithic and particulate stationary phases: Hydrodynamics and adsorption capacity. *J. Chromatogr. A*, **1006**, 207–228.

39 Miyabe, K., Cavazzini, A., Gritti, F., Kele, M., and Guiochon, G. (2003) Moment analysis of mass-transfer kinetics in C_{18}-silica monolithic columns. *Anal. Chem.*, **75**, 6975–6986.

40 Unger, K.K. (1979) *Porous Silica*, Elsevier, Amsterdam, Ch. 5.

41 Kirkland, J.J., Truszkowski, F.A., and Ricker, R.D. (2002) Atypical silica-based column packings for high-performance liquid chromatography. *J. Chromatogr. A*, **965**, 25–34.

42 Kirkland, J.J., Truszkowski, F.A., Dilks, C.H., Jr., and Engel, G.S. (2000) Superficially porous silica microspheres for fast high-performance liquid chromatography of macromolecules. *J. Chromatogr. A*, **890**, 3–13.

43 Desmet, G., Clicq, D., and Gzil, P. (2005) Geometry-independent plate height representation methods for the direct comparison of the kinetic performance of LC supports with a different size or morphology. *Anal. Chem.*, **77**, 4058–4070.

44 Gzil, P., Vervoort, N., Baron, G.V., and Desmet, G. (2004) General rules for the optimal external porosity of LC supports. *Anal. Chem.*, **76**, 6707–6718.

45 Kele, M., and Guiochon, G. (2002) Repeatability and reproducibility of retention data and band profiles on six batches of monolithic columns. *J. Chromatogr. A*, **960**, 19–49.

46 Kucera, P., and Guiochon, G. (1984) Use of open-tubular columns in liquid chromatography. *J. Chromatogr.*, **283**, 1–20.

47 Liu, G., Djordjevic, N.M., and Erni, F. (1992) High-temperature open-tubular capillary column liquid chromatography. *J. Chromatogr.*, **592**, 239–247.

48 Swart, R., Kraak, J.C., and Poppe, H. (1995) Performance of an ethoxyethyl acrylate stationary phase for open-tubular liquid chromatography. *J. Chromatogr. A*, **689**, 177–187.

49 Turowski, M., Yamakawa, N., Meller, J., Kimata, K., Ikegami, T., Hosoya, K., Tanaka, N., and Thornton, E.R. (2003) Deuterium isotope effects on hydrophobic interactions: the importance of dispersion interactions in the hydrophobic phase. *J. Am. Chem. Soc.*, **125**, 13836–13849.

50 Kimata, K., Kobayashi, M., Hosoya, K., Araki, T., and Tanaka, N. (1996) Chromatographic separation based on isotopic chirality. *J. Am. Chem. Soc.*, **118**, 759–762.

51 Eeltink, S., Dolman, S., Swart, R., Ursem, M., and Schoenmakers, P.J. (2009) Optimizing the peak capacity per unit time in one-dimensional and off-line two-dimensional liquid chromatography for the separation of complex peptide samples. *J. Chromatogr. A*, **1216**, 7368–7374.

52 Tolstikov, V.V., Lommen, A., Nakanishi, K., Tanaka, N., and Fiehn, O. (2003) Monolithic silica-based capillary reversed-phase liquid chromatography/electrospray mass spectrometry for plant metabolomics. *Anal. Chem.*, **75**, 6737–6740.

53 McCalley, D.V. (1998) Influence of sample mass on the performance of reversed-phase columns in the analysis of strongly basic compounds by high performance liquid chromatography. *J. Chromatogr. A*, **793**, 31–46.

54 Tock, P.P.H., Duijsters, P.P.E., Kraak, J.C., and Poppe, H. (1990) Theoretical optimization of open-tubular columns for liquid chromatography with respect to mass loadability. *J. Chromatogr. A*, **506**, 185–200.

55 Gilar, M., and Neue, U.D. (2007) Peak capacity in gradient reversed-phase liquid chromatography of biopolymers Theoretical and practical implications for the separation of oligonucleotides. *J. Chromatogr. A*, **1169**, 139–150.

56 Iwasaki, M., Miwa, S., Ikegami, T., Tomita, M., Tanaka, N., and Ishihama Y. (2010) One-Dimensional Capillary Liquid Chromatographic Separation Coupled with Tandem Mass Spectrometry Unveils the Escherichia coli Proteome on a Microarray Scale. *Anal. Chem.*, **82**, 2616–2620.

57 Miyazaki, S., Takahashi, M., Ohira, M., Terashima, H., Morisato, K., Nakanishi, K., Ikegami, T., Miyabe, K., and Tanaka, N. (2010) Monolithic silica rod columns for high-efficiency reversed-phase liquid chromatography. *J. Chromatogr. A*, Submitted.

58 Hara, T., Makino, S., Watanabe, Y., Ikegami, T., Cabrera, K., Smarsly, B., and Tanaka, N. (2010) The performance of hybrid monolithic silica capillary columns prepared by changing feed ratios of tetramethoxysilane (TMOS) and methyltrimethoxysilane (MTMS). *J. Chromatogr. A*, **1217**, 89–98.

59 Miwa, S., Makino, S., Miyamoto, K., Ikegami, T., and Tanaka, N. (2009) Studies on preparation and characterization methods of long monolithic silica capillary columns. Presented at *16th Chromatography Symposium, Nagasaki, May, 2009*.

14
Silica Monolithic Columns and Mass Spectrometry
Keith Ashman

14.1
Introduction

The analysis of drugs, metabolites, proteomes and even newborn screening for various metabolic diseases [1–3] depends heavily on the resolving power of liquid chromatography coupled to electrospray ionization tandem mass spectrometry (ESI LC/MS/MS) to reliably separate, identify, characterize and quantify large numbers of compounds and samples. An essential part in this process is the separation of complex mixtures of analytes by liquid chromatography (LC) prior to their introduction into the mass spectrometer. Due to its high resolution and ease of coupling to (nano-) ESI LC/MS/MS as well as matrix-assisted laser desorption ionization mass spectrometry (LC-MALDI MS), reversed-phase liquid chromatography (RPLC) is the most widely used LC separation method applied to this problem. The main reasons for use of RPLC are the high resolution of the columns but also the need for volatile solvents that can be removed rapidly as ions are generated from molecules in the transition from the liquid to the gas phase that occurs as they enter the mass spectrometer during the electrospray process. When sample complexity increases, for example, in proteomics studies after enzymic cleavage of complex protein mixtures to yield even more complex mixtures of peptides, the peak capacity of standard columns is generally insufficient to achieve suitable numbers of compounds per peak, so that all the components of the sample can be measured by the mass spectrometer. One possible approach to solve this problem is the use of multidimensional LC methods [4], which combine multiple (orthogonal) separation techniques such as ion-exchange, size-exclusion or hydrophilic interaction chromatography (HILIC) with RPLC as a final separation step [5–10]. These methods; however, require sophisticated instrumental setups and often lead to relatively long run times and poor reproducibility. Another approach is to improve the efficiency of the RP column. This can be achieved either by reducing the particle size of the column packing material [11–18] or by increasing the length of the column [19, 20]. Both strategies, however, lead to an increase in column backpressure, which again means that special equipment is required. Due to their high porosity, silica monolithic materials only

Monolithic Silicas in Separation Science. Edited by K.K. Unger, N. Tanaka, and E. Machtejevas
© 2011 WILEY-VCH Verlag GmbH & Co. KGaA, Weinheim
ISBN: 978-3-527-32575-7

cause a relatively small pressure drop per unit of column length and this not only makes it possible to use much longer columns to achieve higher resolution with conventional LC apparatus but also to run at higher flow rates that results in narrower peaks and shorter run times. Silica-based monoliths have several advantages. They have higher loading capacities, mechanical stability and are not subject to pressure changes caused by swelling as the organic solvent content of the eluent changes. This can be a problem with other types of monolithic media [21]. The development of mass spectrometers with faster scanning and data acquisition rates has made it possible to take advantage of the improved chromatographic performance achieved.

In the following pages are described some applications of monolithic silica chromatography media coupled to mass spectrometry including the use of monolithic materials as in line enzymic reactors and affinity columns.

14.2
Offline Chromatography, LC MALDI MS

The offline technique of LC MALDI MS [22], has found application particularly in proteomics experiments, where peptides are separated offline by RP-HPLC and the column effluent mixed with MALDI matrix just before deposition onto a MALDI target plate. In this case rapid high-resolution chromatography is advantageous, since the chromatogram is effectively frozen on the MALDI target plate. The mass-spectrum acquisition is not restricted in the time dimension by peak width, but rather by the number of spots on the target. Hence, narrow peaks and rapid chromatography means less spots, and reduced analysis times. This is in sharp contrast to ESI-MS where the mass spectrometer can only acquire spectra as the compounds being analyzed elute from the column that is, for the duration of a chromatographic peak. A generic workflow for an LC MALDI experiment is shown in Figure 14.1. When applied to the separation of peptides in a proteomics experiment, typically the column effluent is applied to the MALDI target as discrete spots, since with this method it is usual to first scan the entire plate, determine the spot location where each precursor ion is most abundant, before calculating and executing a second scan to acquire MS/MS spectra so that further structural information can be obtained to identify the molecules being analyzed. It is also possible to lay down a continuous trace of sample mixed with matrix by spraying the column effluent onto the MALDI target [23], which results in the entire chromatogram effectively being stored as a continuous trace on the MALDI target. This approach has been applied to the quantitation of small molecules such as drug metabolites by MALDI-MS. The MALDI target can be interrogated multiple times by the mass spectrometer and the chromatogram reconstructed from the mass spectral data. Figure 14.2 shows a comparison between a 75-μm ID 5-μm C18 particle packed column and a 150-mm 100-μm ID Chromolith column (Merck) obtained on a TEMPO LC MALDI system (ABSciex, Concord, Canada). The Chromolith column resolved the mixture of tryptic peptides

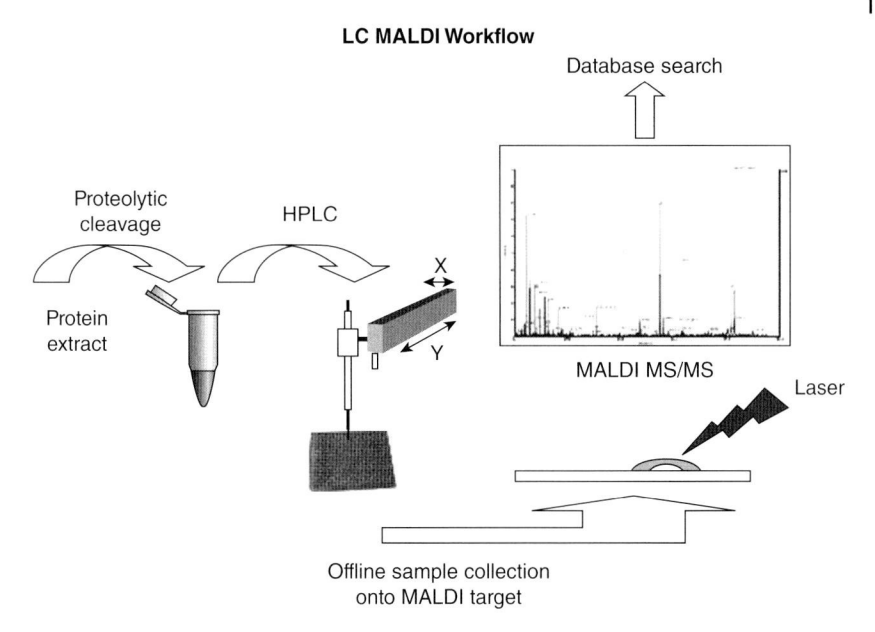

Figure 14.1 A generalized scheme for a liquid chromatography matrix-assisted laser-induced ionization mass spectrometry experiment (LC-MALDI-MS) for proteomics analysis.

significantly better than the packed-bed column, leading to the identification of, in this case, more peptides.

14.3
Online ESI LC/MS/MS for Proteomics and Selected Reaction Monitoring (SRM)

The configuration of a typical ESI LC/MS/MS system is shown in Figure 14.3. There are several types of mass spectrometers currently used in this mode of analysis (see Table 14.1) as well as different methods of ionization of the LC stream [24]. In this type of setup the mass spectrometer functions as a powerful detector. The instrument constantly monitors the total ion current (TIC), which represents the relative abundance of all the ions passing through the analyzer, from which a mass chromatogram, often referred to as a base-peak chromatogram (BPC) can be extracted by displaying only the most abundant ion signal at a given time. This type of display is closely akin to that produced by say an ultraviolet (UV) detector. It is also possible to generate extracted ion chromatograms (XIC) where only signals derived from a specific mass of interest are displayed. In tandem mass spectrometers, which are the bulk of instruments in use today, these primary signals (termed precursor or parent ions) can be used to trigger the machine to obtain MS/MS spectra by fragmenting selected precursor ions [25] to produce daughter or product ions. It is from these fragmentation spectra

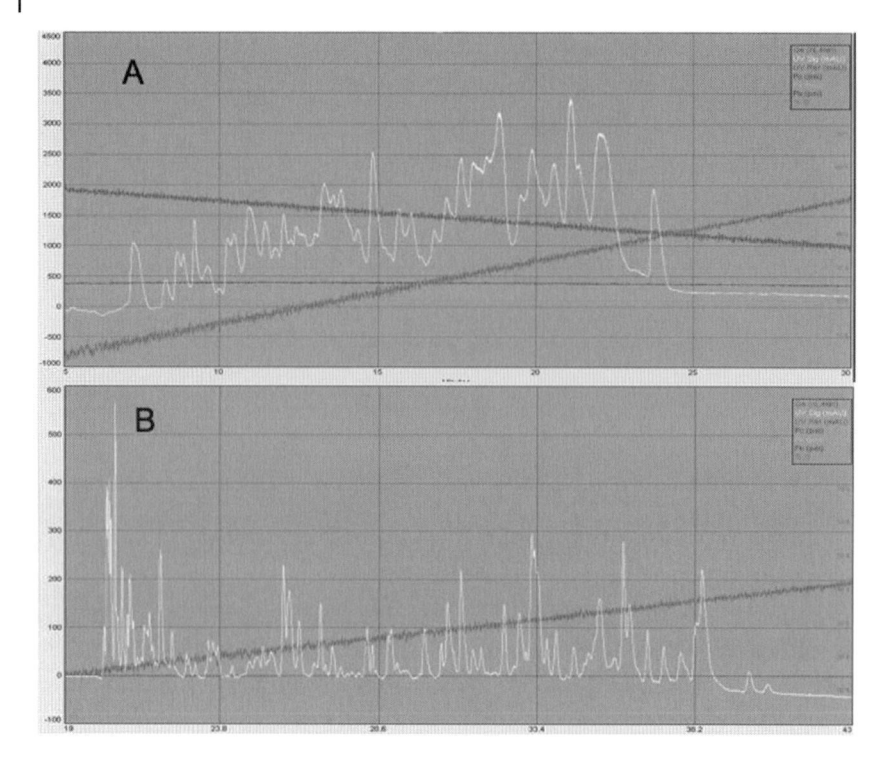

Figure 14.2 A comparison between A. a 150-mm, 75-μm ID C18 packed column and B. a 150-mm, 100-μm ID Chromolith C18 column (Merck). The separation was achieved with linear gradient from 100% A (0.1%TFA in water) 0% B (0.1%TFA in 90% acetonitrile) to 50% A, 50%B over 45 min flowing at 500 nl per minute for the packed column and 2 μl per minute for the monolithic column.

that further structural information can be deduced. So-called shotgun proteomics, based on rapid analysis of RP separated peptides by mass spectrometry has become widely used and the robust nature of monolithic silica columns makes them ideal in this type of application [26] where many complex samples need to be analyzed.

Selected-reaction monitoring (SRM) or multiple-reaction monitoring (MRM) mass spectrometry is the method of choice for the analysis and quantitation of drug metabolites especially by the pharmaceutical industry [27–32] and doping agencies. It is also widely used to analyze naturally occurring metabolites [33], such as those monitored in newborn screening [1] and for the detection of environmental contaminants such as pesticides [34] and antimicrobials [35]. This technique is also beginning to have a significant impact on the analysis of biomarkers for human disease [36, 37]. By combining the separation capacity of RP columns with the mass filtering capability of triple quadrupole (QQQ) and triple quadrupole linear ion-trap hybrid mass spectrometers operating in SRM mode,

Table 14.1 Types of mass spectrometer.

Type of instrument	Type of ionization	Resolution	MS/MS	Major application
Triple Quadrupole, **QQQ**	ESI, MALDI	Low	Yes	Metabolite analysis
Time-of-flight, **TOF**	ESI, MALDI	High	No	Molecular mass profiling
Quadrupole time-of-flight hybrid, **QQTOF**	ESI, MALDI	High	Yes	Proteomics
Ion trap (Paul trap), **IT**	ESI	Low	Yes	Proteomics
Linear ion trap, **LIT**	ESI	Low	Yes	Metabolite analysis, Proteomics
Triple quadrupole linear ion trap hybrid, **qTrap**	ESI	Low	Yes	Metabolite analysis, biomarker analysis
Linear ion trap/orbitrap hybrid **LTQ/Orbitrap**	ESI, MALDI	Very high	Yes	Proteomics
Time-of-flight/time-of-flight, **TOF/TOF**	MALDI	High	Yes	Proteomics
Fourier transform ion cyclotron resonance, **FTICR**	ESI, MALDI	Very High	Yes	Analysis of isotopes, Metabolites, large molecules, e.g. proteins, protein complexes

ESI, electrospray ionization; MALDI, Matrix-assisted laser desorption ionization.

Electrospray LC MS System

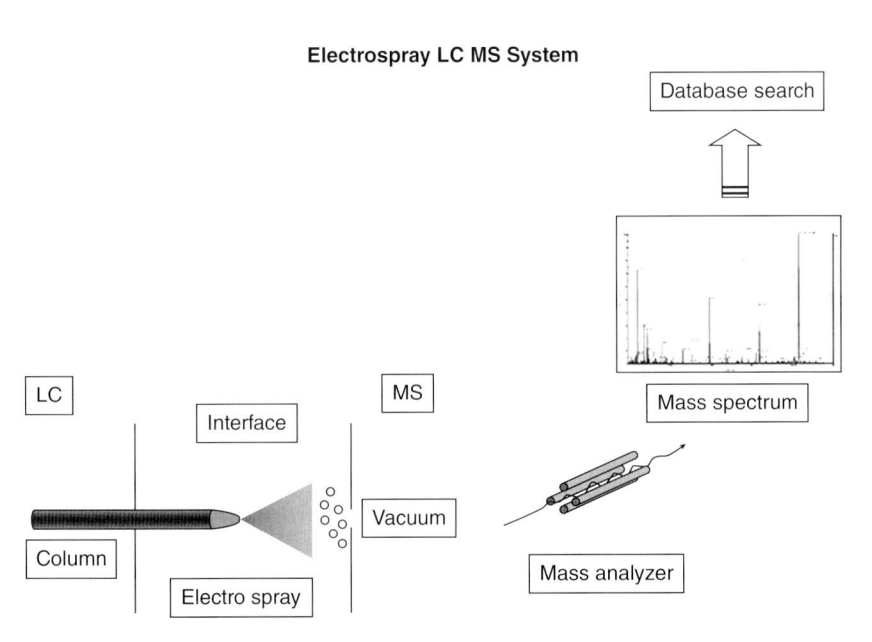

Figure 14.3 A typical layout for a liquid chromatography tandem mass spectrometry system (LC/MS/MS).

it is possible to quantify many proteins in a single analysis from for example cell lysates [38, 39] and blood plasma or serum [40]. This is the most sensitive LC/MS/MS method currently available and has been shown to be able to measure proteins at concentrations as low as 50 copies per cell [38].

14.4
Online Reactors and Affinity Columns Coupled to Mass Spectrometry

A potentially powerful application of silica monolithic columns is the manufacture of online enzyme reactors and affinity columns that can be coupled to mass spectrometers for analysis of the compounds generated or retained by the columns [23, 41–48]. Protein-doped silica monolithic columns have some significant advantages for the generation of affinity and/or enzyme reactors. The sol-gel process is a low-temperature inorganic polymerization reaction that occurs in aqueous solvents under mild pH conditions. Additives can be included along with the buffered protein sample to modify the final properties of the material. For example, polyethylene glycol (PEG) or other polymers, free or tethered sugars can influence material charge, polarity and porosity. These properties allow proteins

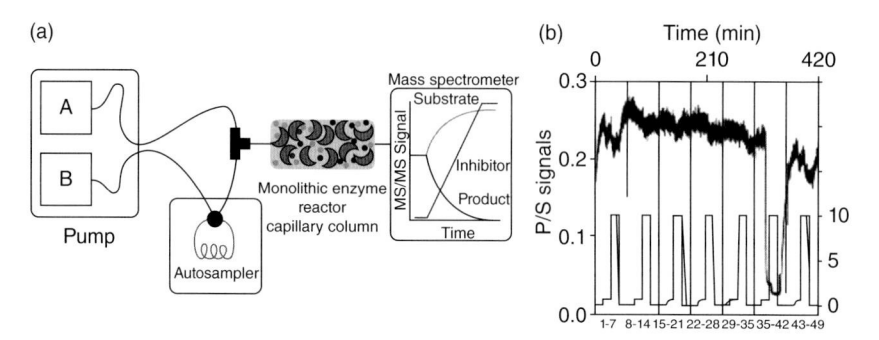

Figure 14.4 (a) Method for inhibitor screening using enzyme reactor chromatography interfaced to mass spectrometry. All mobile phases and samples contain an identical concentration of substrate, while inhibitors are introduced by pump B through an autosampler loop. This configuration allows automated screening of multiple compound mixtures. As the concentration of an inhibitor increases, the enzymes within the column are inhibited. (b) Mixture screening with immobilized enzyme reactor tandem mass spectroscopy (IMER-MS/MS). The figure shows primary screens of 49 compounds against immobilized adenosine deaminase (ADA). The primary screen indicates that a compound in the mixture containing compounds 36–42 reduces the product-to-substrate (P/S) ratio, and thus inhibits ADA. The square injection profiles, shown by the bottom traces in each panel, are programmed by the Eksigent pump and show when the column is exposed to either 500 nmol/l or 10 μmol/l of test mixture. (Reproduced from [26] Hodgson et al., with permission from Analytical Chemistry, **77**, 7152–7519, © 2005).

to be captured without recourse to covalent coupling or other harsh procedures, which can affect their structure and with it their activity. Monolithic silica columns have been used to entrap membrane proteins successfully, so that their binding affinities for small molecules can be measured by frontal affinity chromatography mass spectrometry (FAC-MS/MS) [43]. Immobilized-enzyme reactors (IMER) [41] can be interfaced with mass spectrometers so that both the substrate and products of an enzymic reaction can be quantified without the need to separate them chromatographically prior to MS analysis. It is also possible to screen mixtures of substrates and inhibitors by this approach [43]. This latter technique is outlined in Figure 14.4.

Monolithic silica has also been applied in pipette tip format for sample preparation for mass spectrometry [49], and high-throughput rapid trypsin cleavage of proteins or proteomics [50, 51].

14.5 Conclusion

Despite monolithic silica media being available for some time, they have yet to find wide application in the LCMS context. This situation may change as commercial vendors such as Merck KGaA, Darmstadt, Germany (http://www.merck-chemicals.com/chromolith-hplc-columns/c_oUeb.s1LrkgAAAEWq.AfVhTl) and GL sciences, Tokyo, Japan (http://www.glsciences.com/index.html) introduce more products to the market. Another barrier to their uptake in connection with mass spectrometry is, that by far the largest application of LCMS is the analysis of small molecules such as pesticides and drug metabolites by the pharmaceutical industry [52]. There is generally a lag period before new technologies are adopted in this area, because of the rigorous control of standard operating procedures (SOP) imposed on these types of analysis [53]. The field of protein affinity chromatography stands to benefit enormously from the unique properties of these materials that permit proteins to be captured under mild conditions and their activity monitored by MS. The potential sensitivity gains that could be achieved from the use true nanoelectrospray flow rates (less than 50 nl per minute) requires columns that can operate at such low flow rates [54, 55], with good chromatographic properties and low dead volume (ID less than 50 μm, which is available commercially). Monolithic silica fulfills these criteria and columns as small as 20 μm ID have been made experimentally [54, 56].

The elusive quest for biomarkers of disease, which will ultimately provide the foundation of the much lauded idea of personalized medicine [57, 58] requires technologies that can measure many thousands of compounds across 10 orders of magnitude. Affinity chromatography [59] combined with RP HPLC and mass spectrometry offers one of the most promising technologies likely to solve this highly challenging analytical problem. Again, monolithic silica has great promise to make an important contribution. This author awaits their more widespread use with great interest.

References

1 Shushan, B. (2010) A review of clinical diagnostic applications of liquid chromatography–tandem mass spectrometry. *Mass Spectrom. Rev.*, **29**(6):930–944.

2 Gstaiger, M., and Aebersold, R. (2009) Applying mass spectrometry-based proteomics to genetics, genomics and network biology. *Nat. Rev. Genet.*, **10**, 617–627.

3 Wepf, A., Glatter, T., Schmidt, A., Aebersold, R., and Gstaiger, M. (2009) Quantitative interaction proteomics using mass spectrometry. *Nat. Methods*, **6**, 203–205.

4 Yates, J.R., Ruse, C.I., and Nakorchevsky, A. (2009) Proteomics by mass spectrometry: approaches, advances, and applications. *Annu. Rev. Biomed. Eng.*, **11**, 49–79.

5 Boersema, P.J., Divecha, N., Heck, A.J., and Mohammed, S. (2007) Evaluation and optimization of ZIC-HILIC-RP as an alternative MudPIT strategy. *J. Proteome Res.*, **6**, 937–946.

6 Washburn, M.P., Wolters, D., and Yates, J.R., 3rd (2001) Large-scale analysis of the yeast proteome by multidimensional protein identification technology. *Nat. Biotechnol.*, **19**, 242–247.

7 Dai, J., Shieh, C.H., Sheng, Q.H., Zhou, H., and Zeng, R. (2005) Proteomic analysis with integrated multiple dimensional liquid chromatography/mass spectrometry based on elution of ion exchange column using pH steps. *Anal. Chem.*, **77**, 5793–5799.

8 Delahunty, C., and Yates, J.R., 3rd (2005) Protein identification using 2D-LC-MS/MS. *Methods*, **35**, 248–255.

9 Opiteck, G.J., Lewis, K.C., Jorgenson, J.W., and Anderegg, R.J. (1997) Comprehensive on-line LC/LC/MS of proteins. *Anal. Chem.*, **69**, 1518–1524.

10 Liu, X., Valentine, S.J., Plasencia, M.D., Trimpin, S., et al. (2007) Mapping the human plasma proteome by SCX-LC-IMS-MS. *J. Am. Soc. Mass Spectrom.*, **18**, 1249–1264.

11 Tolley, L., Jorgenson, J.W., and Moseley, M.A. (2001) Very high pressure gradient LC/MS/MS. *Anal. Chem.*, **73**, 2985–2991.

12 MacNair, J.E., Patel, K.D., and Jorgenson, J.W. (1999) Ultrahigh-pressure reversed-phase capillary liquid chromatography: isocratic and gradient elution using columns packed with 1.0-micron particles. *Anal. Chem.*, **71**, 700–708.

13 Shen, Y., Smith, R.D., Unger, K.K., Kumar, D., and Lubda, D. (2005) Ultrahigh-throughput proteomics using fast RPLC separations with ESI-MS/MS. *Anal. Chem.*, **77**, 6692–6701.

14 Nguyen, D.T., Guillarme, D., Rudaz, S., and Veuthey, J.L. (2006) Fast analysis in liquid chromatography using small particle size and high pressure. *J. Sep. Sci.*, **29**, 1836–1848.

15 Nguyen, D.T., Guillarme, D., Heinisch, S., Barrioulet, M.P., et al. (2007) High throughput liquid chromatography with sub-2 micron particles at high pressure and high temperature. *J. Chromatogr. A*, **1167**, 76–84.

16 Guillarme, D., Ruta, J., Rudaz, S., and Veuthey, J.L. (2009) New trends in fast and high-resolution liquid chromatography: a critical comparison of existing approaches. *Anal. Bioanal. Chem.* 2010 Jun;**397**(3):1069–1082.

17 Ikegami, T., Dicks, E., Kobayashi, H., Morisaka, H., et al. (2004) How to utilize the true performance of monolithic silica columns. *J. Sep. Sci.*, **27**, 1292–1302.

18 Ito, S., Yoshioka, S., Ogata, I., Yamashita, E., et al. (2005) Capillary high-performance liquid chromatography/electrospray ion trap time-of-flight mass spectrometry using a novel nanoflow gradient generator. *J. Chromatogr. A*, **1090**, 178–183.

19 van de Meent, M.H., and de Jong, G.J. (2007) Improvement of the liquid-chromatographic analysis of protein tryptic digests by the use of long-capillary monolithic columns with UV and MS detection. *Anal. Bioanal. Chem.*, **388**, 195–200.

20 van de Meent, M.H., and de Jong, G.J. (2009) Potential of long capillary monolithic columns for the analysis of protein digests. *J. Sep. Sci.*, **32**, 487–493.

21 Gu, B., Chen, Z., Thulin, C.D., and Lee, M.L. (2006) Efficient polymer monolith for strong cation-exchange capillary liquid chromatography of peptides. *Anal. Chem.*, **78**, 3509–3518.

22 Mueller, D.R., Voshol, H., Waldt, A., Wiedmann, B., and Van Oostrum, J. (2007) LC-MALDI MS and MS/MS–an efficient tool in proteome analysis. *Subcell. Biochem.*, **43**, 355–380.

23 Kovarik, P., Hodgson, R.J., Covey, T., Brook, M.A., and Brennan, J.D. (2005) Capillary-scale frontal affinity chromatography/MALDI tandem mass spectrometry using protein-doped monolithic silica columns. *Anal. Chem.*, **77**, 3340–3350.

24 Droste, S., Schellentrager, M., Constapel, M., Gab, S., *et al.* (2005) A silica-based monolithic column in capillary HPLC and CEC coupled with ESI-MS or electrospray-atmospheric-pressure laser ionization-MS. *Electrophoresis*, **26**, 4098–4103.

25 Wilm, M. (2009) Quantitative proteomics in biological research. *Proteomics*, **9**, 4590–4605.

26 Guryca, V., Kieffer-Jaquinod, S., Garin, J., and Masselon, C.D. (2008) Prospects for monolithic nano-LC columns in shotgun proteomics. *Anal. Bioanal. Chem.*, **392**, 1291–1297.

27 Xu, R.N., Boyd, B., Rieser, M.J., and El-Shourbagy, T.A. (2007) Simultaneous LC-MS/MS quantitation of a highly hydrophobic pharmaceutical compound and its metabolite in urine using online monolithic phase-based extraction. *J. Sep. Sci.*, **30**, 2943–2949.

28 Xu, R.N., Fan, L., Rieser, M.J., and El-Shourbagy, T.A. (2007) Recent advances in high-throughput quantitative bioanalysis by LC-MS/MS. *J. Pharm. Biomed. Anal.*, **44**, 342–355.

29 Xu, R.N., Vaca, P., Rieser, M.J., and El-Shourbagy, T.A. (2009) Highly sensitive LC-MS-MS analysis of a pharmaceutical compound in human plasma using monolithic phase-based on-line extraction. *J. Chromatogr. Sci.*, **47**, 473–477.

30 Pous-Torres, S., Ruiz-Angel, M.J., Torres-Lapasio, J.R., and Garcia-Alvarez-Coque, M.C. (2009) Performance of a Chromolith RP-18e column for the screening of beta-blockers. *J. Sep. Sci.*, **32**, 2841–2853.

31 Rieux, L., Niederlander, H., Verpoorte, E., and Bischoff, R. (2005) Silica monolithic columns: synthesis, characterisation and applications to the analysis of biological molecules. *J. Sep. Sci.*, **28**, 1628–1641.

32 Heideloff, C., Bunch, D.R., and Wang, S. (2010) A novel HPLC method for quantification of 10 antiepileptic drugs or metabolites in serum/plasma using a monolithic column. *Ther. Drug Monit.*, **32**, 102–106.

33 Bamba, T., and Fukusaki, E. (2009) Separation of hydrophobic metabolites using monolithic silica column in high-performance liquid chromatography and supercritical fluid chromatography. *J. Sep. Sci.*, **32**, 2699–2706.

34 Martinez Vidal, J.L., Plaza-Bolanos, P., Romero-Gonzalez, R., and Garrido Frenich, A. (2009) Determination of pesticide transformation products: a review of extraction and detection methods. *J. Chromatogr. A*, **1216**, 6767–6788.

35 Bogialli, S., and Di Corcia, A. (2009) Recent applications of liquid chromatography-mass spectrometry to residue analysis of antimicrobials in food of animal origin. *Anal. Bioanal. Chem.*, **395**, 947–966.

36 Kitteringham, N.R., Jenkins, R.E., Lane, C.S., Elliott, V.L., and Park, B.K. (2009) Multiple reaction monitoring for quantitative biomarker analysis in proteomics and metabolomics. *J. Chromatogr. B Anal. Technol. Biomed. Life Sci.*, **877**, 1229–1239.

37 Schiess, R., Wollscheid, B., and Aebersold, R. (2009) Targeted proteomic strategy for clinical biomarker discovery. *Mol. Oncol.*, **3**, 33–44.

38 Picotti, P., Bodenmiller, B., Mueller, L.N., Domon, B., and Aebersold, R. (2009) Full dynamic range proteome analysis of S. cerevisiae by targeted proteomics. *Cell*, **138**, 795–806.

39 Picotti, P., Rinner, O., Stallmach, R., Dautel, F., *et al.* (2010) High-throughput generation of selected reaction-monitoring assays for proteins and proteomes. *Nat. Methods*, **7**, 43–46.

40 Anderson, L., and Hunter, C.L. (2006) Quantitative mass spectrometric multiple reaction monitoring assays for major plasma proteins. *Mol. Cell. Proteomics*, **5**, 573–588.

41 Hodgson, R.J., Besanger, T.R., Brook, M.A., and Brennan, J.D. (2005) Inhibitor screening using immobilized enzyme reactor chromatography/mass spectrometry. *Anal. Chem.*, **77**, 7512–7519.

42 Hodgson, R.J., Brook, M.A., and Brennan, J.D. (2005) Capillary-scale monolithic immunoaffinity columns for immunoextraction with in-line laser-induced fluorescence detection. *Anal. Chem.*, **77**, 4404–4412.

43 Lebert, J.M., Forsberg, E.M., and Brennan, J.D. (2008) Solid-phase assays for small molecule screening using sol-gel entrapped proteins. *Biochem. Cell Biol.*, **86**, 100–110.

44 Josic, D., and Clifton, J.G. (2007) Use of monolithic supports in proteomics technology. *J. Chromatogr. A*, **1144**, 2–13.

45 Krenkova, J., and Svec, F. (2009) Less common applications of monoliths: IV. Recent developments in immobilized enzyme reactors for proteomics and biotechnology. *J. Sep. Sci.*, **32**, 706–718.

46 Mallik, R., and Hage, D.S. (2008) Development of an affinity silica monolith containing human serum albumin for chiral separations. *J. Pharm. Biomed. Anal.*, **46**, 820–830.

47 Sakai-Kato, K., and Ishikura, K. (2009) Integration of biomolecules into analytical systems by means of silica sol-gel technology. *Anal. Sci.*, **25**, 969–978.

48 Feng, S., Pan, C., Jiang, X., Xu, S., *et al.* (2007) Fe^{3+} immobilized metal affinity chromatography with silica monolithic capillary column for phosphoproteome analysis. *Proteomics*, **7**, 351–360.

49 Kumazawa, T., Hasegawa, C., Lee, X.P., Hara, K., *et al.* (2007) Simultaneous determination of methamphetamine and amphetamine in human urine using pipette tip solid-phase extraction and gas chromatography-mass spectrometry. *J. Pharm. Biomed. Anal.*, **44**, 602–607.

50 Ota, S., Miyazaki, S., Matsuoka, H., Morisato, K., *et al.* (2007) High-throughput protein digestion by trypsin-immobilized monolithic silica with pipette-tip formula. *J. Biochem. Biophys. Methods*, **70**, 57–62.

51 Ma, J., Liang, Z., Qiao, X., Deng, Q., *et al.* (2008) Organic-inorganic hybrid silica monolith based immobilized trypsin reactor with high enzymatic activity. *Anal. Chem.*, **80**, 2949–2956.

52 Ma, S., and Zhu, M. (2009) Recent advances in applications of liquid chromatography-tandem mass spectrometry to the analysis of reactive drug metabolites. *Chem. Biol. Interact.*, **179**, 25–37.

53 Kuster, M., Lopez de Alda, M., and Barcelo, D. (2009) Liquid chromatography-tandem mass spectrometric analysis and regulatory issues of polar pesticides in natural and treated waters. *J. Chromatogr. A*, **1216**, 520–529.

54 Luo, Q., Shen, Y., Hixson, K.K., Zhao, R., *et al.* (2005) Preparation of 20-micron-i.d. silica-based monolithic columns and their performance for proteomics analyses. *Anal. Chem.*, **77**, 5028–5035.

55 Wilm, M., and Mann, M. (1996) Analytical properties of the nanoelectrospray ion source. *Anal. Chem.*, **68**, 1–8.

56 Luo, Q., Page, J.S., Tang, K., and Smith, R.D. (2007) MicroSPE-nanoLC-ESI-MS/MS using 10-micron-i.d. silica-based monolithic columns for proteomics. *Anal. Chem.*, **79**, 540–545.

57 Aebersold, R., Auffray, C., Baney, E., Barillot, E., *et al.* (2009) Report on EU-USA workshop: how systems

biology can advance cancer research (27 October 2008). *Mol. Oncol.*, **3**, 9–17.

58 Auffray, C., Chen, Z., and Hood, L. (2009) Systems medicine: the future of medical genomics and healthcare. *Genome Med.*, **1**, 2.

59 Schwenk, J.M., Igel, U., Kato, B.S., Nicholson, G., *et al.* (2010) Comparative protein profiling of serum and plasma using an antibody suspension bead array approach. *Proteomics*, **10**, 532–540.

15
Silica Monoliths for Small-Scale Purification of Drug-Discovery Compounds
Alfonso Espada, Cristina Anta, and Manuel Molina-Martín

15.1
Introduction

In today's pharmaceutical business higher demand on compound purity, to remove any ambiguity in compound characterization and screening data, has lead separation sciences to a high level of qualitative and quantitative content [1–5]. Therefore, the role of preparative purification for product isolation is gaining widespread attention in pharmaceutical laboratories [1, 6]. The diversity in chemical structure, the breadth of polarity and the large quantity of material, call for a rugged high-throughput purification solution. In this context, reversed-phase HPLC/UV- or MS-guided collection has become the purification platform of choice to deliver quality leads in medicinal chemistry laboratories [7–10]. The recent advances in automation, detection and method development have made it possible to use HPLC equipment with particle-packed short columns (21–30 × 50–100 mm) containing the classic 5-μm small silica particles for high-throughput purification. However, purification laboratories are still driven to maximize throughput to meet increased sample numbers. Low sample solubility, sample precipitation, low column capacity and solvent evaporation remain the major drawbacks for the purification of new lead compounds. As a result, the purification platform is often considered a long and expensive process.

We recently found that the online dilution (OLD) injection mode addresses those issues without sacrificing chromatography integrity and significantly enhances purification productivity. Particle-packed column sizes of 10–21 × 100 mm operated at flow rates of 5.5 and 25 ml/min, respectively, were successfully integrated into our purification platform. Thus, a refined protocol for processing samples ranging from 10 to 500 mg in a timely fashion was effectively achieved [9]. In an endeavor to further increase sample throughput, the monolithic silica columns were evaluated. Monolithic HPLC silica-based sorbent consists of a single porous-silica rod with a well-defined bimodal pore structure, macropore size of ca. 2 μm and mesopore size of ca. 13 nm. Macropores serve as through-pores of high permeability (due to bed porosity) which allow high-speed separations under conditions of low column backpressure, whereas the presence of mesopores creates an extended surface area of approximately 300 m^2/g for a selective absorption of the analytes [11–13]. Several studies have revealed that the efficiency and performance of monolithic columns, even operated at high flow rates, are equivalent to that of columns packed with 3.5 and 3 µm particles [14–16]. We considered these features ideal for increasing purification speed where higher flow rates as well as higher mass loading are desired. Current applications of monolithic HPLC columns are for analytical purposes. In a preliminary work, we found that crude reaction mixtures in the range from 5 to 25 mg, dissolved in DMSO:MeOH, can be productively purified on a 4.6 × 100 mm column in a single run [17]. These striking results prompted us to take advantage of the first commercially available 10 and 25 × 100 mm column sizes (under the trade name Chromolith® RP-18e) for highly efficient separations. Representative examples of the impact of monolithic silica columns on productivity for the purification of drug-discovery compounds are discussed in this chapter.

15.2
Instrumental and Operating Considerations

15.2.1
Analytical Conditions

Analytical chromatographic separation was carried out on an Agilent 1100 liquid chromatography system equipped with a solvent degasser, quaternary pump, auto sampler, column compartment and a diode-array detector (Agilent Technologies, Waldbronn, Germany). The UV wavelength was set at 214 nm. Electrospray mass spectrometry measurements were performed on a MSD ion-trap mass spectrometer (Agilent Technologies, Palo Alto, CA, USA) interfaced to the HPLC system. MS measurements were acquired in positive ionization mode over the mass range of 100–700 amu. The following ion-source parameters were used: drying gas flow, 12.0 l/min; nebulizer pressure, 60 psig; drying gas temperature, 350 °C; capillary voltage, −3.5 kV. Data acquisition and integration for LC-UV and MS detection were performed using Chemstation software (Agilent Technologies).

Monolithic analytical columns employed in this study were Chromolith Performance RP-18e 4.6 × 100 mm columns (Merck KGaA, Darmstadt, Germany). The acidic mobile phase was water (solvent A) and acetonitrile (solvent B), both containing 0.05% trifluoroacetic acid (TFA). Meanwhile, the alkaline mobile phase was 10 mM or 20 mM ammonium bicarbonate (NH$_4$HCO$_3$, solvent A) and acetonitrile (solvent B). The flow rate prior to the mass spectrometer was 1 ml/min, which was split in a ratio of 3 to 1 to deliver 250 µl min^{-1} into the electrospray interface and 750 µl min^{-1} to the waste reservoir.

15.2.2
Preparative Conditions

The HPLC system for testing the suitability of the monolithic 10×100 mm columns for semipreparative separation was an analytical Agilent 1100 HPLC series equipped with a Gilson 215 injector/collector, a diode-array detector with a preparative flow cell (0.06 mm path length), a quaternary pump to provide a flow rate up to 10 ml/min, a column compartment with a 2-position/10-port valve, and a 12-port solvent valve selector that allows use of mobile phases with different buffers at acid, basic and neutral pH [3]. An auxiliary isocratic pump is employed for the OLD configuration, as reported elsewhere [18]. CC-MODE software package (Agilent Technologies) was used to manage the automated UV-guided collection in this configuration.

The instrumentation for performing separation on a 25×100 mm column was an Agilent 1100 LC-MSD Series Purification System equipped with a Gilson 215 injector/collector, a quaternary pump to provide a flow rate up to 10 ml/min (used as make-up pump), two Agilent 1100 Series preparative pumps to provide a flow rate up to 100 ml/min, a 2-position column compartment, and two 12-port solvent valve selectors that allow use of mobile phases with different buffers at acid, basic and neutral pH. An Agilent Active splitter postcolumn led an adjustable portion of the sample to the UV and MS detectors in a split ratio of 1:4000. The make-up pump was an Agilent 1100 Series quaternary pump. The UV detector was an Agilent 1100 Series diode-array detector with an analytical flow cell (6 mm path length). Electrospray mass spectrometry measurements were performed on a MSD mass spectrometer (Agilent Technologies, Palo Alto, CA, USA) interfaced to the Agilent 1100 HPLC system. MS measurements were acquired simultaneously in both positive and negative ionization modes over the mass range of 100–700 amu. The following ion-source parameters were used: drying gas flow, 10.0 l/min; nebulizer pressure, 35 psig; drying gas temperature, 350 °C; capillary voltage, 2.5 kV for positive and negative mode. Make-up flow was 0.9 ml/min, and the make-up solvent was 80:20 (acetonitrile:water) with 0.1% formic acid and 80:20 (acetonitrile:10 mM NH_4HCO_3 in water) for low and high pH purifications, respectively. The configuration of the HPLC pipework was modified by us to allow the employment of OLD without the need for an additional isocratic pump to carry the sample to the mixer, as reported elsewhere [18]. Automated MS-guided collection through CC-MODE software package was used in this configuration.

The chromatographic separation was performed on Chromolith SemiPrep RP-18e 10×100 mm and Chromolith prep RP-18e 25×100 mm columns (Merck KGaA, Darmstadt, Germany) unless otherwise stated. The mobile phases applied were the same as in the analytical methods. The collection parameters signal slope at the leading and tailing edge of the chromatographic peak as well as the minimum intensity threshold (dependent on the complexity of the mixture, peak spacing and purity requirements) were set for each sample depending on the intensity of the signal of the peak to be collected.

15.3
Preparative Separations and Sample Loading

15.3.1
Semipreparative Monolithic 10 × 100 mm Column

The configuration of the HPLC/UV analytical system was set up for the small scale purification of crude reaction mixtures ranging from 10 to 100 mg using conventional silica C18 10 × 100 mm, 5-μm particle-packed columns. The most favorable flow rate in terms of efficiency (column backpressure, sample loading and purification recovery) for these columns was 5.5 ml/min. Using a gradient time of 8 min (total run time of 10 min), a large number of samples was successfully purified on this system (unpublished results).

A binary mixture of drug compounds (propranolol hydrochloride, pK_a 9.2/Log P 3.1, and nifedipine, pK_a 4.3/Log P 2.97, dissolved in DMSO:MeOH) was used to mimic the behavior of drug-discovery samples (mostly bases and/or compounds with moderate solubility) and to determine the most favorable gradient time and flow rate for the monolithic column. As it is not desirable to collect nonvolatile solvents, the separation between the solvent sample DMSO (fronting peak) and the compound of interest (propranolol.HCl) was also included in these studies. Considering critical LC preparative parameters like resolution (R_s),[1] peak width ($W_{10\%}$) and collection volume (CV) of the target fraction, a fixed gradient time of 4 min with mobile-phase flow rates from 4 to 8 ml/min were investigated for the monolith 10 ID columns. Figure 15.1 shows the relationship of R_s, $W_{10\%}$ and CV versus flow rate. It is well known that flow rate is inversely proportional to $W_{10\%}$ due to longitudinal diffusion. Herein, a significant decrease in $W_{10\%}$ is observed by increasing flow rates. Although the $W_{10\%}$ is slightly increased from 6 to 8 ml/min, the higher flow rate provides the maximum R_s between the fronting and target peaks (DMSO and propranolol.HCl). Taking also into account that CV of propranolol is increased at higher flow rates; 8 ml/min was fixed as the most favorable flow rate for this column on this system.

Chromatograms in Figure 15.2 correspond to the preparative separation of the standard mix on particulate (5-μm particle size) and monolith 10 × 100 mm columns (operated under most favorable condition used by the authors). Because of its high efficiency and stability with acidic mobile phases, Kromasil C18 packing material was selected for this particular test [6]. Although 20 mg of material are very well separated on the particle-packed C18 column in a single run, sample loading on the monolith column is much higher (see chromatograms 2A and 2B). As shown in chromatograms 2B and 2C, lower injection volumes allow increased mass loading without affecting baseline resolution between peaks as a result of narrower peak width. It is worth noting that previous studies indicated

---

1) Resolution is calculated according to the Tangent method (United State Pharmacopoeia) $R_s = 2(T_2 - T_1)/(W_2 + W_1)$. T_1 and T_2 are the retention times of peaks 1 and 2, and W_1 and W_2 are the base width of the peaks 1 and 2.

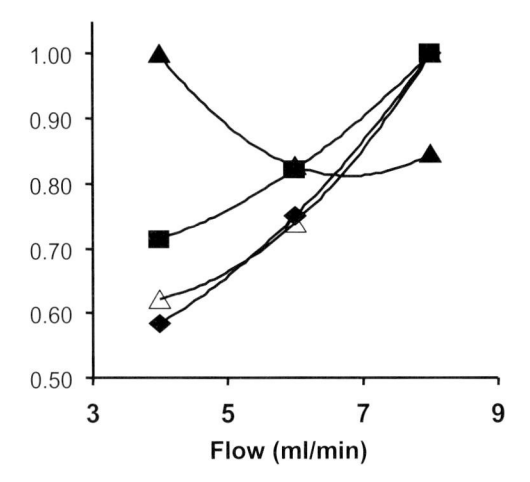

Figure 15.1 Optimization functions for 10 × 100 mm monolithic columns normalized to the maximum measured value. Gradient programming: 2 min at 10% B, 4.5 min gradient from 10 to 90% B, 1.2 min at 95% B. Mobile phase A: 0.05% TFA in water and B: 0.05% TFA in ACN. △: resolution for pair DMSO/propranolol, ■: resolution for pairs propranolol/nifedipine, ▲: $W_{10\%}$ for propranolol, ♦: collected volume for propranolol.

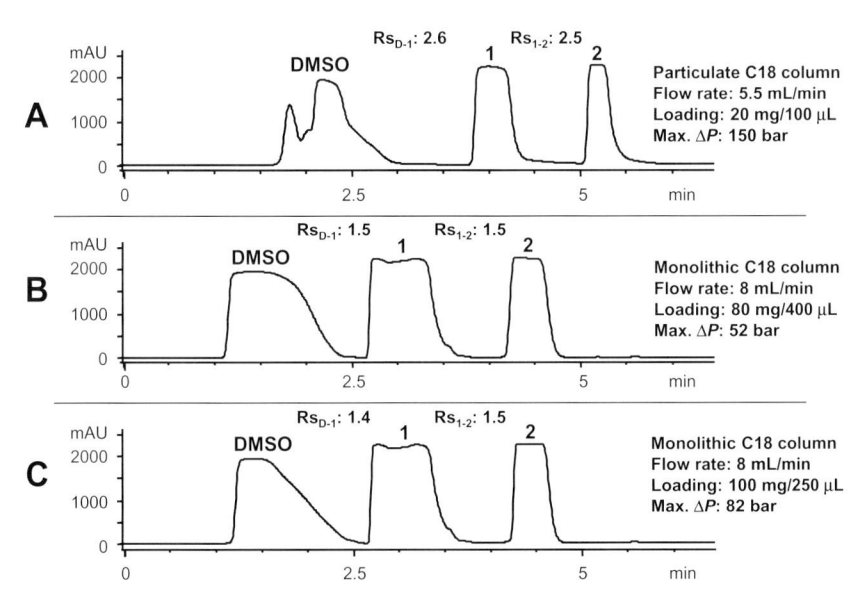

Figure 15.2 Sample loading on monolithic RP-18e *versus* particulate C18 5 μm columns. Gradient programming: 1 min at 10% B, 4 min gradient from 10 to 90% B, 0.2 min up to 95% B, 1.0 min at 95% and 0.3 min back to initial conditions. Mobile phase A: 0.05% TFA in water and B: 0.05% TFA in ACN. Peak 1: propranolol.HCl, Peak 2: nifedipine. R_s: resolution between peaks. Max. ΔP: Maximum system pressure influenced by large mass and DMSO injection loading.

a remarkable peak distortion when mass loading was increased [18]. We noticed that the influence of larger volumes of DMSO on peak distortion is significantly more critical than sample loading [9]. Problems associated with column backpressure and sample precipitation inside the column hinders the separation of more than 20 mg of this binary mixture in the particulate column. Herein, it is important to highlight that the pressure limitation and slow diffusion kinetics at high flow rates limit the use of conventional C18 absorbents for maximum loading capacity [19].

15.3.2
Preparative Monolithic 25 × 100 mm Column

Despite the interest of purifying relatively small amounts of crude material to yield 5–50 mg of pure compounds, purification of larger amounts is also a major requirement in the earlier stage of drug discovery (typically in the range 100–1000 mg). In an effort to further exploit the potential of monoliths, a Chromolith® RP-18e 25 × 100 mm column was set up in an instrument configured to work at flow rates above 10 ml/min.

To optimize gradient time and flow rate for this setup, a mixture containing three standards with different ranges of polarities and chemical functionalities

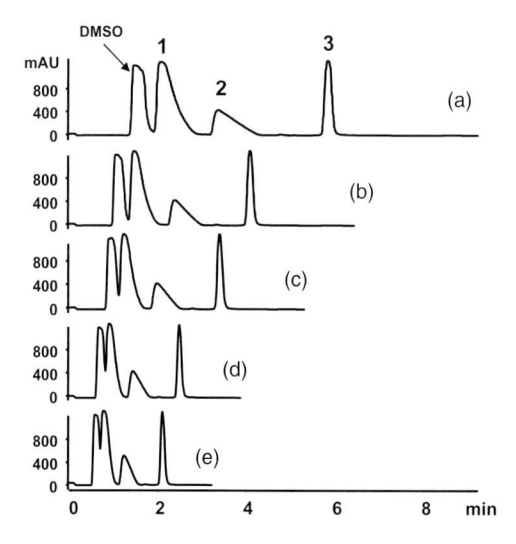

Flow (ml/min)	X (min)	Y (min)	Z (min)	Run time (min)
35	1.4	5.7	2.1	9.2
50	1	4	1.5	6.5
65	0.8	3	1.2	5
80	0.6	2.5	0.9	4
95	0.5	2.1	0.8	3.4

Figure 15.3 Flow optimization for high-throughput purifications on monolithic RP-18e 25 × 100 mm column using focused gradients. Left: chromatograms for mixtures containing 33.3 mg/ml each of Propranolol hydrochloride (1), Verapamil hydrochloride (2) and Niflumic acid (3) in DMSO (250 µl/injection); gradient elution: mobile phase A: 0.05% TFA in water and B 0.05% TFA in MeCN. Flow rates: (a) 35, (b) 50, (c) 65, (d) 80 and (e) 95 ml/min, maximum backpressures: (a) 19, (b) 30, (c) 42, (d) 56 and (e) 70 bars. Right: gradient programming for each of the five injections: X min (30% B), Y min (gradient from 30 to 70% B), and Z min (95% B).

was selected. 250 μl of the mixture (total loading 25 mg) was injected in each run (33 mg/ml propranolol.HCl, 33 mg/ml verapamil.HCl, pK_a 9.0/Log P 3.9 and 33 mg/ml niflumic acid, pK_{a1} 1.7, pK_{a2} 4.7/Log P 4.9 dissolved in DMSO:MeOH 1:1). Mobile phase flow rates from 35 to 95 ml/min were investigated using adjusted gradients to have constant column volumes (see Figure 15.3) [20, 21]. Once the optimum gradient programming had been determined, the mobile-phase flow rate was further tuned for such specific conditions. For simplicity, only parameters involving the peak of interest (verapamil.HCl) were considered. As expected from previous results, the significant decrease in peak width with increasing flow rates makes the collected volume for an individual peak almost unaffected against higher flow rates (see Figure 15.3). However, as the flow rate surpasses 60 ml/min there is a significant loss in resolution due to the closer elution of the peaks of interest. Furthermore, from a practical point of view, and given instrumental and column limitations (backpressures ca. 70 bar, close to 100 bar recommended maximum pressure by the manufacturer for this monolithic column size), it can be difficult to handle flow rates up to 100 ml/min with automated MS-guided collection. A flow rate of 60 ml/min was therefore selected

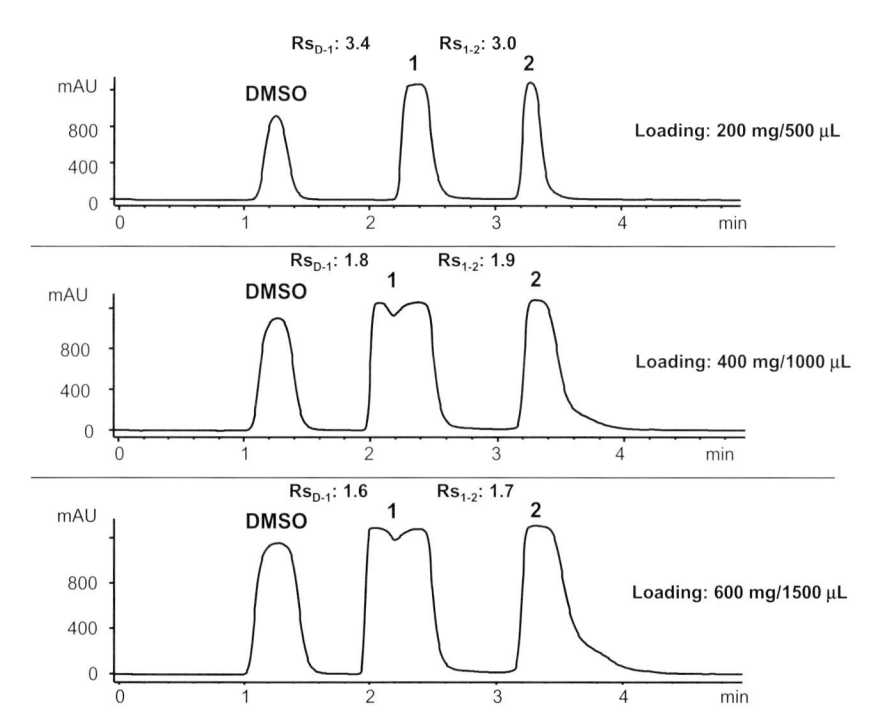

Figure 15.4 Sample loading on monolithic RP-18e 25 × 100 mm column. Flow rate: 60 ml/min. Gradient programming: 0.8 min at 10% B, 3 min gradient from 10 to 90% B, 0.05 min up to 95% B, 1.15 min at 95%. Mobile phase A: 0.05% TFA in water and B: 0.05% TFA in ACN. Peak 1: propranolol.HCl, Peak 2: nifedipine. R_s: resolution between peaks.

as the most favorable to allow the higher throughput without sacrificing separation efficiency. It is important to remark that the 10-mm ID monolithic column was evaluated at flow rates of 15 and 18 ml/min using this system. In terms of chromatographic performance, no substantial improvements were obtained in comparison to the analytical HPLC system.

The effectiveness of the gradient time and mobile phase flow rate were measured by injecting the binary mixture under overloading conditions. Figure 15.4 depicts the chromatograms corresponding to large injection volumes on the 25-mm ID monolithic column. Up to 600 mg of the standard mixture (dissolved in DMSO:MeOH 1:1) were successfully loaded. There is no doubt that R_s for the fronting peak, and peaks 1 and 2, is satisfactory even when overloading is taking place under short run-time conditions, possibly thanks to the high linear velocity of the mobile phase accessible for these columns, which minimizes precipitation issues, and the high surface area intrinsic to monolithic columns. These results confirm that the higher permeability of monoliths permits high flow rates, thus increasing the throughput of the purification platform while maintaining efficiency. In addition, it also supports the use of DMSO as a standard injection solvent with minimal chromatographic impact on these columns, an important consideration when purifying poorly soluble compounds in reversed-phase solvents.

15.4
Purification of Drug-Discovery Compounds

The purification of Lilly proprietary research samples is illustrated in Figure 15.5. Our automated purification approach is divided into three steps: first, compounds of interest (COI) are identified by analytical LC/MS with generic gradient conditions (see UV and extracted ion chromatograms in Figure 15.5, left), secondly, once UV/MS collection parameters are optimized by means of an in-house software program, analytical methods are successfully transferred to preparative LC/UV or LC/MS using in-house automatic tailored gradient algorithm (see preparative chromatograms in Figure 15.5, middle). Finally, fractions containing the COI are evaporated to dryness and analyzed by analytical LC/MS with generic gradient conditions to assess purity (Figure 15.5, right). Chromatograms 5A correspond to the separation of a final reaction mixture on a 10-mm ID column carried out on the analytical HPLC/UV system. The goal of this configuration is to obtain a robust and flexible purification platform to deal with single compounds and libraries in small quantities by using semipreparative monolithic columns. This goal is definitely fulfilled with good quality separation in a very short time, for more than 50 mg of a crude mixture on this column (Figure 15.5a). The CV and recovery of the COI (pK_a 3.5, Log P 3.9, MW 334) were found to be 15 ml and 95% from the semipreparative column. Interestingly, the reproducibility from the analytical (generic gradient) to preparative (tailored gradient) HPLC is very good; most of the undesired peaks around the COI are also detected and identified in

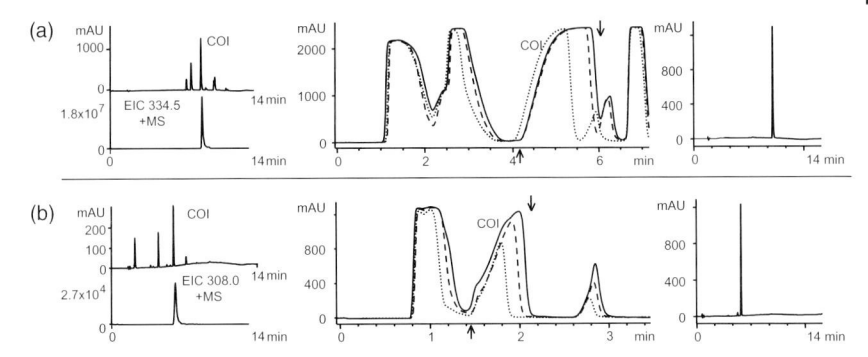

Figure 15.5 Purification of Lilly research samples on Chromolith® RP-18e columns. Left and right: initial (including extracted ion chromatograms) and final analysis, respectively, on 4.6 mm ID Chromolith® RP-18e with 10 min gradient from 10 to 95% B, 2 min at 95%B and 2 min back to initial conditions. Middle: preparative injections (purification) on 10 × 100 mm (a) and 25 × 100 mm (b) columns size. Gradient programming for 10 mm ID column (8.0 ml/min): 2 min at 45% B, 3.5 min gradient from 45 to 50% B, 0.5 min up to 95% B, 1.0 min at 95% and 0.2 min back to initial conditions. Quantities injected (a): 30 (dotted), 40 (dashed) and 50 mg (solid) in 300 μl (DMSO:MeOH 1:1). Gradient programming for 25 mm ID column (60 ml/min): 0.8 min at 30% B, 1.5 min gradient from 30 to 40% B, up to 99% B, 1.2 min at 99%. Quantities injected (b): 200 (dotted), 350 (dashed) and 600 mg (solid) in 260, 460 and 800 μl DMSO. Mobile phase, A: 0.05% TFA in water and B: 0.05% TFA in ACN. Arrows indicate the start (up) and end (down) of the collection.

the preparative chromatogram. In fact, the analyses of the collected fractions after solvent removal to verify the success of the purification, afforded purity levels >95%. Given the dimensions of the 25 × 100 mm column, its loading capacity and the achievable throughput, it is possible to widen the scope of this RP preparative strategy to other purposes such as the purification of intermediates or advanced molecules at a medium scale (up to a few grams), where large amounts of material of very high purity are required to support ADME/toxicological studies. An example corresponding to a Lilly proprietary intermediate (pK_a 3.7, Log P 1.5, MW 307), required for rapid parallel synthesis (RPS) is shown in chromatograms 5B. Considering the invested time for sample handling by the injector/collector prior to chromatographic separation and a maximum loading of 600 mg/0.8 ml DMSO per injection, a throughput of above 4 g/h was reached for this particular sample. This purification strategy yielded 2.6 g of the COI (as TFA salt) with purity higher than 95% following the purification of 5.2 g (quantitative purity of the crude material 41%); a recovery of 90% was achieved.

The separation power and robustness of this purification platform has been demonstrated on the purification of batches containing 5–250 mg/COI with a single injection and one fraction collection. The data shown in Figure 15.6 represents the purification of small batches from multistep RPS and small targeted libraries on 10 and 25 mm ID columns using mobile phase pH of 2.5, 6.8 and 8. Even though the initial purity of some samples was lower than 20%, final purities

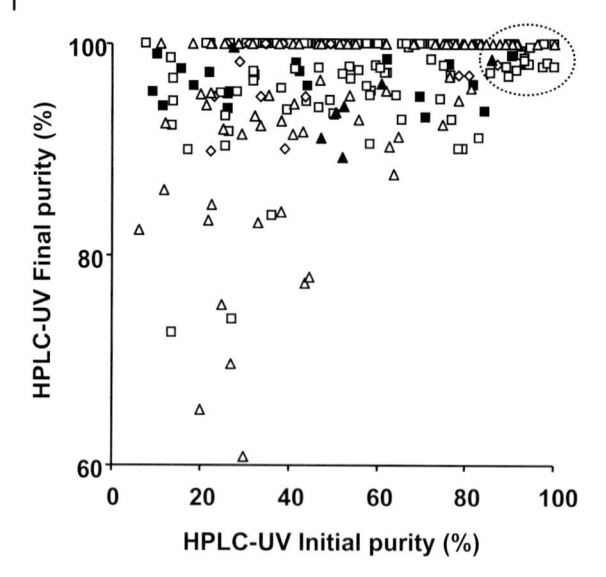

Figure 15.6 Final *versus* initial UV relative purity of individual samples from batch purifications in 10 and 25 mm ID monolithic columns. ■: 10 mm ID column, pH 2.5 (A: 0.05% TFA in water, B: 0.05% TFA in ACN) ($n=27$), ◊: 10 mm ID column, pH 6.8 (A: 10 mM CH$_3$COONH$_4$ in water, B: ACN) ($n=16$), ▲: 10 mm ID column, pH 8.0 (A: 10 mM NH$_4$HCO$_3$ in water, B: ACN) ($n=11$), □: 25 mm ID column, pH 2.5 (A: 0.05% TFA in water, B: 0.05% TFA in ACN) ($n=114$), △: 25 mm ID column, pH 8.0 (A: 20 mM NH$_4$HCO$_3$ in water, B: ACN) ($n=78$).

above 90% were typically achieved. This purification platform facilitates the removal of minor or poorly UV-absorbing impurities. The pool of samples encircled on the right upper corner corresponds to batches of samples purified to remove traces of different solvents and minor impurities resulting from multistep RPS. Because of the low stability of monolith silica sorbent at high pH, most of the samples were purified using 0.05% TFA as a pH modifier. Nevertheless, our data suggest that Chromolith columns can also be used with neutral and alkaline volatile buffers like ammonium acetate and ammonium hydrogen carbonate up to pH 8.

15.5
Conclusions

The examples presented here confirm the suitability of monolithic columns to the automated purification of single compounds and libraries. Higher sample load capacities with excellent chromatographic resolution were observed for the semi and preparative monolithic columns. Further, the high speed of the separation and low pressure drop typical of rod monolithic HPLC columns allow high purification throughput, shortening pre and postpurification procedures by three-fold.

Although, in the very short term, monolithic silica columns are not expected to fully replace particle-packed columns, they have been demonstrated as a suitable alternative for purification of drug-discovery compounds. As monolithic technology is coming of age, RP-HPLC purification platforms using rod monolith columns are benefiting from improved chromatographic attributes like higher resolution, good reproducibility, improved loading capacity, lower backpressures and faster separations.

Acknowledgment

The authors would like to thank Alexander Kraus from Merck KGaA for the generous gift of columns used in this work. We also thank Thomas Castle from Lilly, USA for careful evaluation of the manuscript.

References

1 Zhao, J., Zhang, L., and Yang, B. (2004) Strategies and methods for purifying organic compounds and combinatorial libraries, in *Analysis and Purification Methods in Combinatorial Chemistry*, (ed. B. Yang), John Wiley & Sons, Inc., New Jersey, pp. 255–280.

2 Popa-Burke, I.G., Issakova, O., Arroway, J.D., Bernasconi, P., Chen, M., Coudurier, L., Galasinski, S., Jadhav, A.P., Janzen, W.P., Lagasca, D., Liu, D., Lewis, R.S., Mohney, R.P., Sepetov, N., Sparkman, D.A., and Hodge, C.N. (2004) *Anal. Chem.*, **76**, 7278–7287.

3 Molina-Martin, M., Marin, A., Rivera-Sagredo, A., and Espada, A. (2005) *J. Sep. Sci.*, **28**, 1742.

4 Koppitz, M., Brailsford, A., and Wenz, M. (2005) *J. Comb. Chem.*, **7**, 714.

5 Marín, A., Burton, K., Rivera-Sagredo, A., Espada, A., Byrne, C., White, C., Sharman, G., and Goodwin, L. (2008) *J. Liq. Chromatogr. Relat. Tech.*, **31**, 2–22.

6 Espada, A., Marin, A., and Anta, C. (2004) *J. Chromatogr. A*, **1030**, 43–51.

7 Blom, K.F., Glass, B., Sparks, R., and Combs, A.P. (2004) *J. Comb. Chem.*, **6**, 874–883.

8 Schaffrath, M., Roedern, E.V., Hamley, P., and Stilz, H.U. (2005) *J. Comb. Chem.*, **7**, 546–553.

9 Espada, A., Anta, C., Molina-Martin, M., and Rivera-Sagredo, A. (2006) *Oral Presentation L-06 at Monolith Summer School*, Portoroz, Slovenia.

10 Espada, A., Molina-Martin, M., Dage, J., and Kuo, M.-S. (2008) *Drug Discov. Today*, **13**, 417.

11 Kato, M., Sakai-Kato, K., and Toyo'oka, T. (2005) *J. Sep. Sci.*, **28**, 1893–1908.

12 Al-Bokari, M., Cherrak, D., and Guiochon, G. (2002) *J. Chromatogr. A*, **975**, 275–284.

13 Svec, F. (2005) Short course, Monolith Columns: How to make and use them, at 29th International Symposium on High Performance Liquid Phase Separations and Related Techniques (HPLC 2005), Stockholm, Sweden.

14 Cabrera, K. (2004) *J. Sep. Sci.*, **27**, 843.

15 Leinweber, F.C., and Tallarek, U. (2003) *J. Chromatogr. A*, **1006**, 207.

16 Gritti, F., Piatkowski, W., and Guiochon, G. (2002) *J. Chromatogr. A*, **978**, 81–107.

17 Molina-Martin, M., Anta, C., and Rivera-Sagredo, A. (2007) Espada Poster

presentation P03.01 at 31st International Symposium on High Performance Liquid Phase Separations and Related Techniques (HPLC 2007), Ghent, Belgium.

18 Blom, K.F. (2002) *J. Comb. Chem.*, 4, 295–301.

19 Schulte, M., Lubda, D., Delp, A., and Dingenen, J. (2000) *J. High Resolut. Chromatogr.*, 23, 100–105.

20 Engelhardt, H., and Elgass, H. (1986) *Chromatographia*, 22, 31–39.

21 Dolan, J.W., and Snyder, L.R. (1998) *J. Chromatogr. A*, 799, 21–34.

16
Monolithic Silica Columns in Multidimensional LC-MS for Proteomics and Peptidomics

Egidijus Machtejevas and Eglė Machtejevienė

16.1
Introduction

The state of the organism is reflected to the key process in the living body – protein synthesis, modification and degradation. Proteomics is the large-scale study of gene expression at the protein level, which will ultimately provide direct measurement of protein expression levels and insight into the activity state of all relevant proteins [1]. The ambitious and challenging goal of proteomic research is the comprehensive, qualitative, and quantitative analysis of all proteins expressed by a genome as well as the description of changes occurring at the protein level under the influence of biological stimuli such as diseases or drug treatment. The subproject of proteomics, namely the study of all peptides expressed by a certain cell, organ or organism, is termed peptidomics. The term was introduced in 2001 [2], although, some research groups have been active in this field for quite some years. Proteomic analysis offers a powerful approach to identify disease-associated proteins that can be used as biomarkers for diagnosis and as drug targets for treatment [3].

Peptides often have very specific functions as mediators and indicators of biological processes. They play important roles as messengers, for example, as hormones, growth factors, and cytokines, and thus have a high impact on health and disease. Peptidomics comprises not only peptides, originally synthesized by an organism to perform a certain task, but also degradation products of proteins (degradome). Therefore, proteolytic cleavage of proteins leads to peptides as indicators of protease activity, degradation, and degeneration. Degradome is a very important part of the protein metabolism, and thus also reflects the organism state. The sensitivity of proteomics and peptidomics suffers from the lack of an amplification method, analogous to the polymerase chain-reaction method, to reveal and quantify the presence of low-abundance proteinaceous constituents.

While the human genome sequencing endeavor was dealing with a static system composed of only four building blocks, the following battle of conquering the human proteome was of significantly higher complexity, namely dealing with a dynamically changing system of 20 amino acids (in humans) with a substantial range of post-translational modifications (which are about 100 variations) and huge concentration differences. In conclusion, proteomics bears a much higher degree of complexity than genomics, a fact that was highly underestimated when starting proteomics.

The most characteristic features in the field of proteomics are:

1. chemical constituents of widely different structure (peptides, proteins, and different modifications: sugars, carbohydrates, nucleosides, etc.);
2. extremely large number of constituents (>1 Mio);
3. high diversity in abundance ratio ($1:10^9$);
4. large range of molecular weight (from 100 to several million Da);
5. constituents with relatively small differences in chemical structure for example post-translational modifications (PTMs) of proteins: glycosylated, phosphorylated, etc.; chemically small differences, however biologically highly relevant;
6. number of detected constituents increase exponentially with decreasing concentration/increasing the sensitivity.

The proteome analysis usually includes the following strategies: native protein preseparation, then digestion followed by separation and identification, or alternatively straight digestion, separation and identification by mass spectrometry. Therefore, starting with one protein, after digestion we will end up with approximately 30 to 70 short peptide fragments. Identification of only very few of them will provide sufficient information that protein was present in the sample. Peptidomics does not possess such feature: from the beginning of the analysis to the end we have only one peptide at a certain concentration and we have to identify it. However, when peptides come from the degradome of proteins, then, naturally, peptidomics is in a similar situation to proteomics. The display level is difficult due to the wide range of peptide concentrations that spans over ten orders of magnitude. These challenges motivate the studies to develop reliable analytical platforms. Shortcomings in throughput are due to the absence of technologies that can deliver fast and parallel quantitative analysis of complex protein distributions in an automated fashion. In the future, when proteomics will be more analyzed and understood, and biomarkers identified, straight capture step of biomarkers from complex biosamples might be used.

Although gene sequencing and expression analysis can be performed with high throughput and in an automated manner, major technical problems need to be resolved and new techniques developed before proteomics and peptidomics can become a similar large-scale, highly automated affair. Protein samples of biological origin are by nature highly complex and require sophisticated analytical tools to provide reliable analysis of the components. Proteomics especially challenges the need for robust, automated, and sensitive high-throughput technologies. Most single-dimension separations lack sufficient resolution capability to resolve complex biological matrixes. For example, in human blood serum, 90% of the protein content of serum is composed of 10 basic proteins. The remaining 10% of serum consists of trace amounts of millions of different proteins. Thus, partial purification of proteins is necessary so that proteins in trace amounts can be identified and their exact structural analysis can be performed. To overcome the limited peak capacity and concentration diversities of the analytes utilizing chromatographic separation systems, multidimensional chromatography has been realized by analyte transfer between different separation modes through automated valve switching [4]. Another important prerequisite for the suitability of separation systems for proteomic analysis is the ability to handle very big and very small amounts of biological material [5]. However, the application of several orthogonal LC separation systems also bears the danger of severe sample losses due to adsorption at the separation and capture column and by sample transfer.

LC separation techniques are well suited for the analysis of complex multicomponent samples. However, the analysis of proteome requires well-designed sample-preparation procedures. In the early years of LC attempts were made to inject biological samples directly onto the column [6]. It was quickly realized that this approach let to a rapid loss of column performance. Also, the column selectivity was altered, the column backpressure increased, because of the irreversible adsorption of matrix compounds. Some useful means of sample preparation is required to reduce the sample complexity and to remove components that tend to bond irreversibly.

In a search for new stationary-phase configurations the concept of monolithic columns was explored and investigated in depth. A monolith consists of a continuous rod with a rigid porous polymer or porous silica, composed of micro-and macropores [7–10]. The mobile phase is forced to flow through the macropore channels of the monolith resulting in an enhanced mass transport of the analytes and in improved chromatographic efficiency [11].

The often limited amount of sample and/or the low abundance of proteins of interest have become a driver for miniaturization of LC in the form of capillary separation systems. Capillary separations, generating much higher sensitivity than the analytical type of columns, especially when combined with mass spectrometry [12], often have the drawback of reduced robustness. This is partially due to the limited stability of packed capillary columns and the risk of clogging. New approaches in capillary HPLC, particularly in the field of proteomics, are therefore needed. Monolithic capillary columns made of polymeric [13] or silica-based materials [14] promise to overcome some of the limitations mentioned above, in particular the packing instability. Moreover, the enhanced permeability of monolithic as compared to particulate columns results in a much lower backpressure, allowing the application of higher flow rates [15]. The higher column permeability of monolithic columns as compared to particulate columns is based on the higher macroporosity of the former, which can not be reached by a particulate column. The macropores at a monolithic silica column are typically in the range between 1–2 µm while at a particulate column with 2-µm particle size the interstitial voids have diameters of approximately 0.3 µm (theoretically calculated value). The column pressure drop on a monolithic silica column was shown to be proportional to the size of the macropore diameter at constant mesopore diameter [16].

Capillary liquid chromatography based on silica monolithic stationary phases was used to screen complex peptide libraries by fast gradient elution coupled online to electrospray ionization Fourier transform ion cyclotron resonance mass spectrometry [17]. Minakuchi et al. (1998) has shown that monolithic silica columns posses much higher separation efficiency for polypeptides than conventional columns packed with 5-µm wide-pore silica particles [18].

The most comprehensive study of application of the monolithic silica columns in peptidomics was accomplished by Bischoff et al. [19] analyzing Cytochrome C, elastin, and serum digests. The results from the 100th-run experiment with Cytochrome C digest show good reproducibility of retention times and peak heights at a backpressure that is considerably lower than for packed columns of the same dimensions.

16.2
Liquid Chromatography as a Tool Box for Proteomics

Liquid chromatography (LC) has several important advantages for the proteomic analysis. First, LC offers superior automation possibility and high flexibility [20]. One can vary the selectivity, and every different selectivity (or selectivity influenced by chromatographic conditions) will allow different proteinaceous substances to be caught. Sample volumes in proteomics could vary drastically: the microliter range, for example, blood or plasma to the milliliter range, as for example analyzing urine. Column-size adjustments (mainly inner diameter) could handle the loadability of the system. The sensitivity is boosted by reducing the column internal diameter. The combination of selectivities in multidimensional LC helps to obtain high peak-separation capacities. Quantitation is often problematic dealing with high-complexity samples. Also, in LC-based approaches the handling of extreme abundance differences is quite complex and requires long experience [20]. Automated sample processing in combination with adjusted column dimensions and elution sequences allows the correlation. The limitations are related to biological sample stability at the conditions of the optimum LC analysis. Proteins are prone to aggregation and precipitation especially when the pH, concentration and ion strength are altered. The chromatographic setup contains many tiny capillary connections, column filters/frits–all could be easily clogged if any precipitation occurs. Time of the analysis and the temperature influence the stability of the sample components. Despite its pros and cons LC already has been proven in several cases for automated sample clean up, depletion and enrichment and qualitative analysis. The usefulness of multidimensional LC including efficient sample clean-up procedures has been shown and applied [21].

There are several challenges to be addressed in separation technologies. First, the separation of a complex protein mixture is certainly not an easy task because human body fluids, such as urine or blood, contain several thousand proteins and peptides. Recent publications [22, 23] lead to the impression that the separation of such complex samples–containing at least several thousand different polypeptides–is not possible in a single liquid-chromatographic run, due to the lack of resolving power of one-dimensional LC and limitations in the dynamic range of MS. The initial problem is that a vast number of biomolecules are usually present in a biological sample. It is easy to be misguided by the vast number of publications usually dealing with standard protein digests. Separation of a few digested proteins peptides are shown in Figure 16.1a. It is easy to recognize small

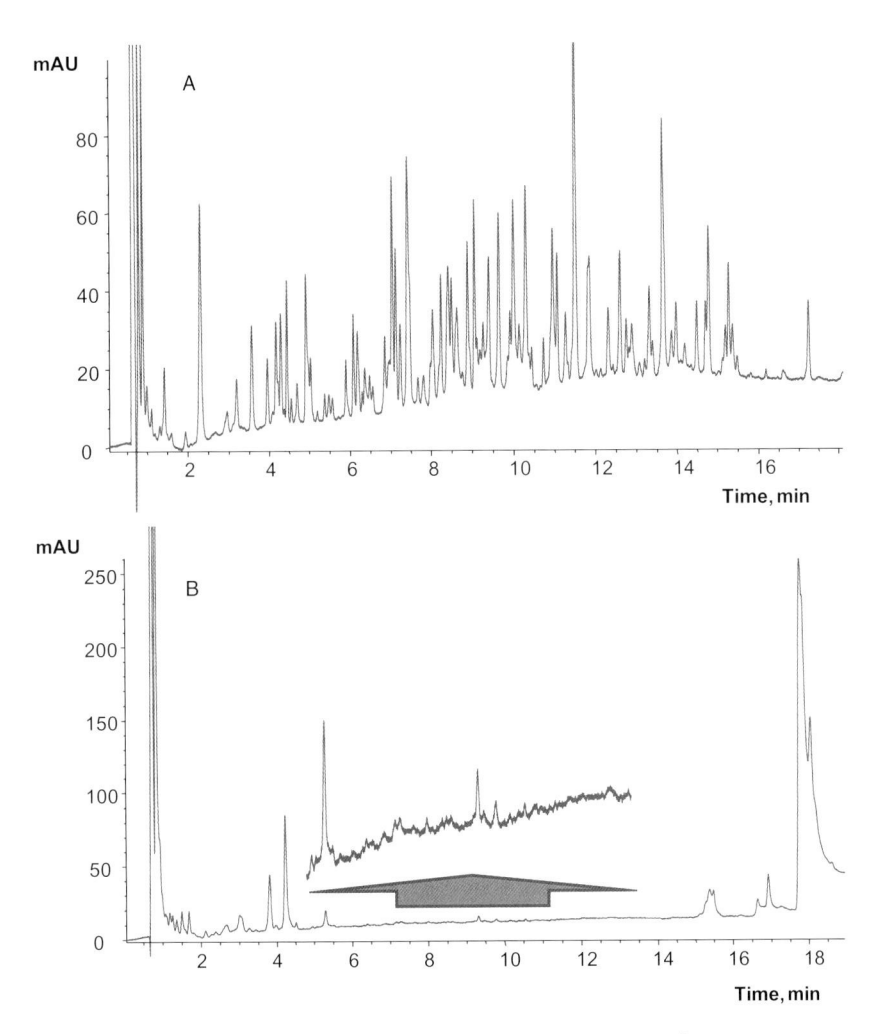

Figure 16.1 Separation of standard protein digest and real biological sample analysis: upper chromatogram–1 µl BSA digest (1 mg/ml), lower chromatogram 8 µl of filtered amniotic fluid. Conditions: column–Chromolith® Performance RP-18e 100 mm × 2 mm ID (Merck KGaA, Darmstadt, Germany); eluents–A: 95% H_2O/5% ACN/0.1% TFA (v/v/v), B: 5% H_2O/95% ACN/0.085% TFA (v/v/v); gradient–from 5% B to 50% B in 20 min; flow rate–0.3 ml/min, detection–UV 214 nm.

differences in dynamic range, and even peak distribution, therefore the conclusion could be drawn that all what we need for successful proteomics analysis is high peak-capacity separation and one dimension then would be sufficient. However, using the same chromatographic conditions and column, also injecting eight times more of a real biosample (amniotic fluid) we do not observe nice and even separation any longer (Figure 16.1b). This is a common situation with all real biosamples.

A two- or three-dimensional approach that employs two or more orthogonal separation techniques or separation methods with different mechanisms of separation will significantly improve the chances of resolving a complex mixture of cell proteins into their individual components. Multidimensional separation typically relies on using two or more independent physical properties of the proteins. Physical properties commonly exploited include size (Stoke's radius), charge, hydrophobicity and biological interaction or affinity. When the targeted properties are independent entities, the separation methods are considered "orthogonal". Multidimensional peptide separation will play an increasingly important role on the way to identify and quantify the proteome.

Multidimensional (multistage, multicolumn) chromatography had been discovered early as a powerful tool to separate complex mixtures. Two of the protagonists were Giddings [24, 25] and Huber [26]. MD-LC is based on coupling columns in an online or offline mode, which are operated in an orthogonal mode, that is, separate the sample mixture by different separation mechanisms. The sample separated on the first column (first dimension) is separated into fractions that can then be further treated independently of each other. The practical consequence is an enormous gain in peak capacity (number of peaks resolved at a given resolution) and the potential of independent optimization of the separation conditions for each fraction. Simultaneously, there is the option of relative enrichment/depletion and peak compression by fractionation.

In MD-LC of proteomics, columns should be combined into a multidimensional system, which allows the separation of proteins under the following conditions: large diversity and abundance of components in chemical structure and composition, small differences in chemical composition, large differences in molecular size and mass, extremely large abundance ratio of $1:10^8$, a wide concentration range facing the fact that the number of constituents increases exponentially with decreasing concentration [27].

To achieve these goals in the desired combination requires a rational design of a separation system with the following selection criteria. First, the choice and sequence of the LC phase system (stationary phase, mobile phase) needs to be tuned to gain a high peak capacity and optimum resolution, to design a system that effectively depletes the high-abundance constituents and enriches the target compounds, and to fractionate the remaining depleted fractions with a high selectivity. In an online mode, the speed of analysis within the dimensions has to be adjusted in such a way that the resolution on the first column is achieved at a low speed. A high-speed analysis in the second dimension is required to resolve the fractions from the first dimension [28]. As a consequence of the high abundance ratio the high-abundance proteins must be depleted and removed most effectively at the beginning of the MD separation to avoid an overload of the subsequent columns, which would compromise their selectivity and reproducibility. In other words, the mass loadability of columns in the multidimensional column train plays a significant role, otherwise displacement phenomena and unwanted protein–protein interactions will take place, which may change the downstream composition of the individual fractions in an irreproducible way [29].

Memory effects of the separation columns in multidimensional LC systems often give rise to cross-contamination between the samples. After the Chromolith® Performance RP column was double overloaded with human hemofiltrate the second gradient run contained 22% of the proteinaceous constituents of the first run. In the sixth blank run the amount declined to 1.3%. In multidimensional LC systems it could easily happen that the second column is overloaded by the sample amount from the first column Therefore, it is absolutely necessary to run at least one blank run in between the analysis of patient samples. However, when the monolithic silica column was double overloaded with a single peptide Angiotensine I, the second run showed only 0.02% of the initial peak. This demonstrates that the chromatographic behavior of the column might differ when standard single peptide or a real complex mixture are applied.

16.3
Selectivity of Columns for MD-LC

While MD-LC MS has found widespread use in the analysis of peptides from natural sources or generated by proteolytic digestion of larger proteins, the method is not suitable for analyzing proteins directly. First, proteins tend to denature under reversed-phase conditions either by stationary-phase- or mobile-phase-induced effects (strongly hydrophobic surfaces, low pH and high organic solvent concentrations) making their quantitative elution rather difficult. Also, measuring the molecular mass of a protein by MS is not sufficient for its unambiguous identification. To circumvent these obstacles the proteins are digested and the separation is performed at the peptide level. One can distinguish two approaches (i) proteins are separated and then digested ("top-down" proteomics [30]); (ii) in "shotgun" proteomics a complex protein mixture is first digested and peptides are then chromatographically resolved ("bottom-up" proteomics [31]). In both cases, separation technologies play a critical role in protein identification and analysis.

Even though in the "shotgun approach" sample complexity is vastly increased, there are an increasing number of reports on the comprehensive analysis of human proteomes using this strategy. The advantage of this strategy is that after digestion, the obtained peptides are more easily separated than large proteins [32–35].

Ion-exchange chromatography (IEC) is most often selected for the first dimension. In principle, there are two options in IEC, either to employ a cation or anion exchanger, which in return dictates the pH working range. Note that either cationic or anionic species are resolved, that is, only a limited number of species from the whole spectrum. The IEC columns are operated via salt gradients with increasing ionic strength. Consequently, the salt load must be removed before the fractions are transferred to the second dimension column. Despite the fact that no IEC columns based on monolithic silica is commercially available, several research samples were tested by the author [unpublished data]. It was found that the performance of monolithic RAM-SCX columns was similar to the RAM-SCX columns packed with particulate silica [36]. However, the monolithic RAM-SCX columns demonstrated much longer lifetime.

RP LC is the method of choice as the second dimension in MD-LC due to its high resolving power, high speed of analysis and its desalting capability [37]. The disadvantage of this approach is that one ends up with an extremely large number of peptides, which need to be resolved. Direct analysis of biofluids without prior digestion is a definitive option in biomarker discovery. Prior digestion gives access to the higher molecular weight proteins, however, at the expense of rendering the mixture much more complex. Assuming that a given biofluid contains 1000 proteins and that each protein will generate approximately 50 proteolytic fragments, we are talking about 50 000, and more, peptides to be resolved. This task can only be approached by MD protein identification technologies [38, 39].

The fractions are subjected to digestion and consecutive reinjection onto a RP column is performed, whereby the separation is based on the hydrophobicity. This is particularly favorable since the mobile phase in the second dimension (RPC) is compatible with the solvent requirements of mass spectrometry. The restrictions associated with this method lie in the limited size of proteins that can be investigated (MW < 20 000 Daltons) and the insolubility or incomplete separation of very hydrophobic peptides. All peptide-containing fractions are then investigated by mass spectrometry to generate a peptide map [40]. This approach has already been found to be sufficient to deal with smaller subsets of the proteome (i.e., several hundred proteins) [41]. These studies also clearly demonstrate that this methodology is not yet suitable for the analysis of a whole proteome due to its enormous complexity. Therefore, preselection of the protein from a given tissue or a preseparation seems mandatory. For example, for the analysis of human urine solid-phase extraction (C-18 packings) to trap peptides, followed by IEX chromatography in the first dimension collecting 30 fractions and analysis of the collected fractions by RP LC (C-18) in the second dimension [42] was successfully employed. A similar procedure was used for the separation of proteins and peptides in human plasma filtrate [43] and plasma [44].

It is most common to use RPC as the second dimension. The term RPC stands for a number of columns with different degrees of hydrophobicity. The most commonly applied phases are n-octadecyl bonded silicas (RP-18 columns). Various dimensions of RP monolithic silica columns are available (see Table 16.1). An intrinsic feature of RP columns is their desalting property: salts are eluted at the front of the chromatogram, when running a gradient elution with an acidic buffer/acetonitrile mobile phase with increasing acetonitrile content. The hydro-

Table 16.1 Overview on commercial monolithic silica columns.

Column ID, mm	0.05	0.1	0.2	2	3	4.6	10	25
Column length, cm	15	5[a], 15, 30	5[a], 15	0.5[a], 2.5, 5, 10	0.5[a], 2.5, 5, 10	0.5[a], 1[a], 2.5, 5, 10	10	10
Commercially available functionalities	RP-18e	RP-8e, RP-18e	RP-18e	RP-18e	RP-18e	Si, RP-8e, RP-18e	RP-18e	Si, RP-18e

a) Guard/trap columns.

phobic surface of the RP packing and the hydrophobic eluent are not favorable with respect to providing a biocompatible environment for proteins: they may change their conformation or denature, which may be seen by the appearance of broad peaks, splitting of peaks, etc. RP columns possess a much lower mass loadability than IEC columns (10 mg of protein per gram of packing as compared to 100 mg in IEC).

An advantage of RPC is the fact that the eluents are compatible to MS, provided volatile buffers such as ammonium acetate are employed.

In case of an online MD-LC system, the speed of analysis in the second dimension should be as high as possible [26]. This, however, conflicts with the requirement of high resolution or high peak capacity. The highest peak capacity in gradient elution RPC is obtained with a shallow gradient at relatively low flow rates. Thus, a compromise between the desired peak capacity and the gradient time is inevitable. Often, gradient times of several hours are applied for the analysis of peptides from protein digests. Due to the possibility of using high flow rates, these columns have distinct advantages over packed columns for the separation of very complex mixtures in multidimensional chromatography systems because they allow shortening of the often extensive run times.

16.4
Dimensions of Columns in MD-LC

In order to end up with enough material for MS-based identification, in some cases one has to start with large sample amounts or large volumes. In MD-LC this means one employs large ID columns at the beginning and ends up with max 2 mm ID or capillary columns of 50 to 200 μm ID. Corresponding to the column ID the volume flow-rate changes from ml/min to nl/min. At constant column length, the amount of packing in the columns decreases. For example, a 10-cm 4-mm ID column contains approximately one gram of packing, a 100 mm × 100 μm capillary column contains approximately 10 μg of packing.

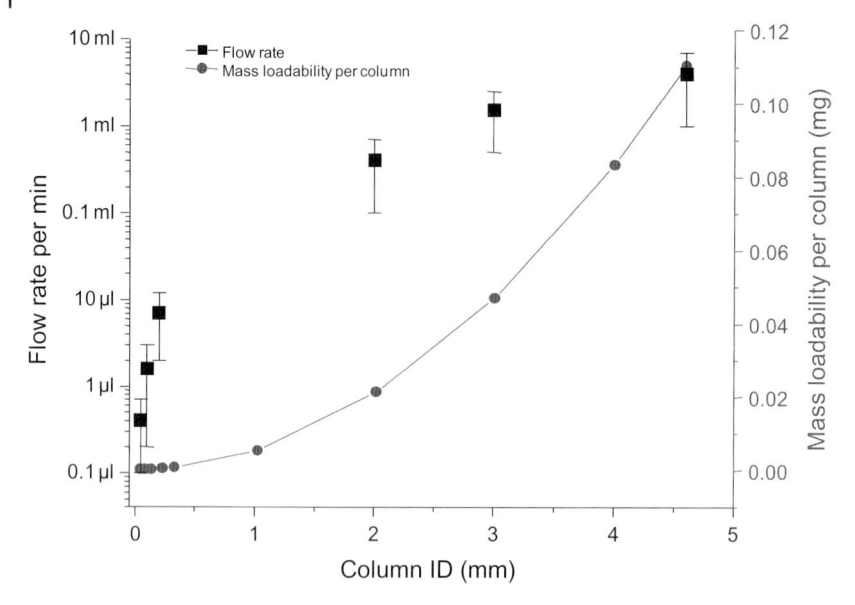

Figure 16.2 Estimation of monolithic silica columns flow rates and mass loadability per column. Columns 2, 3 and 4.6 mm ID are 10 cm long; columns 50 µm, 100 µm and 200 µm are 15 cm long.

Column miniaturization from mm to µm ID has two consequences apart from flow-rate reduction: the injection volume decreases as well as the mass loadability. The diminution of flow-rate and injection volume in relation to column ID and an example of the mass loadability for peptides of RP columns of gradated ID is displayed in Figure 16.2.

In MD-LC all columns except SEC columns are operated under gradient elution conditions. Usually one starts with eluent A and adds increasing amounts of eluent B with a stronger eluting solvent (linear gradient). Other gradient operations are step gradient or pulse injection of the stronger eluent. Depending on the mode of gradient operation one will obtain different results at equal sampling rate for the fractionation on the first-dimension column. Usually, approximately 5 to 40 fractions of equal volume are collected from the first dimension. Each fraction generates approximately another 50 fractions on the second-dimension column. The optimal number of fractions from the second column is determined by the time to be invested analyzing fractions by mass spectrometry, also investigating spectra, and number of peptides one would like to see. Seek to obtain a compromise for the required amount of useful data in a reasonable time frame. Note that the evaluation of MS data, and the system automation are also still under development [40].

Either the fractions from the first dimension can be stored on a trap column (see Table 16.1) of suitable dimension or they can directly be deposited on top of the second-dimension column. The direct deposition approach is simpler, and requires less equipment; however, usually the process time is drastically increased. The use of trap columns is mandatory when the flow rates in the first- and second-dimension columns differ by an order of magnitude (see Figure 16.2). Trap columns are also employed as a powerful desalting tool. When selecting trap columns some precautions have to be taken. A trap column is placed between the first- and the second-dimension column. The first-dimension column is much bigger than the trap column. Therefore, proteinaceous samples should be split into a certain number of fractions, not only to fit the orthogonality condition, but also to match the first column to trap column loadability ratio. The trapped material is then transported to the analytical column in the second dimension. It should be emphasized that the analytical columns should not be overloaded by the material coming from the trap column.

16.5
Monolithic Silica Columns

Commercially available monolithic silica columns used in proteomics analysis are listed in Table 16.1. One of the major advantage of those columns is the low column backpressure that allows one to select a desired flow rate at a multidimensional separation platform for proteomics (see Figure 16.2). The combination of different size columns fulfils the injection volume requirement for various samples. The possibility to vary the flow rate over a large range up to very high linear flow velocities combined with the robustness of the monoliths reduces considerably the "down times" during washing and re-equilibration of the column [45].

It is important to notice that comparing a particulate and a silica monolithic guard column showed that the particulate column was clogging much faster than the monolithic column (Figure 16.3). 120 injections of plasma (50 µl each) led to an increase of approximately 6 bar at the particulate column backpressure, while at the monolithic column the backpressure rise was only approximately 1 bar. The lifetime of the short silica monolithic columns used as a trap column or as a guard column heavily depends on the type and the volume of the biofluid injected. Figure 16.4 demonstrates the decrease of the column efficiency applying filtrated urine and plasma. After injecting plasma the column performance dropped drastically when half the column volume was injected (Figures 16.4b and d). For urine, the column stability was at least 20 times higher (Figures 16.4a and c). This definitely depends on the sample complexity. When the intensities of four peptides are monitored over the total injection volume, (see Figure 16.5) it is clear that the intensity stability over the total injection volume depends on the monitored compound. Comparing the lifetime of a monolithic silica columns and particle-packed columns of the same dimension under the same conditions the monolithic silica column lifetime is double that of the particulate-packed column. This is not a surprise, as any particle-packed column contains particles and frits to maintain the particles in the column. The pore diameters of the interstitial voids between

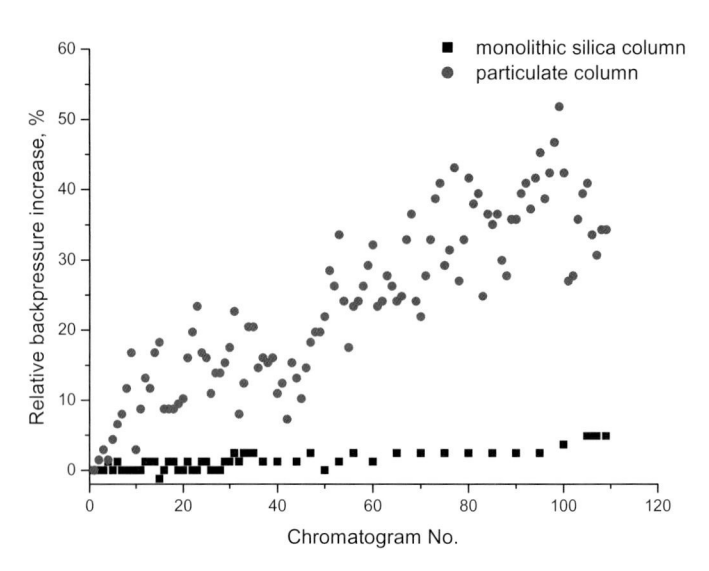

Figure 16.3 Changes in backpressure of the particulate (25 μm particles) and the monolithic silica columns injecting filtrated human plasma sample. Particulate column: 25 × 4 mm ID, RAM-SCX. Monolithic silica column: 5 × 4.6 mm ID, Chromolith® guard column. Sample human plasma, 50 μl per injection.

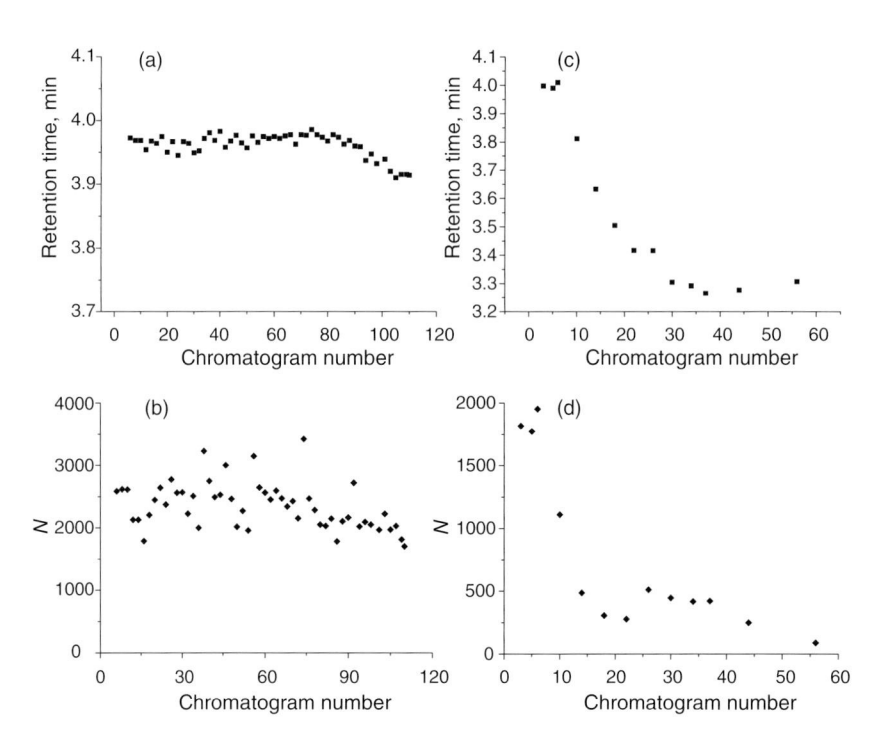

Figure 16.4 Changes in monolithic column performance injecting filtrated human urine (a, b) and plasma (c, d) samples. Monolithic silica column: Chromolith® RP-18e Guard 5 × 4.6 mm ID. Samples: human urine 100 μl per injection, human plasma 50 μl per injection.

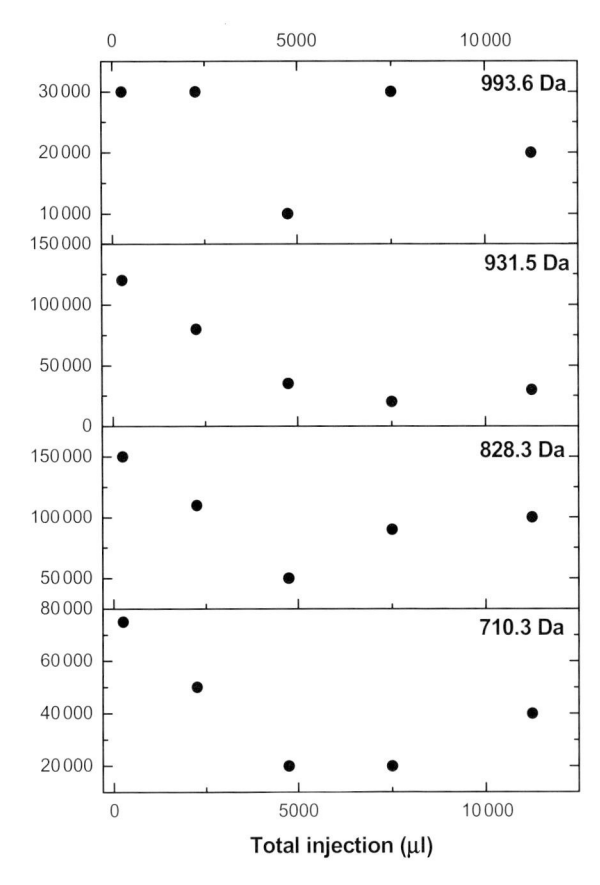

Figure 16.5 Measurements of overall long term MD-LC-MS system reproducibility by monitoring four peptide signal intensities. Sample: filtered amniotic fluid. Columns: first dimension – 5 cm × 4.6 mm ID SCX, Chromolith® RP-18e Guard 5 mm length and 3 mm ID was used as a trap column, second dimension – Chromolith® Performance RP-18e 10 cm × 2 mm ID column.

the particles are much smaller than the particle diameter itself. For example, if the column is packed with 2-µm particles the space between particles is about 0.3 µm. A monolithic silica column made as a continuous rod does not contain frits and the flow-through pores (macropores) are about 2 µm in diameter.

Capillary separations, although delivering much improved sensitivity, especially when combined with mass spectrometry, often have the drawback of reduced robustness. This is partially due to the limited stability of packed capillary columns and the risk of clogging (same aspect as discussed above). The reproducibility of packed capillaries requires special skills and experience. New approaches to capillary LC, particularly in the field of proteomics, are therefore needed. Monolithic capillary columns made of polymeric [46] or silica-based materials promise to overcome some of the limitations mentioned above, in particular that of packing stability. Furthermore, due to their higher porosity they have less tendency to become clogged. Usually, microcolumns for LC are fabricated by packing beads with a controlled range of diameters and pore sizes. To obtain a better efficiency, columns have been packed with particles of sub-2 µm in diameter [47] generating a column backpressure in excess of 400 bar. As a result, particle sizes in the range of 3 to 5 µm are a good compromise between column efficiency and pressure drop. Moreover, it was demonstrated that the recently developed monolithic-type HPLC columns could be operated at high flow rates while maintaining a high efficiency. In this context, Fort *et al.* [48], investigated the reproducibility of the preparation of the first columns, getting reproducibilities higher than with particle-based columns. Because of their capacity to perform fast separations monolithic silica columns can be used for fast screening methods and applications in multidimensional LC systems. Conditioning and regeneration of these monolithic columns can be done in a short time interval when compared with the corresponding capillary packed columns, thus making more effective use of costly LC-MS equipment. Monolithic silica columns can be easily integrated in an automated system to perform unattended runs. These columns are flexible and they show a good performance at both low (1.5 µl/min) and high (4.5 µl/min) flow rates. Such a flow-rate range is highly compatible with MALDI plate-spotting strategy. The fractions could be spotted directly, then the flow of a 4-µl/min allows up to 8 fractions to be spotted. If the flow rate is set to 3 µl/min, and an equal flow of MALDI matrix solution is added postcolumn (7 mg/ml recrystallized α-cyanohydroxycinnamic acid, 2 mg/ml ammonium phosphate, 0.1% trifluoroacetic acid, 80% acetonitrile) the combined eluent is automatically spotted onto a stainless steel MALDI target plate every 6 s (0.6 µl/spot), a total of 370 spots are obtained per original SCX fraction [49]. Haffey demonstrated a similar approach and obtained 3828 MALDI-TOF spots from the 12 SCX fractions [50]. Such a separation strategy offers enormous discrimination power and imposing peak capacity.

16.6
Applications of Monolithic Silica in Proteomics – A Brief Survey

The most extensive study was performed by Bischoff's group. In order to evaluate the performance and robustness of the newly developed monoliths silica capillary columns, a tryptic digest of Cytochrome C was analyzed 100 times over a time period of 100 h. The results show good reproducibility of retention times and peak heights at a backpressure that is considerably lower than for packed columns of the same dimensions. Increasing the column length and thus the overall efficiency in terms of plate numbers is an option that is not easily available with packed columns due to backpressure limitations. Although modern mass spectrometers may be able to deal with a small number of components simultaneously (e.g., automatic precursor ion selection of 5 ions), ion suppression effects may interfere with obtaining quantitative data, which are very relevant in comparative proteomics studies. The gain in resolution when going from a 15- to a 50-cm capillary as well as the increased retention of earlier eluting peptides has been demonstrated [19]. The sample loadability in terms of total protein amount retained by the column was also calculated by injecting increasing amounts of protein until it appears in the flow-through. For a 0.1-mm ID 150-mm length column 5 µg was the maximum loading capacity. Silica monolithic capillary columns show excellent separation performance for peptides and proteins, similar to columns packed with 2-µm spherical particles. Peak widths at half the height are of the order of 0.2 min, and the elution of low-abundance compounds is not deteriorated in the presence of overloaded compounds, as observed during the chromatographic separation of BALF samples where some components were in a much higher concentration [19].

A year later, Regnier's group, demonstrated the advantage of the flow-rate variation possibility for 4.6-mm ID Chromolith® Performance columns. It was concluded that silica monolith reversed-phase chromatography columns show little loss in the resolution of peptides ranging up to several thousand in molecular weight as the mobile-phase velocity is elevated from the conventional 2.5–25 mm/s [51]. Moreover, at 25 mm/s with a 100-mm length column, the operating pressure did not exceed 150 bar. This is well within the pressure limit of most commercial LC instruments. The separation of a tryptic digest of cytochrome C in 6 and 60 min seemed almost identical. Resolution at 25 mm/s linear velocity was 77% of that at 2.5 mm/s. It was concluded that the fact that peptide separations could be achieved 10 times faster than with a conventional packed column with moderate loss in resolution could have a major impact on analytical throughput in proteomics.

A shotgun approach was used for the analysis of the proteome of A. thaliana after depletion of Rubisco, a high abundance protein found in plant leafs [52]. The first dimension was based on SCX, whereas the second dimension was performed using RPLC. The silica monolith (0.1 mm) was coupled to the SCX column by means of a 10-port valve and two C18 trap columns to desalt the fractionated sample prior to injection on the monolith. Individual RPLC runs were in excess of 2 h. Using this setup, a total of about 3500 MS/MS spectra were acquired during each run, enabling about 300 unique proteins to be identified.

An interesting study was performed by Guryca *et al.* to provide a side-by-side comparison of monolithic nano-LC columns used in reversed-phase chromatography of proteins tryptic digests [53]. They compared PepMap (LC Packings, Amsterdam, The Netherlands, 3 µm 100 Å, ID 75 µm, 15 cm), Chromolith® CapRod® (Merck, Darmstadt, Germany; silica monolith-C18, ID 100 µm, 15 cm) and PS-DVB (LC Packings; polystyrene monolith, ID 100 µm, 5 cm) columns (all C18 modification), in terms of the number of peptides identified and also with respect to their chromatographic characteristics. In terms of performance the peak shapes obtained on Chromolith® CapRod® and PepMap columns appeared to be very similar, and the peak widths for both columns were in the range 0.3–0.4 min. The PS-DVB column exhibited somewhat disappointing performance that could be attributed, to the mobile-phase composition used. However, it was concluded that generally the performance of both silica-based columns was superior to that of monolithic PS-DVB [53]. Also a similar finding was observed comparing peptide identification power. Comparing column throughput Chromolith® CapRod®

column was superior at 5.0 μl/min in contrast to flow rates of up to 0.8 μl/min for PS-DVB column and to 0.5 μl/min for the particulate (PepMap) column. Moreover, it was found that, for short gradients, the number of identifications is not affected by the flow rate (3–10 mm/s). The results shown demonstrate the greater potential of monolithic compared to particle-based columns, as higher flows can be utilized, enabling the number of identifications per unit time to be significantly increased.

The benefits of implementing monolithic silica columns into the multidimensional separation platforms were demonstrated by Machtejevas [54]. First, monolithic silica columns offer a high variability of flow-rate adjustments, which is particularly useful in the setup of multidimensional LC MS system to adjust for different column sizes. Also, the implementation of the monolithic silica columns into the multidimensional LC MS system (see Figure 16.6) met the requirement of high reproducibility as with particulate columns. However, in terms of column robustness and usage flexibility monolithic silica columns were superior to packed particulate columns for the following reasons: (i) one can cut the top end column easily when damaged, (ii), there is no change in the column permeability as a result of pressure fluctuation (iii); the end of the capillary can be directly connected to the MS [54]. The overall analysis of amniotic fluid strategy could be described as follows: online sample preparation/peptide extractions of filtered amniotic fluid was achieved using restricted access technology with strong cation-

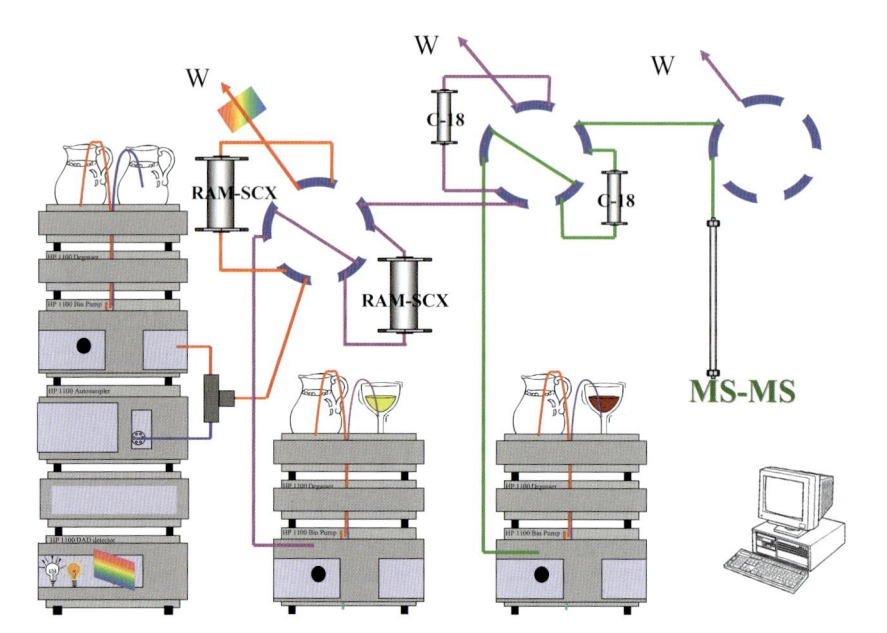

Figure 16.6 Multidimensional chromatographic platform configuration: 5 cm × 4 mm I.D RAM-SCX column, two Chromolith® RP-18e Guard 2 mm ID columns used as traps and analytical Chromolith® CapRod® RP-18e 15 cm long × 200 μm ID coupled directly to ion trap MS.

exchange groups hidden in the 6-nm pores (with such sample pretreatment we obtained a peptide profile that was in a molecular weight range between 700 to 4500 Da); at the second dimension reverse-phase trap columns were used to desalt fractions coming out from the RAM-SCX column; the third step is to separate trapped peptides using monolithic silica capillary column RP-18e directly connected to the mass spectrometer. The system runs in a fully automated way, by a scale-down strategy, gaining in sensitivity. In all peptide displays (Figure 16.7),

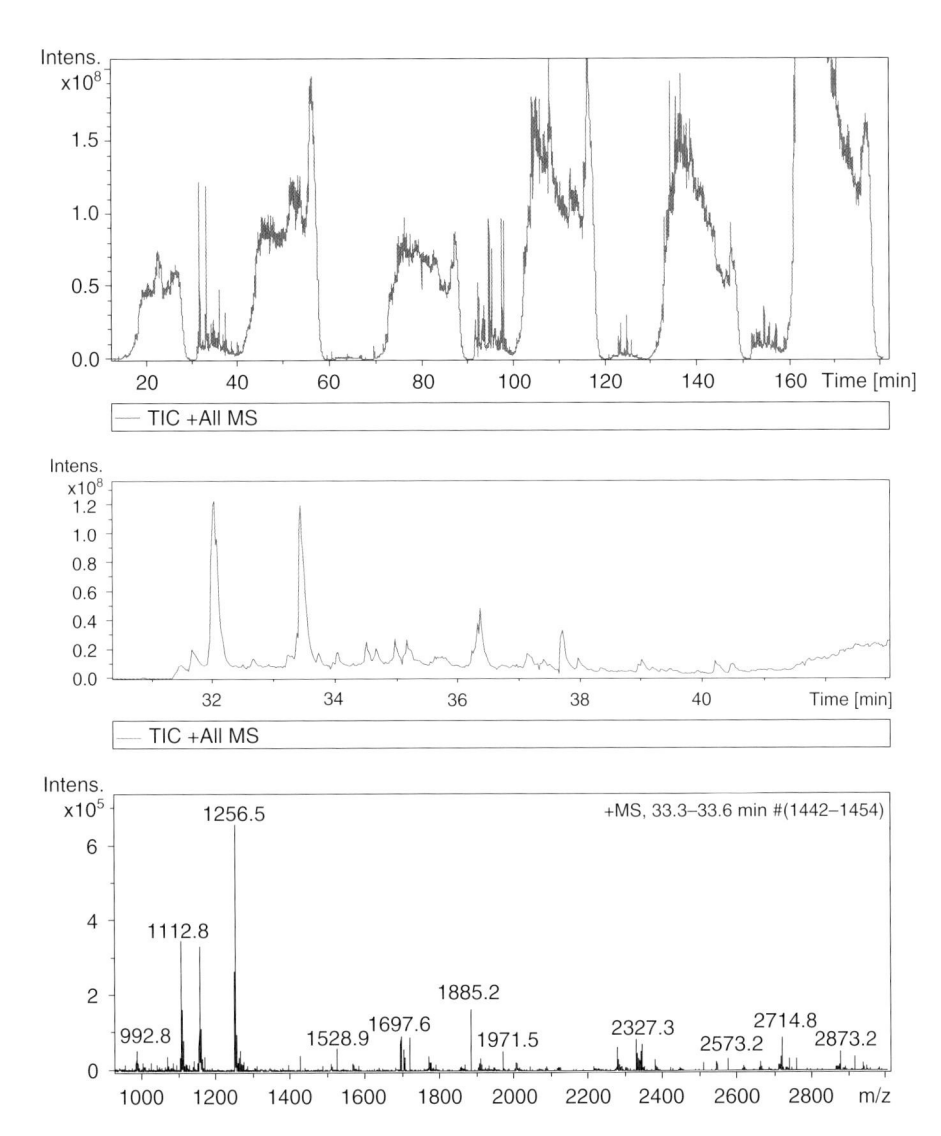

Figure 16.7 Analysis example of 100-µl filtered amniotic fluid separation using HPLC setup described in Figure 16.6.

between 1000 and 4000 mass spectrometric signals appeared, which correspond to 500–2000 individual peptides.

16.7
Summary and Conclusions

The benefits of monolithic silica columns used in proteomic analysis are:

1. Monolithic silica columns offers superior low backpressure compared to particulate-packed columns, therefore high variability of flow-rate adjustments is possible, which is particularly useful in the setup of a long-term robust multidimensional LC MS system.
2. The monolithic silica columns exhibit superior long-term stability and data reproducibility analyzing various proteinaceous samples. The lifetime of the monolithic silica columns is roughly double compared to particle-packed columns.
3. Much higher flow rates allows speeding up the overall analysis: fast separation, washing and re-equilibration.

References

1 Pandey, A., and Mann, M. (2000) Proteomics to study genes and genomes. *Nature*, **405**, 837–846.
2 Clynen, E., De Loof, A., and Schoofs, L. (2003) The use of peptidomics in endocrine research. *Gen. Comp. Endocrinol.*, **132**, 1–9.
3 He, Q.Y., Chen, J., Kung, H.F., Yuen, A.P., and Chiu, J.F. (2004) Identification of tumor-associated proteins in oral tongue squamous cell carcinoma by proteomics. *Proteomics*, **4**, 271–278.
4 Link, A.J. (2002) Multidimensional peptide separations in proteomics. *Trends Biotechnol.*, **20** (12S), 8–13.
5 Machtejevas, E., Unger, K.K., and Ditz, R. (2006) Multidimensional column Liquid Chromatography (LC) in proteomics – where are we now? in *Proteomics in Drug Research* (eds M. Hamacher, K. Marcus, K. Stühler, A. van Hall, B. Warscheid, and H.E. Meyer), Wiley-VCH Verlag GmbH, Weinheim, Germany, pp. 89–111.
6 Westerlund, D. (1987) Direct injection of plasma into column liquid chromatographic systems. *Chromatographia*, **24**, 155–164.
7 Hansen, L.C., and Sievers, R.E. (1974) Highly permeable open-pore polyurethane columns for liquid chromatography. *J. Chromatogr.*, **99**, 123–133.
8 Liaoa, J.-L., Zhanga, R., and Hjertén, S. (1991) Continuous beds for standard and micro high-performance liquid chromatography. *J. Chromatogr.*, **586/1**, 21–26.
9 Svec, F., and Frechet, J.M. (1992) Continuous rods of macroporous polymer as high-performance liquid chromatography separation media. *J. Anal. Chem.*, **64**, 820–822.
10 Minakuchi, H., Nakanishi, K., Soga, N., Ishizuka, N., and Tanaka, N. (1996) *Anal. Chem.*, **68**, 3498.
11 Meyers, J.J., and Liapis, A.I. (1999) Network modeling of the convective flow and diffusion of molecules adsorbing in monoliths and in porous particles packed in a chromatographic column. *J. Chromatogr. A*, **852**, 3–23.
12 Strittmatter, E.F., Ferguson, P.L., Tang, K., and Smith, R.D. (2003) Proteome analyses using accurate mass and elution time peptide tags with capillary

LC time-of-flight mass spectrometry. *J. Am. Soc. Mass Spectrom.*, **14**, 980–991.

13 Chirica, G.S., and Remcho, V.T. (2001) Novel monolithic columns with templated porosity. *J. Chromatogr. A*, **924**, 223–232.

14 van Nederkassel, A.M., Aerts, A., Dierick, A., Massart, D.L., and Vander Heyden, Y.J. (2003) Fast separations on monolithic silica columns: method transfer, robustness and column ageing for some case studies. *Pharm. Biomed. Anal.*, **32** (2), 233–249.

15 Vervoort, N., Gzil, P., Baron, G.V., and Desmet, G.A. (2003) A correlation for the pressure drop in monolithic silica columns. *Anal. Chem.*, **75**, 843–850.

16 Skudas, R., Grimes, B.A., Machtejevas, E., Kudirkaite, V., Kornysova, O., Hennessy, T.P., Lubda, D., and Unger, K.K. (2007) Impact of pore structural parameters on column performance and resolution of reversed-phase monolithic silica columns for peptides and proteins. *J. Chromatogr. A*, **1144/1**, 72–84.

17 Leinweber, F.C., Schmid, D.G., Lubda, D., Wiesmueller, K.H., Jung, G., and Tallarek, U. (2003) Silica-based monoliths for rapid peptide screening by capillary liquid chromatography hyphenated with electrospray ionization Fourier transform ion cyclotron resonance mass spectrometry. *Rapid Commun. Mass Spectrom.*, **17**, 1180–1188.

18 Minakuchi, H., Ishizuka, N., Nakanishi, K., Soga, N., and Tanaka, N. (1998) Performance of an octadecylsilylated continuous porous silica column in polypeptide separations. *J. Chromatogr. A*, **828**, 83–90.

19 Barroso, B., Lubda, D., and Bischoff, R. (2003) Applications of monolithic silica capillary columns in proteomics. *J. Proteome Res.*, **2**, 633–642.

20 Neverova, I., and Van Eyk, J.E. (2005) Role of chromatographic techniques in proteomic analysis. *J. Chromatogr. B*, **815**, 51–63.

21 Machtejevas, E., and Unger, K.K. (2008) Proteomics Sample Preparation in Sample Preparation for HPLC–Based Proteome *Analysis* (ed. J. von Hagen), Wiley-VCH Verlag GmbH, Weinheim, Germany, pp. 245–264.

22 Issaq, H.J., Chan, K.C., Janini, G.M., Conrads, T.P., and Veenstra, T.D. (2005) Multidimensional separation of peptides for effective proteomic analysis. *J. Chromatogr. B*, **817**, 35–47.

23 Westermeier, R., Naven, T., and Hēpker, H.-R. (eds) (2008) *Proteomics in Practice. A Guide to Successful Experimental Design*, 2nd edn, Wiley-VCH Verlag GmbH, Weinheim, Germany, p. 482.

24 Giddings, J.C. (1984) Two-dimensional separations: concept and promise. *Anal. Chem.*, **56**, 1258–1270.

25 Giddings, J.C. (1995) Sample dimensionality: a predictor of order-disorder in component peak distribution in multidimensional separation. *J. Chromatogr. A*, **703**, 3–15.

26 Huber, J.F.K., and Lamprecht, G. (1995) Assay of neopterin in serum by means of two-dimensional high-performance liquid chromatography with automated column switching using three retention mechanisms. *J. Chromatogr. B*, **666**, 223–232.

27 Sandra, K., Moshir, M., D'hondt, F., Verleysen, K., Kasa, K., and Sandra, P. (2008) Highly efficient peptide separations in proteomics Part 1. Unidimensional high performance liquid chromatography. *J. Chromatogr. B*, **866**, 48–63.

28 Wagner, K., Miliotis, T., Marko-Varga, G., Bischoff, R., and Unger, K.K. (2002) An automated online multidimensional HPLC system for protein and peptide mapping with integrated sample preparation. *Anal. Chem.*, **74**, 809–820.

29 Willemsen, O., Machtejevas, E., and Unger, K.K. (2004) Enrichment of proteinaceous materials on a strong cation-exchange diol silica restricted access material (RAM): protein-protein displacement and interaction effects. *J. Chromatogr. A*, **1025**, 209–216.

30 Wolters, D.A., Washburn, M.P., and Yates, J.R., III (2001) An automated multidimensional protein identification technology for shotgun proteomics. *Anal. Chem.*, **73**, 5683–5690.

31 Regnier, F., Amini, A., Chakraborty, A., Geng, M., Ji, J., Riggs, L., Sioma, C., Wang, S., and Zhang, X. (2001) Multidimensional chromatography and the signature peptide approach to proteomics. *LC-GC*, **19**, 200–213.

32 Gevaert, K., Van Damme, J., Goethals, M., Thomas, G.R., Hoorelbeke, B., Demol, H., Martens, L., Puype, M., Staes, A., and Vandekerckhove, J. (2002) Chromatographic isolation of methionine-containing peptides for gel-free proteome analysis. *Mol. Cell. Proteomics*, **1**, 896–903.

33 Griffin, T.J., Gygi, S.P., Ideker, T., Rist, B., Eng, J., Hood, L., and Aebersold, R. (2002) Complementary profiling of gene expression at the transcriptome and proteome levels in Saccharomyces cerevisiae. *Mol. Cell. Proteomics*, **1**, 323–333.

34 MacCoss, M.J., McDonald, W.H., Saraf, A., Sadygov, R., Clark, J.M., Tasto, J.J., Gould, K.L., Wolters, D., Washburn, M., Weiss, A., Clark, J.I., and Yates, J.R., III (2002) Shotgun identification of protein modifications from protein complexes and lens tissue. *Proc. Natl. Acad. Sci. USA*, **99**, 7900–7905.

35 McDonald, W.H., Ohi, R., Miyamoto, D.T., Mitchison, T.J., and Yates, I.J.R., III (2002) Comparison of three directly coupled HPLC MS/MS strategies for identification of proteins from complex mixtures: single-dimension LC-MS/MS, 2-phase MudPIT, and 3-phase MudPIT. *Int. J. Mass Spectrom.*, **219**, 245–251.

36 Machtejevas, E., Denoyel, R., Meneses, J.M., Kudirkaite, V., Grimes, B.A., Lubda, D., and Unger, K.K. (2006) Sulphonic acid strong cation-exchange restricted access columns in sample cleanup for profiling of endogenous peptides in multidimensional liquid chromatography Structure and function of strong cation-exchange restricted access materials. *J. Chromatogr. A*, **1123**, 38–46.

37 Liu, H., Lin, D., and Yates, J.R., III (2002) Multidimensional separations for protein/peptide analysis in post-genomic era. *Biotechniques*, **32** (4), 898–902.

38 Washburn, M.P., Wolters, D., and Yates, J.R. (2001) Large-scale analysis of the yeast proteome by multidimensional protein identification technology. *Nat. Biotechnol.*, **19**, 242–247.

39 Pang, J.X., Ginnanni, N., Donge, A.R., Hefta, S.A., and Opiteck, G.J. (2002) Biomarker discovery in urine by proteomics. *J. Proteome Res.*, **1**, 161–169.

40 Schulz-Knappe, P., Zucht, H.-D., Heine, G., Jürgens, M., Hess, R., and Schrader, M. (2001) Peptidomics: the comprehensive analysis of peptides in complex biological mixtures. *Comb. Chem. High Throughput Screen.*, **4**, 207–217.

41 Hille, J.M., Freed, A.L., and Wätzig, H. (2001) Possibilities to improve automation, speed and precision of proteome analysis: a comparison of two-dimensional electrophoresis and alternatives. *Electrophoresis*, **22**, 4035–4052.

42 Heine, G., Raida, M., and Forssmann, W.-G. (1997) Mapping of peptides and protein fragments in human urine using liquid chromatography–mass spectrometry. *J. Chromatogr. A*, **776**, 117–124.

43 Raida, M., Schulz-Knappe, P., Heine, G., and Forssmann, W.-G. (1999) Liquid chromatography and electrospray mass spectrometric mapping of peptides from human plasma filtrate. *J. Am. Mass Spectrom.*, **10**, 45–54.

44 Richter, R., Schulz-Knappe, P., Schrader, M., Standker, L., Jurgens, M., Tammen, H., and Forssmann, W.-G. (1999) Composition of the peptide fraction in human blood plasma: database of circulating human peptides. *J. Chromatogr. B*, **726**, 25–35.

45 Rieux, L., Niederlander, H., Verpoorte, E., and Bischoff, R. (2005) Silica monolithic columns. *Synthesis*, **28**, 1628–1641.

46 Svec, F., Tennikova, T., and Deyl, Z. (eds) (2003) *Monolithic Materials: Preparation, Properties and Applications*, Elsevier, Amsterdam, The Netherlands.

47 Szabolcs, F., Jeno, F., and Katalin, G. (2009) Characterisation of new types of stationary phases for fast liquid

chromatographic applications. *J. Pharm. Biomed. Anal.*, **50** (5), 703–712.
48 Kele, M., and Guiochon, G. (2002) Repeatability and reproducibility of retention data and band profiles on six batches of monolithic columns. *J. Chromatogr. A*, **960**, 19–49.
49 Fort, P.E., Freeman, W.M., Losiewicz, M.K., Singh, R.S.J., and Gardner, T.W. (2009) The retinal proteome in experimental diabetic retinopathy: up-regulation of crystallins and reversal by systemic and periocular insulin. *Mol. Cell. Proteomics*, **8**, 767–779.
50 Haffey, W.D., Mikhaylova, O., Meller, J., Yi, Y., Greis, K.D., and Czyzyk-Krzeska, M.F. (2009) iTRAQ Proteomic identification of pVHL-dependent and -independent targets of Egln1 prolyl hydroxylase knockdown in renal carcinoma cells. *Adv. Enzyme Regul.*, **49** (1), 121–132.
51 Xiong, L., Zhang, R., and Regnier, F.E. (2004) Potential of silica monolithic columns in peptide separations. *J. Chromatogr. A*, **1030**, 187–194.
52 Wienkoop, S., Glinski, M., Tanaka, N., Tolstikov, V., Fiehn, O., and Weckwerth, W. (2004) Linking protein fractionation with multidimensional monolithic reversed-phase peptide chromatography/mass spectrometry enhances protein identification from complex mixtures even in the presence of abundant proteins. *Rapid Commun. Mass Spectrom.*, **18**, 643–650.
53 Gury a, V., Kieffer-Jaquinod, S., Garin, J., and Masselon, C.D. (2008) Prospects for monolithic nano-LC columns in shotgun proteomics. *Anal. Bioanal. Chem.*, **392**, 1291–1297.
54 Machtejevas, E., Andrecht, S., Lubda, D., and Unger, K.K. (2007) Monolithic silica columns of various formats in automated sample clean-up/multidimensional liquid chromatography/mass spectrometry for peptidomics. *J. Chromatogr. A*, **1144**, 97–101.

17
Silica Monoliths in Solid-Phase Extraction and Solid-Phase Microextraction

Zhi-Guo Shi, Li Xu, and Hian Kee Lee

17.1
Introduction

Monolithic materials have been attracting intensive attention as alternative stationary phases for high-performance liquid chromatography (HPLC), capillary electrochromatography (CEC) and as extraction sorbents, due to their fast dynamic transport, good loading capacity, low backpressure, etc. Basically, monolithic materials are grouped into two matrices: polymer- and silica-based. The former has been known for almost five decades and has been widely used in different areas, while the latter has just gained popularity recently.

In 1992, Nakanishi *et al.* pioneered the preparation of silica monolith via a sol-gel process [1]. The synthetic process may be briefly described as follows: Silica precursors (e.g., tetramethyl orthosilicate [TMOS], or tetraethyl orthosilicate [TEOS]) are hydrolyzed with the assistance of an acid or base catalyst in the presence of water-soluble high molecular mass polymers such as polyacrylic acid, polyethylene oxide, polyethylene glycol (PEG), etc. The mixture reacts in mild conditions to form a gel. The rate of phase separation and sol-gel transition in the process jointly determine the porous structure of the gel. Under favorable conditions, a structure of interconnected silica skeletons and interwoven through-pores may be obtained. Generally, the gel should be subjected to hydrothermal treatment and calcination to get a high-strength, rod-like monolith. The as-obtained rod typically possesses 0.5–8 μm through-pores and 0.3–5 μm silica skeletons with 2–20 nm mesopores in them [2]. Silica monoliths have been commercialized by chemical or chromatographic supply companies such as Merck, Phenomenex and Conchrom Trenntechniken. Both reversed and normal phases are commercially available. They have found applications in separation science, especially as the stationary phases for HPLC and CEC, for fast separation or preparative purposes.

However, as far as sample pretreatment is concerned, although much attention has been paid to polymeric monoliths, the silica monolith has been more or less neglected. The popularity of polymeric monoliths in sample pretreatment applications may be ascribed to their facile preparation and less likelihood of shrinkage

Monolithic Silicas in Separation Science. Edited by K.K. Unger, N. Tanaka, and E. Machtejevas
© 2011 WILEY-VCH Verlag GmbH & Co. KGaA, Weinheim
ISBN: 978-3-527-32575-7

during post-treatment process. Nevertheless, silica monoliths can circumvent some disadvantages associated with polymeric ones, such as swelling and decomposition under the influence of temperature and/or organic solvents. It possesses huge potential for applications at extreme conditions, especially at high temperature and/or in various organic solvents. It has been gaining more attention in the sample pretreatment area recently [3–5].

Another favorable property of silica lies in its active surface, which can be facilely derivatized. This implies that a greater diversity of desirable functional groups can be obtained through surface modification, leading to more selective, more efficient, and thus sensitive extraction.

Shrinkage, however, cannot be completely avoided when preparing silica monoliths. It occurs even more seriously in the case of a large-diameter monolith. Therefore, in some cases, silica monolith cannot be directly obtained via *in situ* preparation. To prepare a silica monolith column for HPLC, it is generally first synthesized in a mold of a wider dimension than the column. During the drying treatment, the monolith will therefore shrink to fit the dimension of the column (before the experiment, careful design is therefore necessary based on the expected degree of shrinkage, such that the most suitable mold is selected). The resulting silica monolith is then removed from the mold, and encased in the column for use. Possibly due to this necessity, most applications of silica monoliths for sample pretreatment are at the microscale, that is, in-capillary microextraction or microfluidic extraction. Despite the fact that the monolith in smaller devices still tends to shrink more or less during the post-heat treatment, this shrinkage can be minimized or prevented by covalent attachment of the monolith to the capillary or tubing walls. Hence, it can be directly synthesized *in situ*, which may ensure reliable and reproducible performance of the material for the subsequent sample pretreatment.

Bare-silica monoliths have been used for the purification of DNA, peptide and nucleic acid in biological samples [6–14], and hybrid or derivatized silica monoliths have been applied to analytes of environmental or biological concern [14–25]. They can be used in online or offline systems. To provide an overview of the current status of silica monolith and to encourage the further development of these materials in the sample pretreatment field, here we focus on their applications in solid-phase extraction (SPE) and solid-phase microextraction (SPME).

17.2
Extraction Process

The sequence of steps in extraction using a silica monolith is basically the same as that of other silica-based sorbents for SPE. Typically, the experiment is carried out as follows: (1) Preconditioning. The monolith is initially flushed with organic solvent (e.g., methanol) and water in that order, respectively; (2) Sample loading. The sample solution is then percolated through the column for a prescribed time at a certain flow rate; (3) Washing. The monolith is washed with water for several minutes to remove any possible impurity; (4) Elution. The eluent (in the case of the online preconcentration system for HPLC or micro-HPLC, the mobile phase is usually used as the eluent) is pumped through the monolith at a specified flow rate for a prescribed time. For an online preconcentration system, the eluent can be directed into the subsequent analytical system through a column-switching valve or split-flow system. For offline application, the eluent is collected in a sample vial, ready for the subsequent analysis; (5) Washing. After extraction, the monolith is regenerated by further rinsing with the eluent or a stronger elution solvent. For each extraction process, the above five steps are repeated.

17.3
Extraction Platforms

17.3.1
Online Extraction

A valve-switching configuration is desired when online extraction is performed prior to HPLC, micro-HPLC or nano-LC separation. Generally, one or two six-port switching valves (depending on the sample-loading techniques) are necessary to achieve this online extraction process. The monolithic extraction column of an appropriate length is directly connected to the valve instead of a sample loop. When the valve is in the sample-loading position, a certain volume of the sample solution is introduced to the monolith by several means. The Surveyor autosampler is the most convenient automatic means of delivering the sample solution to the sorbent [15]. An external syringe pump may also be used to assist the delivery of the sample solution [16]. In some cases, direct manual loading of sample solution is also possible [17].

After sample loading, the valve is switched to the "inject" position. The elution solution, normally the mobile phase, is pumped through the monolith to desorb the analytes from the silica monolith into the analytical column. After analysis, the extraction column is rinsed with the proper solution before the next injection. In this way, the online purification/preconcentration and separation can be achieved automatically [15–17, 20–25].

Apart from the above online sample pretreatment in conjunction with HPLC, micro-HPLC and nano-LC, another common application of silica monoliths for extraction is in a microfluidic configuration. Basically, a silica monolith is prepared *in situ* in the channel of the chip. To deliver the sample flow and the elution solution, a microsyringe pump is necessary to accomplish the extraction process. Since the consumption of the sample solution is in the microliter range, microfluidic-based extraction is especially suitable for purification and enrichment of biological samples, targeting such analytes as DNA [6–11] and RNA [12].

Alternatively, Wen *et al.* [13] directly used a commercial CE system to accomplish the entire extraction process. The capillary containing a silica monolith of the desired length was fixed in the CE instrument. Pressure was used to load the sample solution, wash the column and elute the analyte (DNA in this case). Online monitoring of the DNA was carried out by laser-induced fluorescence. In this study, CE was employed only as a platform to automate the capillary extraction process; there was no separation involved.

17.3.2
Offline Extraction

In addition to the online configuration, offline extraction is also very popular for its simple setup and manipulation. In this mode, capillary extraction or in-tube microextraction is the most frequently used configuration. The extraction monolith is first prepared in a capillary of the desired length and width. A syringe pump is then used to deliver the required liquid phases during the extraction process. After extraction, the elution solution is collected for subsequent analysis, for example, capillary electrophoresis (CE) [18] and inductively coupled plasma mass spectrometry (ICP-MS) [19].

To simplify the extraction process, a monolithic silica-filled extraction tip was prepared by Miyazaki *et al.* [14]. The silica monolith was tightly attached to the pipette wall by supersonic adhesion. When the pipette was used to withdraw the sample solution, adsorption of the analytes took place on the silica tip. The eluent could then be taken up through the pipette tip in a similar manner to desorb the analytes. The operation of the whole process was claimed to be simple and convenient. Nevertheless, a special instrument was needed to effect the supersonic adhesion. This is generally not available in a normal analytical laboratory.

17.4
Applications

17.4.1
SPE and SPME

Currently, silica monoliths can be prepared by two means: sol-gel entrapped particle process or direct sol-gel process. Their applications in SPE and SPME are discussed below and also summarized in Table 17.1.

17.4.1.1 Silica Monolith from Entrapped Particles
Landers' group contributed much to the purification of DNA, nucleic acids and proteins using silica-based materials [6–9, 13, 26]. In their early study, various silica beads were investigated for the purification of DNA, exhibiting high recovery efficiency, up to 80% [26]. However, this silica bead-packing approach entailed some difficulty of bead retention especially in the microchip chamber; and the reproducibility was unsatisfactory, which may be a result of gradually increasing backpressure during repeated use.

Table 17.1 SPE and SPME applications of silica monoliths.

Monolith	Extraction format	Analytes	Sample matrix	Detection	LOD (range)	Reference
Silica	Offline capillary/ microchip	DNA	Standard	Fluorescence	ng	[6]
Silica	Offline microchip	Human genomic DNA bacterial DNA	Human whole blood colony samples and spores	Fluorescence fluorescence	ng ng	[7]
Silica	Offline microchip	Human genomic DNA bacterial DNA viral DNA	Human whole blood bacteria human spinal fluid	Fluorescence fluorescence fluorescence	ng ng ng	[8]
Silica	Offline microchip	DNA	Human whole blood	Fluorescence	ng	[9]
Polymer monolith impregnated with silica particles	Offline microchip	DNA	Standard	UV	ng	[10]
Silica C_{18}	Offline microchip offline microchip	DNA aromatic hydrocarbons, aromatic phenols, aromatic carboxylic acids	Lysate solution standard	Fluorescence HPLC-UV	ng poor	[11]
Polymer monolith impregnated with silica particles	Offline microchip	Viral RNA	Infected mammalian cells	Polymerase chain reaction- fluorescence	ng	[12]
Silica	Online capillary	DNA	Human whole blood	LIF	ng	[13]
C_{18} silica titania- coated silica	Offline extraction tip	Proteins proteins phosphopeptide	Digestion digestion digestion	HPLC-UV HPLC-UV LC-MS	– – –	[14]

Table 17.1 Continued

Monolith	Extraction format	Analytes	Sample matrix	Detection	LOD (range)	Reference
Chromolith Flash RP-18e (Merck KGaA)	Online SPE	16 phthalate metabolites	Human urine	LC-MS/MS	0.11–0.90 ng/ml	[15]
C_{18}	Online capillary	3 phthalates	Laboratory distilled water/ tap water	HPLC-UV	0.21–0.87 ng/ml	[16]
C_{18}	Online capillary	Pesticides	standard	HPLC-UV	–	[17]
SO_3H-silica	Offline capillary	Anaesthetics	Urine	CE-UV	6.6–36.7 ng/ml	[18]
Amino-silica	Offline capillary	Aluminum	Rainwater, fruit juice	ETV-ICP-MS	1.6 ng/l	[19]
C_{18}	Online capillary	BSA tryptic digest	Digestion	LC-UV	0.53–1.32 ng/µl	[20]
Chromolith SpeedROD RP-18e	Online	Isoquinoline drug (GW328713X)	Plasma	MS	5 ng/ml	[21]
Chromolith Flash RP-18e	Online	Drugs	Plasma	LC-MS	10–25 ng/ml	[22]
Chromolith Flash RP-18e	Online	Drugs	Plasma	LC-MS	–	[23]
Chromolith Flash RP-18e	Online	Benzodiazepines	Whole blood	LC-MS	2.5–80 µg/L (LOQ)	[24]
C_8	Online capillary	PAHs	Standard	HPLC-UV	2.4–8.1 ng/ml	[25]

To implement a more practical operation, the same group studied a silica xerogel and "monolithic" silica bead [6]. Both of these materials were based on the sol-gel technology, with the difference that the former was only formed from the silica sol, while the latter actually employed the silica sol to gel the silica beads. In this preliminary experiment, the authors found the performance of the silica xerogel was even worse than that of the silica-bead-packing format, despite the fact that the silica xerogel could be conveniently and satisfactorily held in the microchip channel. They ascribed the observation to the mechanical instability and poor porous structure of the xerogel. Compared to a silica monolith, a silica xerogel lacks the macroporous structure even though it has a monolithic appearance.

On the other hand, monolithic silica beads provided a workable system. The technology of immobilizing particles by sol-gel means was first developed by Zare's group to avoid tedious preparation of frits associated with the packed capillary [27]. The packing material was assembled with a sol-gel matrix. The dried gel served as a "glue" to bond the packing material together and to the capillary wall when the sol solution was polymerized. The silica beads, in this study, were immobilized in a capillary by the matrix formed by TEOS sol. The ultimate material exhibited a monolithic configuration, whose porous structures relied on both silica beads and the sol-gel structure. The extraction recovery of DNA was more than 80% and the relative standard deviation (RSD) was only 3.05%. The authors also developed the material into a microchip-based online purification platform.

In a later thorough study by the same group, the microchip SPE system containing the immobilized silica beads was used to purify human genomic DNA from whole blood and bacterial DNA from colony samples and spores. Fast extraction, less than 15 min, could be obtained. They also demonstrated the potential application of this microchip format in a portable device [7]. The silica monolith from direct sol-gel means was also attempted by the group to extract DNA [8, 9, 13] (see below).

The particle-entrapped monolith was also examined by Zhang *et al.* [20]. This strategy was aimed at eliminating cracking or shrinkage of the column bed commonly associated with silica monoliths in large-bore tubing. The authors evaluated the preparation conditions and characteristics, such as permeability and mechanical strength in detail. C_{18}-bonded silica particles were gelled using methyltriethoxysilane as precursor in 320- and 530-µm ID capillaries. The postcolumn drying procedure was regarded as the crucial step to produce a crack-free column. Low-pressure nitrogen was passed through the column during drying. The permeability of the particle-entrapped monolith was slightly lower than that of a packed column, as a result of decreasing interstitial voids between the particles. However, this slightly lower permeability showed no serious influence on its application.

The performance of the aforementioned C_{18} particle-entrapped column was evaluated for the online enrichment of bovine serum albumin tryptic digest followed by micro-HPLC determination. The extraction performance between 320- and 530-µm ID was almost the same in terms of the analytical signal intensity. However, the separation efficiency was decreased compared with direct injection (no extraction), and was worse in the case of the 530-µm ID column. This is possibly due to the increasing dead volume between the precolumn and separation column with the increasing inner diameter of the former. A larger-bore column allowed a fast loading flow rate, and was expected to have a greater mass-loading capacity. However, the experimental result suggested almost the same mass loading capacity between the 320- and 530-µm ID precolumns. The authors claimed that it was as a result of the decreasing separation efficiency in the case of the larger-diameter precolumn. Both of these different internal-diameter precolumns exhibited satisfactory recovery, reproducibility and chemical stability. Finally, the authors suggested the potential of the large-bore particle-entrapped monolithic precolumns for LC-tandem mass spectrometry (MS/MS) and LC-LC-MS/MS applications.

Similarily, Mitra *et al.* [11] immobilized micro- and nanosized silica particles in poly(dimethylsiloxane) (PDMS) channels. The 3-µm silica particles or 5-µm C_{18} particles were packed into the channel tightly, with the two ends sealed with a PDMS prepolymer and crosslinking mixture. Afterwards, the sol consisting of methyltrimethoxysilane (MTMS), propyltrimethoxysilane (PTMS) and trifluoroacetic acid, was injected into the channel to gel the silica particles under thermal treatment. MTMS and PTMS were found to play a crucial role to obtain a stable, porous and uniform bed in this effort. However, it was not the case as far as nanosilica was concerned. Obviously, it could be problematic to directly pack the silica nanoparticles in the channel. Hence, silica nanoparticles were first dispersed into the TEOS sol solution by sonication. These sol-containing silica nanoparticles were then injected into the channels to form the bed. In this case, the sol of MTMS and PTMS did not work well. A very dense bed with lower interparticle porosity resulted, leading to very low flow rate. This may be due to the fact that silica nanoparticles take part in the sol-gel reaction themselves, and react with the PDMS surface to give a tighter bed, while larger particles could compensate with greater interparticle porosity.

Both of the above micro- and nanosized silica particle-immobilized beds were examined for the purification of genomic DNA from crude *E. coli* cell lysate solution. Higher capacity was obtained with silica nanoparticles, whereas higher extraction efficiency could be achieved with the microsized silica particles. The observation could be possibly ascribed to the high surface area of nanoparticles. When the authors tried to capture chemical species on a C_{18} bed, they failed. They explained that these chemical species are nonpolar, and have the tendency to associate with the PDMS surface rather than the C_{18}. Therefore, they concluded that microchip SPE on PDMS substrates was more suited for polar analytes, but not for nonpolar ones. However, this assumption seems unreasonable. Both PDMS and C_{18} are well known for their properties to adsorb nonpolar species. In this experiment, C_{18} particles were gelled by MTMS and PTMS. This may imply that the surface of the resulting monolithic C_{18} particles is based on MTMS and PTMS, and no longer on C_{18}. The material exposed to the sample solution actually comprises the functional groups on MTMS and PTMS. These are methyl and propyl groups, which possess much less hydrophobic affinity than C_{18}. This may be an explanation for the failure to extract nonpolar species, as observed by the authors.

The same strategy as reported in Landers' work [6] was adopted by Bhattacharyya and Klapperich [10]. Instead of inorganic microfabricated devices (i.e. silicon, glass, or quartz), a thermoplastic polymer of cyclic polyolefin, was used to fabricate a plastic microfluidic chip. This approach is more attractive because of the suitability for disposable applications and low cost. The plastic channel was first pretreated via a benzophenone-initiated surface photopolymerization process, to allow the adhesion of the monolith to it. Afterwards, monomers, porogens, initiator and silica microspheres (0.7 µm) were introduced into the channel and reacted *in situ* by UV irradiation. In this way, a porous poly(methyl methacrylate-co-ethylene dimethacrylate) monolith impregnated with silica particles was obtained. In the subsequent experiments, the monolith was demonstrated to be effective for repetitive DNA extractions across different channels. It exhibited comparable extraction efficiency with that reported in previous work [7, 8]. However, when extractions were repeated on the same channel, the extraction efficiency was reduced gradually with repeated use. This could be a result of the breakdown of the monolithic structure. Nevertheless, since the aim of this work was to prepare disposable devices, the stability of the single device was not of primary concern.

In a recent effort, porous poly(butylmethacrylate-co-ethylene dimethacrylate)-embedded silica particles were used for the isolation of viral RNA in a microfluidics platform [12]. Even though this application did not use a silica monolith directly, it involved the exploitation of the porous structures of monolithic materials to derive the benefits of mass transfer and the silica surface to bind RNA. It could be expected that the direct usage of silica-based monoliths for the same application should be more straightforward and easier to handle.

17.4.1.2 Silica Monolith from Direct Sol-Gel Strategy

Precolumns and Tips A company, Merck KGaA pioneered the commercialization of silica monoliths as LC and guard column stationary phases. Later, other companies such as Phenomenex and Conchrom also provided similar products. Normally, the commercialized monolithic silica affords through-pores of about 1 µm and mesopores of about 13 nm, which ensure favorable mass transfer as well as a large surface area during the separation/extraction process. Connectors with zero dead volume are commercially available, which offer the convenience of lining the precolumn to various detection systems for the purpose of online preconcentration/cleanup. Applications of commercialized silica monoliths aiming at sample pretreatment have been widely explored.

A Merck product, Chromolith SpeedROD RP-18-e column was used as a precolumn before MS to purify and enrich an isoquinoline pharmaceutical candidate in human plasma [21]. Through a column-switching valve, the eluent was controlled to be delivered to the waste port or the MS inlet port as desired. The monolithic precolumn showed excellent robustness after almost 300 direct injections of diluted plasma. Satisfactory sensitivity was obtained, with a limit of quantification (LOQ) of 5 ng/ml for the isoquinoline.

The same precolumn was also employed for concentrating phthalate metabolites [15] by a column-switching strategy. Under the control of an HPLC autosampler, the extraction and separation can be automatically performed. The total time for these two steps was 27 min. Limits of detection (LODs) were in the low ng/ml range. The established method was used to screen human urine samples, to investigate human exposure to phthalates [15].

Three drugs and their metabolites in human plasma [22, 23] were monitored by a Chromolith Flash precolumn and LC-MS. Endogenous components such as proteins and salts were rapidly removed from the human plasma, while the analytes of interest were entrapped by the monolithic support. In this way, direct injection of a biological sample into an LC-MS system could be realized. More than 100 samples could be handled in the same precolumn without any noticeable decrease in performance [22, 23].

Bugey and Staub [24] compared the performance of silica monolith and restricted access material sorbents in the online preconcentration of benzodiazepines in whole blood samples. They found that a silica monolith gave better validation data and was more robust.

As mentioned above, Miyazaki *et al.* proposed the use of monolithic silica held in a pipette tip for extraction [14]. A bare-silica monolith or a modified silica monolith was first synthesized in a mold, and then fixed into a 200-µl pipette tip by supersonic adhesion. With the proper purification steps, such as equilibration, sample loading, tip rinsing and sample elution, a C_{18}-bonded silica monolithic tip could be used to retain peptides and proteins due to their strong affinity to the hydrophobic silica surface, while salts, detergents and other hydrophilic contaminants from the sample were removed. However, high molecular mass proteins were not as efficiently recovered as the smaller ones. This may be a result of the presence of 20-nm mesopores in the monolithic silica, which provided some difficulties for sufficient infiltration of such large molecules as proteins [14].

Additionally, a bared monolithic silica tip was investigated for peptide and protein purification. Based on the principal that dissociated silica provides silanol groups and these groups can adsorb basic compounds, negative-charged and acid compounds would pass through the tip.

More interestingly, a titania-coated monolithic silica extraction tip was studied to isolate phosphorylated proteins with high selectivity and sensitivity [14]. This is a result of favorable affinity of titania to phosphorous compounds. The method was applied to the detection of phosphopeptide from ß-casein tryptic digestion followed by LC-MS analysis.

This work demonstrated one obvious advantage of monolithic materials, which is low pressure drop originating from the highly porous structure of the monolith. The extraction operation requires no extra equipment, for example, pump or vacuum system. Aspiration and dispensation could be achieved by normal manual operation as one would operate a conventional pipettor. This format of SPE can predictably be extended after suitable modification of silica surface for more diverse applications.

Capillaries and Chips As the fabrication of silica monolith in a microdevice can overcome the problem relating to shrinkage, most of the in-house prepared silica monoliths are based on capillary or chip modes. Accordingly, the extractions carried out in these devices are regarded as capillary in-tube microextraction or microchip SPE, respectively.

The C_{18}-bonded monolithic silica column was first developed by Shintani et al. for in-tube SPME coupled to micro-HPLC [17]. Micro-HPLC is a miniaturized version of HPLC that is more compatible with capillary in-tube SPME. The coupling was realized by placing the silica monolith in a capillary leading to the sample loop position of a six-port injector. Manual injection could be applied because the monolithic material has a low pressure drop. Meanwhile, a large volume of sample solution, 50 μl instead of the usual 0.1 μl associated with normal micro-HPLC, was injected to enhance the sensitivity. Compared to the open-tubular column, the silica monolith exhibited higher sample capacity. It provided 50 times higher sensitivity for thiuram, mecoprop, iprodione, pencycuron and bensulide pesticides in a complex matrix.

Similarly, a C_{18} silica monolith was used as an online sample-enrichment system for phthalates prior to micro-HPLC [16]. It was synthesized by derivatizing the silica monolith with dimethyloctadedylchlorosilane. The LOD enhancement was as high as 2000-fold. However, in that report, the method was only evaluated for laboratory distilled water and tap water samples, not more challenging real aqueous matrices.

As mentioned previously, Landers' group has contributed much to the area of purification of biological samples based on silica materials. The use of monolithic silica for microchip SPE application was proposed by this group in 2006 [8]. Thanks to the development of the synthetic technologies, macroporous silica monolith can be conveniently prepared in situ via the sol-gel approach in the microchip channel. The addition of PEG to the synthesis is crucial to form macroporous structures by inducing phase separation and pore formation. The resulting macropores (of the order of 1 μm) should be more favorable to DNA extractions. DNA extraction efficiencies for simple systems (less complex matrices) were ~85%, while efficiencies for the extraction of human genomic DNA from blood were ~70%. However, it was observed that blockage of pores occurred in repeated extractions on a single device.

In more recent efforts by the same authors [9, 13], an organic silica-based monolith was attempted for the DNA purification in a microdevice. The 3-(trimethoxysilyl)propyl methacrylate (TMSPM) monolith was fabricated by UV-induction. The advantage of UV-induced photopolymerization is that it enables the accurate placement of monolithic matrixes within the architecture of microscale devices. Compared to the pure inorganic sol-gel monolith, gel shrinkage and cracking is lessened. To further enhance the extraction performance of this TMSPM monolith, a derivatization of its surface was adopted. TMOS was chosen to modify the monolith, due to its superior reactivity to TEOS. A continuous network of silicon dioxide was formed after TMOS modification, which was supposed to increase the number of silica binding sites on the monolith surface. In subsequent experiments, the TMOS-derivatized TMSPM monolith was found to enhance the relative column capacity by ~60-fold and the relative extraction efficiency by ~40% when compared to the underivatized material for the extraction of prepurified genomic DNA and whole blood. One possible negative result from the derivatization may be the reduced pore size. The derivatization procedure herein was actually a coating step and would occupy more void space with the increasing amount of TMOS (as observed in this study). On this basis, therefore, the coating procedure should be controlled carefully, ensuring acceptable extraction efficiency and mass transfer. Otherwise, the mass transfer would be sacrificed and the backpressure might be too high. The comparison between the TMOS-derivatized TMSPM monolith and a commercial SPE was also evaluated. The former was demonstrated to have higher extraction efficiency than the Qiagen spin column for the extraction of prepurified human genomic DNA [13]. However, this was only efficient in the case of low-volume samples. As for larger-volume samples, the Qiagen spin column, which is designed to handle larger volumes, exhibited better performance.

This TMOS-derivatized monolith was also implemented in a microfluidic device for a two-stage purification of DNA [9]. A precolumn, packing with C_{18} beads, was used to remove proteins in the presence of blood. With this cleanup step, although some hydrophilic proteins passed through the precolumn, 70% protein could be trapped onto it. More than 97% DNA was shown to pass through the precolumn, and be available for further enrichment. In this way, interferences from the interaction of proteins and the monolith could be reduced significantly. The binding capacity of the monolith of DNA was therefore increased. The overall extraction efficiency was about 69%.

The above examples demonstrated the applications of the chemical-grafting silica monoliths via traditional chemical-bonding methods. However, this type of silica-based material suffers from several shortcomings. First, this procedure involves a multistep synthesis. There is no doubt that too many steps used for the manufacture of the monolithic column can lead to more variability in their performance. Secondly, the loading of the functional groups into the silica matrix may be low. To address these issues, a hybrid silica-based monolith prepared by one-pot sol-gel technology may be an alternative. Organic functional groups should generally be more evenly distributed in the structure of the inorganic matrix, facilitating the attainment of better performance in applications. The ratio of the functional groups to the silica matrix may be adjustable across a particular range to provide optimum performance.

A hybrid organic–inorganic octyl silica monolithic column was for the first time proposed for the online in-tube SPME coupling with micro-HPLC by Zheng *et al.* [25]. The authors claimed that the hybrid organic–inorganic materials had superior advantages over chemical-grafting silica materials. Silica–carbon (Si–C) or silica–nitrogen (Si–NH–C) bonds, instead of Si–O–Si–C, result from the one-step hybrid synthetic strategy, which are more resistant to extreme conditions, for example, high pH. Additionally, the combination of the functional monomer and the matrix monomer as the precursor source of the material can afford a flexible choice of the degree of the functionality. Their ratio can be adjusted in a certain range, leading to the different porous structures and functional properties. The functional monomer is normally in greater proportion than the chemical-grafting silica material. The preparation of such hybrid silica monolith is a more straightforward and direct approach, involving a single step. This simplifies the synthetic process, saves time and avoids uncertainties that may result from multistep reaction and workup.

Zheng et al. [25] used their C_8 silica monolith for the enrichment of polycyclic aromatic hydrocarbons (PAHs) followed by micro-HPLC-UV determination. Around 300- to 450-fold enhancement could be obtained. Intraday and interday reproducibilities on the same column were satisfactory, with RSDs of less than 7.4 and 8.1%, respectively. The column-to-column RSDs were between 1.3% and 8.0%. These results imply that the hybrid material possesses stable structure and property.

In a recent work, Xu and Lee presented an organic–silica hybrid monolith containing mercapto groups for in-tube SPME applications [18]. The hybrid monolith was successfully prepared by sol-gel technology. Synthetic conditions were carefully investigated, including the solvent category, PEG content, reactant ratios (TMOS/MPTS), temperature and catalysts. The resulting hybrid monolith showed bimodal porous structures with high sulfur content (3.05%) and large specific surface area (467 m^2/g). Due to the high reactivity, mercapto groups can be favorably modified further to afford various functionalities. For example, in this case, they were conveniently oxidized by hydrogen peroxide to produce sulfonic acid groups, implying a high potential for extraction applicability based on ion-exchange principles. Hence, the monolithic material was employed for in-tube microextraction, considering four anaesthetics as model compounds. Through studies on the extraction parameters, such as loading solution, loading time, elution solution and elution time, optimized extraction performance was accomplished by loading the sample solution for 800 s at a flow rate of 1.5 ml/h and eluting by 0.25% ammonia solution (80% methanol) at a flow rate of 0.5 ml/h for 600 s. The procedure was successfully applied to a spiked human urine sample for analyte extraction and concentration. The LODs achieved were between 6.6 and 36.7 µg/l. The hybrid silica monolith may also be subjected to other derivatization approaches for more analytical (or even nonanalytical such as catalytic) applications.

The use of silica-base monolithic material for extractions has been extended to inorganic analytes by Zheng and Hu [19]. An N-(2-aminoethyl)-3-aminopropyltrimethoxysilane (AAPTS)-silica monolith was prepared by a sol-gel method in a capillary. It was used as an offline preconcentration device prior to electrothermal vaporization (ETV)-ICP-MS for aluminum (Al) fractionation. Labile monomeric Al could be retained on the AAPTS-silica monolith in the pH range of 4–7, while the total monomeric Al could be adsorbed on the monolith in the pH range of 8–9. Therefore, a two-step capillary microextraction with the sample solution maintained under different pH conditions was employed to extract labile and total monomeric Al, respectively. The eluent was subjected to ETV-ICP-MS determination separately. In this manner, the mount of labile and total monomeric Al could be measured. The nonlabile monomeric Al was then calculated by subtraction. The developed method was applied to determine the fraction of Al in rainwater and fruit juice. The authors claimed that the AAPTS-silica monolith possessed good pH resistance. However, the use of an elution solution containing 1 M hydrochloric acid seems to be problematic in terms of stability and durability of the silica material.

The aforementioned hybrid materials have been especially investigated as sorbents for in-tube SPME. However, the sol-gel approach to hybrid materials is limited because it requires careful control of synthetic conditions to obtain the defined monolithic configuration in the presence of dissimilar precursors. This implies that the synthetic conditions to prepare hybrid monolithic structures containing different functionalities have to be optimized individually.

17.4.2
Other Applications of Silica Monolith

Being similar to extraction applications, adsorption is another field to which silica monoliths may be applied. Liu *et al.* [28] studied in detail the adsorption kinetics of Erioglaucine (a triphenylmethane dye) on four different hybrid gel monoliths. The authors prepared hybrid gel monoliths using bis(trimethoxysilylprolyl)amine (TSPA) or a mixture of TSPA and n-propyltriethoxysilane, bis(trimethoxysilyl) hexane or TEOS as precursors. Parameters influencing adsorption, including temperature, pH and ionic strength, were examined. It was found that adsorption was promoted by hydrophobic attraction, hydrogen bonding, and electrostatic attraction under acidic condition. On the other hand, adsorption was much reduced under basic conditions due to the suppression of electrostatic attraction. The stability of silica monoliths was generally satisfactory. No obvious shrinkage occurred under different pH conditions or ionic strength, but a slight shrinkage could be observed after about 165 h of use.

The application of silica monolith to the adsorption of small organic molecules was investigated by Štandeker *et al.* [29]. Different degrees of hydrophobicity of the silica aerogels were obtained by varying the ratios of MTMS to TMOS or trimethylethoxysilane to TMOS in the synthesis. Toluene, benzene, ethylbenzene, chloroform xylene, chlorobenzene, 1,2-dichloroethane and trichloroethylene were used as model analytes. The aerogels showed superior adsorption capacity for the target organic molecules when compared to granulated active carbon.

17.5
Conclusion and Outlook

The current review has focused on the use of silica monoliths for sample pretreatment, specifically covering SPE and SPME applications. Silica monoliths have the following merits:

1. They possess a bimodal structure, interlaced through-pores and skeletons with mesopores in them. This highly porous structure leads to high surface area, which is beneficial to the surface modification processes, and extraction

capacity. Moreover, it allows a favorable mass transfer and low backpressure during use. As a result, less time is needed to condition the column, and high flow rates can be applied to deliver sample solution and to elute analytes. Moreover, manual sample introduction is also possible.

2. Modification of the silica surface is easily accomplished. Although it seems most applications so far are concentrated on bare or C_{18} silica, the future will likely see greater diversity of functionality being imposed on the silica surface. Silica can be derivatized by silylation. Greater selectivity and sensitivity can thus be obtained. Additionally, silica containing functional groups may be made to undergo further derivatization. For example, the successful use of mercapto-group-incorporated silica monolith for in-tube microextraction was described above. After simple oxidation, mercapto groups were converted to sulfonic acid groups, which were effective in extracting anaesthetics analytes in human urine. Potentially, the mercapto-group-incorporated silica monolith can be further derivatized to provide more functionalities, owing to the high reactivity of mercapto groups.

3. Online sample pretreatment involving monoliths is easy to handle. Unlike particle packed columns, no frits are required and the material can be synthesized *in situ*. This affords convenience when connecting the column to the analytical instrument.

4. Silica material has good resistance to organic solvent and elevated temperatures. No obvious deformation or swellings is observed in most organic solvents. These properties imply potential applications under extreme conditions.

5. Biocompatibility of silica material is generally regarded as satisfactory. This is important when the material is used for biological sample preparation.

Silica monoliths do have some disadvantages, such as shrinkage of the skeleton during the drying step and irreversible adsorption. Nevertheless, these materials represent useful approaches to fast, selective, sensitive, and automatic extraction, particularly automatic sample pretreatment procedure at the microscale.

References

1 Tanaka, N., Ishizuka, N., Hosoya, K., Kimata, K., Minakuchi, H., Nakanishi, K., and Soga, N. (1993) *J. Chromatogr. A*, **14**, 50–51.
2 Minakuchi, H., Nakanishi, K., Soga, N., Ishizuka, N., and Tanaka, N. (1996) *Anal. Chem.*, **68**, 3498–3501.
3 Kato, M., Sakai-Kato, K., and Toyo'oka, T. (2005) *J. Sep. Sci.*, **28**, 1893–1908.
4 Svec, F., and Geiser, L. (2006) *LC-GC North Am.*, **24**, 22–27.
5 Svec, F. (2006) *J. Chromatogr. B*, **841**, 52–64.
6 Wolfe, K.A., Breadmore, M.C., Ferrance, J.P., Power, M.E., Conroy, J.F., Norris, P.M., and Landers, J.P. (2002) *Electrophoresis*, **23**, 727–733.

7 Breadmore, M.C., Wolfe, K.A., Arcibal, I.G., Leung, W.K., Dickson, D., Giordano, B.C., Power, M.E., Ferrance, J.P., Feldman, S.H., Norris, P.M., and Landers, J.P. (2003) *Anal. Chem.*, **75**, 1880–1886.

8 Wu, Q., Bienvenue, J.M., Hassan, B.J., Kwok, Y.C., Giordano, B.C., Norris, P.M., Landers, J.P., and Ferrance, J.P. (2006) *Anal. Chem.*, **78**, 5704–5710.

9 Wen, J., Guillo, C., Ferrance, J.P., and Landers, J.P. (2007) *Anal. Chem.*, **79**, 6135–6142.

10 Bhattacharyya, A., and Klapperich, C.M. (2006) *Anal. Chem.*, **78**, 788–792.

11 Karwa, M., Hahn, D., and Mitra, S. (2005) *Anal. Chim. Acta*, **546**, 22–29.

12 Bhattacharyya, A., and Klapperich, C.M. (2008) *Sens. Actuators B Chem.*, **129**, 693–698.

13 Wen, J., Guillo, C., Ferrance, J.P., and Landers, J.P. (2006) *Anal. Chem.*, **78**, 1673–1681.

14 Miyazaki, S., Morisato, K., Ishizuka, N., Minakuchi, H., Shintani, Y., Furuno, M., and Nakanishi, K. (2004) *J. Chromatogr. A*, **1043**, 19–25.

15 Kato, K., Silva, M.J., Needham, L.L., and Calafat, A.M. (2005) *Anal. Chem.*, **77**, 2985–2991.

16 Lim, L.W., Hirose, K., Tatsumi, S., Uzu, H., Mizukami, M., and Takeuchi, T. (2004) *J. Chromatogr. A*, **1033**, 205–212.

17 Shintani, Y., Zhou, X., Furuno, M., Minakuchi, H., and Nakanishi, K. (2003) *J. Chromatogr. A*, **985**, 351–357.

18 Xu, L., and Lee, H.K. (2008) *J. Chromatogr. A*, **1195**, 78–84.

19 Zheng, F., and Hu, B. (2008) *Spectrochim. Acta B*, **63**, 9–18.

20 Gu, X., Wang, Y., and Zhang, X. (2005) *J. Chromatogr. A*, **1072**, 223–232.

21 Plumb, R., Dear, G., Mallett, D., and Ayrton, J. (2001) *Rapid Commun. Mass Spectrom.*, **15**, 986–993.

22 Souverain, S., Rudaz, S., and Veuthey, J.-L. (2003) *Chromatographia*, **57**, 569–575.

23 Veuthey, J.-L., Souverain, S., and Rudaz, S. (2004) *Drug Monit.*, **26**, 161–166.

24 Bugey, C.S. (2007) *J. Sep. Sci.*, **30**, 2967–2978.

25 Zheng, M.-M., Lin, B., and Feng, Y.-Q. (2007) *J. Chromatogr. A*, **1164**, 48–55.

26 Tian, H., Hühmer, A.F.R., and Landers, J.P. (2000) *Anal. Biochem.*, **283**, 175–191.

27 Dulay, M.T., Kulkarni, R.P., and Zare, R.N. (1998) *Anal. Chem.*, **70**, 5103–5107.

28 Liu, H., You, L., Ye, X., Li, W., and Wu, Z. (2008) *J. Sol-Gel Sci. Technol.*, **45**, 279–290.

29 Štandeker, S., Novak, Z., and Knez, Ž. (2007) *J. Colloid Interface Sci.*, **310**, 362–368.

Index

a

α,α′-azobisisobutyronitrile (AIBN)   37, 40
additives   23–26
affinity columns   278, 279
aging *see* coarsening
AIBN *see* α,α′-azobisisobutyronitrile
alkaloids   196, 199
alkylbenzenes   256–259, 262, 264, 265
alumina   23, 51
ammonium peroxydisulfate (APS)   40
amphiphilic organic molecules   216
amylase tris(3,5-dimethylphenyl carbamate)   236, 238, 240
analyte hydrophobicity   163–166
anodic alumina   51
antibiotics   199
APS *see* ammonium peroxydisulfate
aqueous alkaline silicate *see* water-glass
axial dispersion   147, 148, 151

b

band broadening
– characteristic reference lengths   111–113
– definition of terms used   121, 122
– eddy-dispersion and longitudinal diffusion   107, 108, 120
– general plate-height model   106–120
– Giddings coupling theory   116–120
– Kataoka model   111, 116, 119
– Knox model   116–118
– mass transfer kinetics   105–107, 110, 111
– modeling   105–125
– nonporous skeleton case   115, 116, 118
– penetration model   111, 116, 119
– porous skeleton case   116–120
– prediction   114–120
– tetrahedral skeleton model   106, 111, 112, 114–120
bare-silica monoliths   213, 215–217

Barrett–Joyner–Halenda (BJH)   54, 55, 66, 73, 74
basic compounds
– activity of monoliths towards basic solutes   175–179
– endcapping   173, 178, 179, 186, 187
– hybrid capillary silica monoliths   183–186
– overload   180
– particle-packed columns   176, 177
– performance of silica monoliths   173–188
– reproducibility of analyses   174
– reversed-phase HPLC   173–188
– silanol groups   173, 174
– van Deemter plots   180–183, 185
benzene isotopologues   257–259
BET isotherm   64–66, 146
beta blockers   190
bimodal pore structure   49, 50, 60, 210–212, 332, 333
biomarkers of disease   279, 297
BJH *see* Barrett–Joyner–Halenda
blind ink-bottle-shaped pores   52
bulk monolith preparation   23–26

c

caffeines   190
capillary columns
– basic compounds   183–186
– future developments   267, 268
– gradient mode   259–261
– high-efficiency separations   249–261, 268, 269
– high-speed separations   249–254, 261–266, 268, 269
– hybrid silica monoliths   183–186, 250, 251, 254, 268
– isocratic mode   256–259
– isotopic separations   257–259

Monolithic Silicas in Separation Science. Edited by K.K. Unger, N. Tanaka, and E. Machtejevas
© 2011 WILEY-VCH Verlag GmbH & Co. KGaA, Weinheim
ISBN: 978-3-527-32575-7

– particle-packed columns 253, 255, 256, 260, 261, 266–268
– performance characteristics 254–256, 259–261, 265–267
– phase ratios 261–265, 267–269
– preparative methods 31, 250–252
– properties 252–254
capillary electrochromatography (CEC)
– basic compounds 186
– chiral separation techniques 231, 233–235
– solid-phase extraction/microextraction 319
capillary electrophoresis (CE) 231
capillary liquid chromatography (CLC) 231
catalytic converters 1
CE *see* capillary electrophoresis
CEC *see* capillary electrochromatography
cellulose tris(3,5-dimethylphenyl carbamate) 236, 238, 240–242
ceramic monolith supports 1, 2
cetyltrimethylammonium bromide (CTAB) 214, 219, 220
cetyltrimethylammonium chloride (CTAC) 214, 216–219
characteristic reference lengths 111–113
characteristic wavelength 86–89
chemical cooling 14, 15
chiral separation techniques 231–248
– covalently bonded chiral selectors 237–244
– drug molecules 198–199
– *in situ* copolymerization 234, 235
– monolithic chiral stationary phases 231–248
– particle-packed columns 232, 233, 243, 245
– physically adsorbed chiral selectors 236, 237
– polymer-based materials 233–235
– quality control techniques 199–201
– silica monoliths 235–244
– state-of-the-art and future directions 245
chiral stationary phases (CSP) 231–248
chord lengths 90, 91
chromatographic separations
– band broadening 105–125
– general plate-height model 106–120
– laboratory methods for separation columns 29–31
– morphology and pore structure 3
– polymer-coated materials 38–41

– principles and development 2–5
– *see also* individual techniques
CLC *see* capillary liquid chromatography
closed pores 51, 52
coarsening stage 17, 18, 21, 26
cocontinuous domain structures
– capillary columns 252, 262, 263, 268
– chromatographic separations 3
– microscopic characterizations 81
– phase separation 17, 18
COI *see* compounds of interest
column coupling 196
column efficiency
– basic compounds 177, 178, 182, 183, 186, 187
– capillary columns 249–261, 266, 268, 269
– chiral separation techniques 232
– fast ion chromatography 209, 210, 223
– high performance liquid chromatography 135–137, 147–151
column impedance 137, 138, 150–152
column permeability
– fast ion chromatography 211
– high performance liquid chromatography 132–135, 140–142, 147, 149, 152
compounds of interest (COI) 292
confined geometries 95–102
confocal laser scanning microscopy 71
contrast-variance enhancement (CVE) 84
controlled-pore glass (CPG) 54–57
covalently bonded chiral selectors 237–244
covalently bonded ion-exchange groups 221, 222
CPG *see* controlled-pore glass
cross-contamination 303
CSP *see* chiral stationary phases
CTAB *see* cetyltrimethylammonium bromide
CTAC *see* cetyltrimethylammonium chloride
CVE *see* contrast-variance enhancement
cytochrome C 300, 310, 311

d

DAHMA *see* 3-diethylamino-2-hydroxypropyl methacrylate
Darcy equation 132, 134
DDAB *see* didodecyldimethylammonium bromide
deformations in confined geometries 95–102

didodecyldimethylammonium bromide (DDAB)   208, 209, 213–215, 217, 220
3-diethylamino-2-hydroxypropyl methacrylate (DAHMA)   43, 44, 215, 216, 222
diffusion coefficients   109, 110, 111, 121
dilute and shoot method   200
N,N-dimethylacetamide (DMA)   100–102
dipole–dipole interactions   161, 167–170
dipole–induced-dipole interactions   161, 167, 168, 170
DMA *see* N,N-dimethylacetamide
domain formation by phase separation   15–19, 31
Dorsey–Foley equation   177, 178, 181–183
drug discovery compounds   292–295
drug molecules
– analysis times   190–196
– chiral drug separation   196–199
– column coupling   194–196
– flow programming   194–196
– particle-packed columns   189–198
– quality control techniques   189–205
– silica monoliths   189–201
drying and heat treatment   28

e

eddy-dispersion   107, 108, 120, 148, 151
electron-pair donor/electron-pair acceptor (EPD/EPA) interactions   161, 167, 169
electrospray ionization (ESI)   40, 41, 273, 275–278, 300
enantioseparation *see* chiral separation techniques
endcapping   173, 178, 179, 186, 187, 251
entrapment   56–60, 322–327
environmental applications   1, 2, 223
EPD/EPA *see* electron-pair donor/electron-pair acceptor
epoxy-clad columns   29
ESI *see* electrospray ionization
estrogens   190
external porosity   131, 132, 138, 139, 146, 152
extracted ion chromatograms (XIC)   275

f

FAC *see* frontal affinity chromatography
fast Fourier transformation (FFT)   86, 88
fast ion chromatography (FIC)   207–230
– advantages of silica monoliths   210–212
– amphiphilic organic molecules   216
– analytical applications   223–225
– anion separations   215–220, 222, 224
– bare-silica monoliths   213, 215–217
– cation separations   215–218
– column efficiency   209, 210
– conditions and parameters   210
– covalently bonded ion-exchange groups   221, 222
– fundamental principles   207
– future developments   225, 226
– historical development   207–210
– ion-pair separation   219, 220
– modified silica monoliths   216–221
– operational parameters   221–223
– particle-packed columns   211
– polymer-coated materials   215, 216
– pore structure   210–213
– reversed-phase   217–221
– surfactant-coated materials   213–215, 219–221, 223
– types and properties of silica monolith columns   212–216
FFT *see* fast Fourier transformation
fibrous gel skeletons   89, 90, 93, 94
FIC *see* fast ion chromatography
flow optimization   290
flow programming   194–196
flow-through pores
– characterization   49–53, 55–61, 62, 70, 74, 75
– fast ion chromatography   210, 211
– *see also* macroporous silica monoliths
FLUENT program   133
fragmentation of domains   17
frontal affinity chromatography (FAC)   279

g

gamma-alumina   2
gas chromatography (GC)   231
Gaussian curvatures   91–94
general plate-height model   106–120
Giddings coupling theory   116–120
gradient mode   259–261
gyroid models   87–89

h

Hagen–Poisseuille equation   67, 68, 70
Happel's equation   135
heat treatment   28
height equivalent to a theoretical plate (HETP)   135–138, 142–144, 147–152, 254–257, 259–261
helium pycnometry   130
heterogeneous catalysis   1, 2
HETP *see* height equivalent to a theoretical plate

hexamethyldisilazane (HMDS)   178, 179, 186
high-efficiency separations   249–261, 268, 269
high performance liquid chromatography (HPLC)
– analysis times   191–193
– axial dispersion   147, 148, 151
– basic column properties   128, 129
– chiral drug separation   198–201
– chiral separation techniques   231, 232
– column efficiency   135–137, 147–151
– column impedance   137, 138, 150–152
– column permeability   132–135, 140–142, 147, 149, 152
– definition of terms used   153, 154
– drug molecules   189–200
– external and internal porosities   131, 132, 138, 139, 146, 152
– flow programming   194–196
– high-speed, high-efficiency separations   249–252
– historical development   127, 128
– ion-exchange   42, 43
– kinetic properties   147–151
– loadability/saturation capacities   146
– mass-transfer kinetics   129, 136, 148–150, 151
– mobile-phase velocity   137, 149
– particle-packed columns   127, 128, 132, 138–140, 142, 143, 149, 152
– performance characteristics   127–156
– preparative methods   12, 24, 26, 27, 31
– principles and development   3, 4
– proteomics and peptidomics   299
– quality control techniques   189–200
– quantitative structure–retention relationships   163–166, 168
– radial homogeneity   142–144
– retention factors   144–146
– small-scale purification   285–295
– solid-phase extraction/microextraction   319
– thermodynamic properties   144–146
– through-pore size/velocity   129, 131, 133, 134, 138–144
– total column porosity   129–131, 134, 138
– two-dimensional   194–196
– see also reversed-phase HPLC
high-speed separations   249–254, 261–266, 268, 269
HILIC see hydrophilic-interaction chromatography
HMDS see hexamethyldisilazane
human genome sequencing   297, 298

hybrid monoliths
– basic compounds   183–186
– high-speed/high-efficiency separations   250, 251, 254, 268
– microscopic characterizations   95, 96
– solid-phase extraction/microextraction   330, 331
hydrogen bonding   161, 170
hydrophilic-interaction chromatography (HILIC)
– basic compounds   186
– fast ion chromatography   221
– mass spectrometry   273
– polymer-coated materials   38–42
– quality control techniques   200
hydrophilicity   234
hydrophobicity   163–166
hysteresis curves   56–60, 64–66

i

ICP-MS see inductively coupled plasma mass spectrometry
IEC see ion-exchange chromatography
imaging and image analysis   50, 53, 70–75, 84, 85
immobilized enzyme reactors (IMER)   278, 279
impedance   137, 138, 150–152
in situ copolymerization   234, 235
inductively coupled plasma mass spectrometry (ICP-MS)   219
industrial applications   223
interconnected pores   50
internal porosity   131, 132, 139
inverse size-exclusion chromatography (ISEC)   50, 53, 60–64, 67, 73, 74
ion-exchange chromatography (IEC)
– multidimensional LC-MS   303–305, 310–313
– polymer-coated materials   42, 43
– quality control techniques   200
– see also fast ion chromatography
ion-pair separation   219, 220
ISEC see inverse size-exclusion chromatography
isocratic mode   256–259
isoflavones   196
isotopic separations   257–259

k

Kataoka model   111, 116, 119
Kelvin equation   54
Knox model   116–118, 136, 148
Kozeny–Carman equation   67, 68, 70, 75, 112, 132, 133

l

laboratory methods for separation columns   29–31
Landauer–Davis (LD) expression   109, 110
Langmuir isotherm   64, 65, 146
laser scanning confocal microscopy (LSCM)   81, 82, 83–101
– characteristic wavelength   86–89
– chord lengths   90, 91
– curvature distributions   93, 94
– deformations in confined geometries   95–102
– fundamental parameters   84–94
– high performance liquid chromatography   133, 140, 142
– imaging and image analysis   84, 85
– macropore size and skeleton thickness   89, 90, 93, 94
– mean curvatures and Gaussian curvatures   91–94
– MTMS-formamide system   96–99
– MTMS-methanol system   96, 99–102
– organic–inorganic hybrid monoliths   95, 96
– pore structure   85, 86
– preparative methods   95, 96
– surface areas   86, 91
– three-dimensional observations   81, 84–86, 95–102
– wetting transition   97–99
LC *see* liquid chromatography
LC-MS *see* liquid chromatography–mass spectrometry
LD *see* Landauer–Davis
LFER *see* linear free-energy relationships
limits of detection (LOD)   223, 328, 331
limits of quantification (LOQ)   327
linear free-energy relationships (LFER)   159–161
linear solvation energy relationships (LSER)   161, 169–171
lipophilicity   163–166
liquid chromatography (LC)
– band broadening   105–125
– chiral separation techniques   231
– multidimensional LC-MS   297–317
– polymer-coated materials   40, 41
– proteomics and peptidomics   297–317
– quality control techniques   200
– small-scale purification   292
liquid permeability (LP)   50, 53, 67–70, 74, 75
liquid-liquid extraction (LLE)   200
LLE *see* liquid-liquid extraction
loadability   146, 233
– capillary columns   259–261
– multidimensional LC-MS   306
– small-scale purification   288–292
LOD *see* limits of detection
longitudinal diffusion   107, 108
LOQ *see* limits of quantification
LP *see* liquid permeability
LSCM *see* laser scanning confocal microscopy
LSER *see* linear solvation energy relationships

m

McGowan's characteristic volume   162, 169
macroporous silica monoliths
– characterization   49–53, 57–61, 70, 74, 75
– deformations in confined geometries   95–102
– fundamental parameters   84–94
– microscopic characterizations   81–103
– preparative methods   12, 17–25, 30, 31, 82, 95, 96
– *see also* flow-through pores
magnetic resonance imaging (MRI)   71, 72
MALDI *see* matrix-assisted laser desorption ionization
marching cube algorithm (MCA)   81, 86
MAS *see* 3-(methacrylamidopropyl)trialkoxysilanes
mass spectrometry (MS)   201, 219, 273–283
– affinity columns   278, 279
– ESI techniques   273, 275–278
– MALDI technique   273–275, 277
– multidimensional LC-MS   297–317
– offline techniques   273–275, 277
– online techniques   273, 275–279
– proteomics   275–278
– proteomics and peptidomics   297–317
– selected reaction monitoring   275–278
– small-scale purification   285, 287, 292
– solid-phase extraction/microextraction   326, 328
– standard operating procedures   279
– tandem techniques   273, 275–279, 326
mass-transfer kinetics
– band broadening   105–107, 110, 111
– basic compounds   182
– fast ion chromatography   221
– high performance liquid chromatography   129, 136, 148–151
– pore structure   61–63
matrix-assisted laser desorption ionization (MALDI)   273–275, 277, 310

MCA *see* marching cube algorithm
MCM-41   51, 55
MCM-48   55
mean curvatures   91–94
medicinal plants   190
memory effects   303
mercapto-group-incorporated silica monoliths   333
mercury intrusion-extrusion porosimetry (MP)   50, 53–61, 67, 72–75, 81, 130, 132
mesoporous silica monoliths
– characterization   49–56, 58–61, 62, 63, 73, 74
– chromatographic separations   3
– fast ion chromatography   213
– high performance liquid chromatography   129, 131
– preparative methods   12, 21–23, 25, 27, 28
3-(methacrylamidopropyl)trialkoxysilanes (MAS)   37, 40, 41
3-methacryloxypropyltrimethoxysilane (MOP)   37, 41, 43, 44
method of moments   106
methyltriethoxysilane (MTES)   22, 24
methyltrimethoxysilane (MTMS)
– basic compounds   183–186
– capillary columns   250, 251, 254, 257, 268
– microscopic characterizations   95, 96
– preparative methods   22, 24, 28, 29
– solid-phase extraction/microextraction   326, 332
micellar liquid chromatography (MLC)   165
microporous materials   50, 51, 55, 56, 131
microscopic characterizations   81–103
– characteristic wavelength   86–89
– chord lengths   90, 91
– curvature distributions   93, 94
– deformations in confined geometries   95–102
– fundamental parameters   84–94
– imaging and image analysis   84, 85
– laser scanning confocal microscopy   81–84
– macropore size and skeleton thickness   89, 90, 93, 94
– mean curvatures and Gaussian curvatures   91–94
– MTMS-formamide system   96–99
– MTMS-methanol system   96, 99–102
– organic–inorganic hybrid monoliths   95, 96

– pore structure   85, 86
– preparative methods   82, 95, 96
– surface areas   86, 91
– three-dimensional observations   81, 84–86, 95–102
– wetting transition   97–99
MLC *see* micellar liquid chromatography
mobile-phase velocity   137, 149
molecular modeling   166–169
Monte Carlo simulations   56
MOP *see* 3-methacryloxypropyltrimethoxysilane
morphology
– characterization   61
– chromatographic separations   3
– domain formation by phase separation   16, 19, 20
MP *see* mercury intrusion-extrusion porosimetry
MRI *see* magnetic resonance imaging
MRM *see* multiple-reaction monitoring
MS *see* mass spectrometry
MS/MS *see* tandem mass spectrometry
MSC *see* multisyringe chromatography
MTES *see* methyltriethoxysilane
MTMS *see* methyltrimethoxysilane
multidimensional LC-MS   297–317
– applications   310–314
– column dimensions   305–307
– memory effects and cross-contamination   303
– monolithic silica columns   307–310
– proteomics and peptidomics   297–317
– selectivity of columns   303–305
– toolbox for proteomics   300–303
multiparameter QSRR   161–163
multiple-reaction monitoring (MRM)   276–278
multisyringe chromatography (MSC)   199, 253, 261, 262
multivariate analysis   161, 166

n

n-octadecyl grafted chains   58–60
nano-liquid chromatography (nano-LC)   231
NG *see* nucleation growth
nitrogen oxides (NO_x)   1, 2
nitrogen sorption   50, 53, 55, 64–67, 73, 74
NLDFT *see* nonlocal density functional theory
non-equilibrium theory   106
nonlocal density functional theory (NLDFT)   55, 56, 66, 67, 73, 74

nonporous skeleton case 115, 116, 118
NO$_x$ *see* nitrogen oxides
nucleation growth (NG) 15, 16, 31
nucleic bases 43, 44, 193
nucleosides 211
nucleotides 43, 44

o

obstruction factors 109
octadecyl methacrylate (ODM) 37, 38
octadecyldimethyl-N,N-diethylaminosilane (ODS-DEA) 35–38, 178, 179, 251
ODM *see* octadecyl methacrylate
ODS-DEA *see* octadecyldimethyl-N,N-diethylaminosilane
offline MALDI technique 273–275
OLD *see* online dilution
oligosaccharides 40, 41
online dilution (OLD) mode 285, 287
online extraction method 200
open tubular columns (OTC) 128
opened pores 51, 52
organic–inorganic hybrid monoliths 95, 96, 330, 331
organic monolithic materials *see* polymer-based materials
Ostwald ripening 21, 22
OTC *see* open tubular columns
overload 180

p

p-styrenesulfonic acid (pSSA) 43, 44
PAH *see* polynuclear aromatic hydrocarbons
palladium heterogeneous catalysis 2
parallel-pore model (PPM) 62, 63
particle-entrapped monoliths 56–60, 322–327
particle-packed columns
– basic compounds 176, 177
– chiral separation techniques 232, 233, 243, 245
– fast ion chromatography 211
– high-speed, high-efficiency separations 253, 255, 256, 260, 261, 266–268
– performance characteristics 127, 128, 132, 138–140, 142, 143, 149, 152
– quality control techniques 189–196, 200
– small-scale purification 285, 286, 295
PCA *see* principle components analysis
PDMS *see* poly(dimethylsiloxane)
peak asymmetry 173, 175–180, 184
peak capacity *see* column efficiency

PEEK *see* poly(ether-ether-ketone)
penetration model 111, 116, 119
PEO *see* poly(ethylene oxide)
peptides 259–261, 297–317
permeability
– fast ion chromatography 211
– high performance liquid chromatography 132–135, 140–142, 147, 149, 152
– *see also* liquid permeability
pH gradient methods 165
phase ratios 261–265, 267–269
phase retention factor 107, 108
phase separation
– additives 23
– arresting transient structures 16–18
– domain formation 15–18, 31
– macropore control 19–21
– mesopore control 21, 22
– polymerization-induced 14, 15
physical cooling 14, 15
physically adsorbed chiral selectors 236, 237
platinum heterogeneous catalysis 2
PMA *see* poly(methacrylate)
PNM *see* pore-network model
Poiseuille's pressure drop law 112
polycyclic aromatic hydrocarbons (PAH) 331
poly(dimethylsiloxane) (PDMS) 326
poly(ether-ether-ketone) (PEEK) 3, 178, 179, 187, 211–213, 251, 252
poly(ethylene oxide) (PEO) 21, 23–27, 87–89
polymer-based monoliths
– characterization 64
– chiral separation techniques 233–235
– chromatographic separations 2
– solid-phase extraction/microextraction 319, 320, 323
polymer blends 87–89
polymer-coated materials 35–46
– advantages 43–45
– chemical modification processes 35, 36
– fast ion chromatography 215, 216
– hydrophilic-interaction chromatography 38–42
– ion-exchange chromatography 42, 43
– retention factors 36, 37
– reversed-phase chromatography 35–38
– separation of oligosaccharides 40, 41
– separation of tocopherol isomers 38, 39
polymerization-induced phase separation 14–17

poly(methacrylate) (PMA) 209
polynuclear aromatic hydrocarbons (PAH) 256, 257
polysaccharide-based CSPs 236, 238, 240–242
poly(styrene-divinylbenzene) (PS-DVB) 209
polyurethane foams 2
pore coarsening *see* coarsening
pore-free silica glasses 11, 12
pore-network model (PNM) 62–64
pore structure
– bimodal 49, 50, 60, 61, 210–212, 332, 333
– characterization 49–80
– chromatographic separations 3
– comparison of methods 73–75
– connectivity 52
– fast ion chromatography 210–213
– flow-through pores 49–53, 57–61, 62, 70, 74, 75
– general characteristics of porous materials 50–53
– high performance liquid chromatography 129, 130
– imaging and image analysis 50, 53, 70–75
– inverse size-exclusion chromatography 50, 53, 60–64, 67, 73, 74
– liquid permeability 50, 53, 67–70, 74, 75
– mass transfer kinetics 61–63
– mercury intrusion-extrusion porosimetry 50, 53–61, 67, 72–75
– mesopores 49–56, 58–63, 73, 74
– microporous materials 50, 51, 55, 56
– microscopic characterizations 85, 86
– nitrogen sorption 50, 53, 55, 64–67, 73, 74
– opened/closed pores 51, 52
– preparative methods 11, 12
– small-scale purification 285, 286
– solid-phase extraction/microextraction 332, 333
– *see also* through-pore size/velocity
porous skeleton case 116–120
post-translational modifications (PTM) 298
PPM *see* parallel-pore model
preparative methods 11–33
– additives to induce phase separation 23–25
– arresting transient structures 17, 18
– background and concepts 12–21
– bulk monolith preparation 23–27
– capillary columns 30, 250–252
– domain formation by phase separation 15–18, 31
– drying and heat treatment 28, 29
– epoxy-clad columns 29
– gelation and aging 26–28
– historical development 11, 12
– hydrophilic-interaction chromatography 38–42
– ion-exchange chromatography 42, 43
– laboratory methods for separation columns 29–31
– macroporous silica monoliths 12, 16–25, 30, 31, 82
– mesoporous silica monoliths 12, 20, 21, 25, 27, 28
– polymer-coated materials 35–46
– polymerization-induced phase separation 14–18
– pore structure 11, 12
– reproducibility 29
– reversed-phase chromatography 35–38
– silica source and catalyst 21, 22
– sol-gel processes 11–13, 17–21
preparative-scale separations 231, 243, 288–292
principle components analysis (PCA) 161, 166
propyltrimethoxysilane (PTMS) 326, 331
protein affinity chromatography 278, 279
proteomics 275–278, 297–317
PS-DVB *see* poly(styrene-divinylbenzene)
pSSA *see* *p*-styrenesulfonic acid
PTM *see* post-translational modifications
PTMS *see* propyltrimethoxysilane
purification *see* small-scale purification

q

QQQ *see* triple quadrupole
QSRR *see* quantitative structure–retention relationships
quality control (QC) techniques
– analysis times 190–196
– chiral drug separation 198
– column coupling 196
– drug molecules 189–205
– flow programming 194–196
– particle-packed columns 189–198, 201
– silica monoliths 189–201
quantitative structure–retention relationships (QSRR) 159–172
– analyte hydrophobicity 163–166
– fundamental principles 159–163
– linear free-energy relationships 159–161
– linear solvation energy relationships 161, 169–171

– methodology and goals   159, 160
– monolithic columns   169–171
– multiparameter approaches   161–163
– structural descriptors from calculation chemistry   166–169

r

radial homogeneity   142–144
rapid parallel synthesis (RPS)   293, 294
retention factors
– capillary columns   257
– high performance liquid chromatography   144–146
– polymer-coated materials   36, 37
– quantitative structure–retention relationships   160–162
reversed-phase (RP) HPLC   129–131, 144–147
– activity of monoliths towards basic solutes   175–179
– basic compounds   173–188
– endcapping   173, 178, 179, 186, 187
– fast ion chromatography   217–221
– hybrid capillary silica monoliths   183–186
– mass spectrometry   273
– multidimensional LC-MS   301, 304–306, 311
– overload   180
– performance of silica monoliths   173–188
– polymer-coated materials   35–38
– quality control techniques   191, 192, 199
– quantitative structure–retention relationships   163–166, 168
– reproducibility of analyses   174
– silanol groups   173, 174
– small-scale purification   285, 292–295
– van Deemter plots   180–183, 185
rhodium heterogeneous catalysis   2
RP see reversed-phase
RPS see rapid parallel synthesis

s

sample loading see loadability
SAS see solvent-accessible molecular surface area
saturation capacities   146
SAX see strong anion exchange
SBA-15   55
scanning electron microscopy (SEM)   37, 38
– capillary columns   252, 263
– column properties   132, 133, 138
– macroporous silica monoliths   81, 93
– pore structure   70–72

SCX see strong cation-exchange
SD see spinodal decomposition
sedimentation   16, 17
selected reaction monitoring (SRM)   275–278
self-similar coarsening see coarsening stage
SEM see scanning electron microscopy
sequential-injection chromatography (SIC)   199
SFC see super/subcritical fluid chromatography
share-flask methods   163, 164
Sherwood numbers   110, 111
Shirasu porous glass (SPG)   11
SIC see sequential-injection chromatography
silanol groups
– basic compounds   173, 174
– phase separation   19–21
– reversed-phase HPLC   173, 174
silsesquioxanes   21
skeleton thickness   89, 90, 93, 94
small-scale purification   285–296
– analytical conditions   286
– drug discovery compounds   292–295
– instrumental and operational factors   286, 287
– particle-packed columns   285, 286, 295
– preparative conditions   287
– preparative separations and sample loading   288–292
sol-gel processes
– chromatographic separations   2, 3
– preparative methods   11–13, 18–21
– solid-phase extraction/microextraction   327–332
solid-phase extraction/microextraction (SPE/SPME)   319–334
– applications   322–332
– capillaries and chips   328–332
– entrapped particles   322–327
– extraction platforms   321, 322
– extraction process   320, 321
– future developments   332, 333
– offline extraction   322–324
– online extraction   321–324
– polymer-based monoliths   319, 320, 323
– precolumns and tips   327, 328
– quality control techniques   199
– silica monoliths   319, 320, 322–332
– sol-gel processes   327–332
solvent pycnometry   130
solvent-accessible molecular surface area (SAS)   166, 167
SPE see solid-phase extraction; solid-phase extraction/microextraction

SPG *see* Shirasu porous glass
spheroidization   16, 17
spinodal decomposition (SD)   15–18, 31, 86–89
SPME *see* SPE *see* solid-phase extraction/microextraction
SRM *see* selected reaction monitoring
statistical modeling   73
strong anion exchange (SAX)   42, 43
strong cation-exchange (SCX)   42, 43, 304, 310–313
structural descriptors from calculation chemistry   166–169
super/subcritical fluid chromatography (SFC)   231
surface areas   86, 91
surfactant-coated materials   213–215, 217–221, 223
synthesis *see* preparative methods

t

tailing   173, 175–180, 184
tandem mass spectrometry (MS/MS)   273, 275–279, 326
TEM *see* transmission electron microscopy
tetraethoxysilane (TEOS)   21, 23, 82, 329, 330, 332
tetrahedral skeleton model (TSM)   106, 111, 112, 114–120, 133
tetramethoxysilane (TMOS)
– basic compounds   183–186
– capillary columns   250–252, 254, 257, 261–264, 267, 268
– preparative methods   21, 23–26, 29, 36
– solid-phase extraction/microextraction   329, 330, 332
thin-layer chromatography (TLC)   163
three-way catalysis   2
through-pore size/velocity
– band broadening   107, 108, 112–114
– capillary columns   253, 262, 269
– high performance liquid chromatography   129, 131, 133, 134, 138–144
titania   21, 24
titania-coated monolithic silica   198, 328, 329
TLC *see* thin-layer chromatography
TMCS *see* trimethylchlorosilane
TMOS *see* tetramethoxysilane
TMSI *see* trimethylsilylimidazole
tocopherols   38, 39
total column porosity   129–131, 134, 138, 139

transmission electron microscopy (TEM)   55
trimethylchlorosilane (TMCS)   179, 186
trimethylsilylimidazole (TMSI)   178, 179
triple quadrupole (QQQ) mass spectrometry   276, 277
TSM *see* tetrahedral skeleton model
two-dimensional HPLC   194–198
two-phase systems   3

u

ultraperformance liquid chromatography (UPLC)   192
ultraviolet (UV) detection   292–294
uphill diffusion   16
UPLC *see* ultraperformance liquid chromatography
UV *see* ultraviolet

v

van Deemter plots   41, 42, 115, 117, 120
– basic compounds   180–183, 185
– capillary columns   255
– chiral separation techniques   234
– fast ion chromatography   212, 223
– reversed-phase HPLC   180–183, 185
volatile organic compounds (VOC)   1
Vycor® process   11, 54

w

wall-coated open tubular columns (WCOT)   128
Washburn equation   54, 55, 130
water-glass   22
WAX *see* weak anion exchange
WCOT *see* wall-coated open tubular
WCX *see* weak cation-exchange
weak anion exchange (WAX)   42, 43
weak cation-exchange (WCX)   42, 43
wetting transition   97–99
Wilson–Geankoplis correlation   111, 116, 119

x

X-ray diffraction (XRD)   55
xanthines   167, 168
xerogels   324, 325
XIC *see* extracted ion chromatograms

z

zeolites   50, 51
zirconia   21, 23
zone-retention factor   107, 108